2-6-70

ANCILLARY TECHNIQUES OF GAS CHROMATOGRAPHY

ANCILLARY TECHNIQUES
OF
GAS CHROMATOGRAPHY

EDITED BY
LESLIE S. ETTRE

Encyclopedia of Industrial Chemical Analysis
Port Chester, New York

AND
WILLIAM H. McFADDEN

International Flavors & Fragrances (U.S.)
Union Beach, New Jersey

WILEY-INTERSCIENCE

A DIVISION OF JOHN WILEY & SONS
NEW YORK-LONDON-SYDNEY-TORONTO

Preface

If one investigates the trend of gas chromatography, there is observed a definite indication toward use of instrumental analytical methods or chemical manipulations for qualitative analysis of the original sample. A special characteristic of this trend is that the coupling of the different instrumental or chemical methods with gas chromatography are usually carried out in *one unified system.* Thus, all these manipulations or methods can be regarded in this respect as *ancillary techniques of gas chromatography.*

All these ancillary techniques are well known in chemistry and physics. Infrared spectroscopy, mass spectrometry, proton magnetic resonance spectroscopy, or thin layer chromatography are well-established instrumental methods of analysis and many books discuss their various aspects in detail. Similarly, pyrolysis, hydrogenation, and oxidation are common techniques used in organic chemistry for over a hundred years, and the study of chemical catalytic reactions is a discipline in itself. It is not our intention to discuss the principles of these methods and techniques. Our aim is to summarize the available information concerning their *connection* or, as it is commonly called, their *interfacing*, with gas chromatography.

A few years ago, when one of us (L.S.E.) was involved in editing the book *The Practice of Gas Chromatography*, this trend was already recognized, and two special chapters entitled "Reaction Gas Chromatography" and "Ancillary Systems" were included. To this date, these two chapters possibly represent the most thorough summary of these subjects. However, as soon as that book was published a great need was felt for an even more detailed discussion of the individual systems interfacing gas chromatography with the various ancillary methods.

The principal aim in the present book is not simply to give a literature survey of all the techniques dealing with a particular interfaced system, but rather to summarize the aspects of a particular system, discuss its pros and cons, describe the most important instrumental arrangements, and discuss what kind of information can be obtained through its use. It should be emphasized again that we do not want to teach gas chromatography or any of the chemical or physical techniques coupled with the gas chromatograph. The subject is how individual techniques are *interfaced with gas chromatography.*

v

When deciding on the scope of the present book, we definitely felt that methods used in sample preparation, such as derivative formation (ester preparation, formation of silyl esters, etc.), are standard chemical procedures and do not belong to our treatise. On the other hand, special and selective detectors which are used mainly not "to detect" but rather "to identify"—even if they are part of the "standard" gas chromatograph—definitely belong to the scope of this book. These detectors are therefore discussed separately in the last chapter.

When dealing with the combination of an ancillary method with gas chromatography, the scientists involved often argue as to whether in such a system, gas chromatography serves the other method, or vice versa. Using the combination of gas chromatography and mass spectrometry as an example, it is always debated whether the gas chromatograph serves as the sampling system for the mass spectrometer, or the latter should be considered as the detector of the gas chromatograph. It is not our intention to resolve this debate; however, it was felt that it would serve the book best if the two editors looked at the systems from the two ends. One of us (L.S.E.) is a gas chromatographer; the other (W.H.M.) is mainly experienced in the instrumental analytical methods used as ancillary techniques.

As with *The Practice of Gas Chromatography* (which we consider a basic monograph supplemented by the present volume), we have asked a number of individual investigators who are well experienced in the various interfaced systems to contribute by writing a chapter on the subject of their specialty. We are most grateful for their cooperation. If this book attains the stated objectives, the credits are due to their efforts.

We sincerely hope that the present volume will contribute to the further growth of the application of integrated systems which couple gas chromatography with other physical and chemical methods. The progress of the past ten years has often exceeded the wildest dreams of the organic chemist. We hope that this volume plants a few ideas for the next decade.

LESLIE S. ETTRE
WILLIAM H. MCFADDEN

January 1969

Acknowledgments

The illustrations and tables in this book, aside from those constructed or obtained by the authors, come from a variety of sources. Acknowledgment is made to the following individuals, publications, and corporations whose courtesy made the use of this material possible.

Academic Press, New York, N.Y.: Fig. 2-3.
Acta Chimica Academiac Scientiarum Hungaricae (Akadémiai Kiadó, Budapest, Hungary): Fig. 2-4.
American Chemical Society, Washington, D.C.: Fig. 10-24.
The Analyst (Society for Analytical Chemistry, London, England): Figs. 3-9, 7-8.
Analytical Chemistry (American Chemical Society, Washington, D.C.): Figs. 3-1, 3-7, 3-12, 3-13, 3-14, 4-7, 4-11, 4-12, 4-13, 4-14, 4-16, 4-17, 4-18, 5-3, 5-8, 5-10, 5-14, 5-23, 5-24, 6-29, 9-1, 9-2, 9-3, 9-4, 9-5, 9-6, 9-7, 9-8, 9-9, 9-11, 10-8, 10-9, 10-11, 10-12, 10-18, 10-19, 10-20, 10-22; Tables 3-1, 3-2, 3-8, 4-1.
Annals of the New York Academy of Sciences (The New York Academy of Sciences, New York, N.Y.): Fig. 10-26.
Arkiv for Kemi (K. Svenska Vetenskapsakademier, Stockholm, Sweden): Fig. 5-7b.
Barnes Engineering Co., Stamford, Conn.: Fig. 6-12a.
Beckman Instruments, Inc., Fullerton, Calif.: Figs. 6-10, 6-12b, 6-17.
Bibliographisches Institut, Mannheim, Germany: Figs. 8-4, 8-8, 8-16, 8-20.
Biochemical Journal (The Chemical Society, London, England): Figs. 10-15, 10-16, 10-17.
Carle Instruments Co., Anaheim, Calif.: Figs. 6-5, 6-14.
Chemistry & Industry (Society of Chemical Industry, London, England): Figs. 7-5, 7-6, 7-7.
Dohrmann Instruments Co., Mountain View, Calif.: Figs. 10-1, 10-2.
Forster Verlag, Zurich, Switzerland: Fig. 8-21.
Hayes, J. M., NASA Ames Research Center, Moffett Field, Calif.: Fig. 5-13.
Industrial & Engineering Chemistry (American Chemical Society, Washington, D.C.): Figs. 2-9, 2-16, 2-17.

Institute of Petroleum, London, England: Fig. 6-16.

Journal of Agriculture & Food Chemistry (American Chemical Society, Washington, D.C.): Fig. 10-25.

Journal of the American Chemical Society (American Chemical Society, Washington, D.C.): Figs. 2-7, 2-8.

Journal of the American Oil Chemists' Society (American Oil Chemists' Society, Chicago, Ill.): Fig. 5-6.

Journal of the Association of Official Agricultural (Analytical) Chemists (Association of Official Analytical Chemists, Washington, D.C.): Figs. 4-3, 4-6, 4-8, 10-6, 10-7.

Journal of Chromatography (Elsevier Publishing Co., Amsterdam, The Netherlands): Figs. 2-10, 9-10.

Journal of Dairy Research (Cambridge University Press, London, England): Fig. 5-22.

Journal of Gas Chromatography (Preston Technical Abstracts Co., Evanston, Ill.): Figs. 3-2, 3-3, 3-5, 3-8, 3-18, 4-1, 10-4, 10-5; Tables 3-3, 3-4, 3-5, 3-6, 3-7.

Lipids (American Oil Chemists' Society, Chicago, Ill.): Fig. 5-26.

Magyar Kémikusok Lapja (Magyar Kémikusok Egyesülete, Budapest, Hungary): Figs. 2-12, 2-13.

McCloskey, J. A. Baylor University College of Medicine, Houston, Tex.: Figs. 5-2, 5-9, 5-17, 5-18.

Microchemical Journal (Interscience, New York, N.Y.): Fig. 3-4.

Mikrochimica Acta (Springer Verlag, Vienna, Austria): Figs. 3-15, 4-9, 4-19, 4-20.

Nature (Macmillan (Journal) Ltd., London, England): Fig. 3-11.

G. Newnes Ltd., London, England: Fig. 5-7a.

Oil & Gas Journal (The Petroleum Publishing Co., Tulsa, Okla.): Fig. 2-11.

Perkin-Elmer Corporation, Norwalk, Conn.: Figs. 5-16, 6-4, 6-8.

Pharmaceutica Acta Helvetiae (Société Suisse de Pharmacie, Geneva, Switzerland): Fig. 3-6.

Plenum Press, New York, N.Y.: Figs. 10-3, 10-10, 10-21.

Spex Industries, Inc. Metuchen, N.J.: Fig. 6-31.

Varian Aerograph, Walnut Creek, Calif.: Figs. 10-13, 10-14.

Varian Associates, Palo Alto, Calif.: Fig. 5-12.

Warner & Swasey Co., Flushing, N.Y.: Figs. 6-23, 6-24, 6-25.

Zeitschrift für analytische Chemie (Springer Verlag, Heidelberg, Germany): Figs. 3-10, 3-16, 3-17.

Contributors

MORTON BEROZA, *Entomology Research Division, Agricultural Research Service, U.S. Department of Agriculture, Beltsville, Maryland*

LESLIE S. ETTRE, *Encyclopedia of Industrial Chemical Analysis, Wiley-Interscience Division of John Wiley & Sons, Inc., Port Chester, New York*

STANLEY K. FREEMAN, *Research & Development Division, International Flavors & Fragrances (U.S.), Union Beach, New Jersey*

GORDON E. HALL, *Physical Chemistry Section, Unilever Research Laboratories, Colworth House, Sharnbrook, Bedford, England*

MAY N. INSCOE, *Entomology Research Division, Agricultural Research Service, U.S. Department of Agriculture, Beltsville, Maryland*

RUDOLF KAISER, *Badische Anilin & Soda Fabrik A.G., Ludwigshafen am Rhein, German Federal Republic*

CHARLES T. MALONE, *Corporate Research Department, The Coca Cola Company, Atlanta, Georgia*

WILLIAM M. McFADDEN, *Research & Development Division, International Flavors & Fragrances (U.S.), Union Beach, New Jersey*

ROBERT W. McKINNEY, *Washington Research Center, W.R. Grace & Company, Clarksville, Maryland*

CHARLES MERRITT, JR., *Pioneering Research Division, U.S. Army Natick Laboratories, Natick, Massachusetts*

PAUL STEINGASZNER, *High Pressure Research Institute, Budapest, Hungary*

J. THROCK WATSON, *Institut de Chimie, Université de Strasbourg, Strasbourg, France*

Contents

CHAPTER 1

Principles:
What Are "Ancillary Techniques?"

Leslie S. Ettre, *Encyclopedia of Industrial Chemical Analysis,*
Wiley-Interscience Division of J. Wiley & Sons, Port Chester,
New York

I. Introduction

It is a common expression today to say that gas chromatography is the most widely used modern analytical method. Indeed, the number of gas chromatographs presently used in the whole world probably surpasses the number of instruments of any other single analytical method.

In essence, gas chromatography is a *separation technique*—the best separation technique available. Its peculiarity is that while the mixture to be separated is introduced as a single slug, it is introduced into a continuously moving gas flow and at the end of the column (in which the separation takes place), the individual fractions (i.e., the components of the original sample) emerge separated in time. Due to this discontinuity in the continuous process, many manipulations can be carried out with the

1

separated pure fractions as they emerge from the column. Similarly, if manipulations of the original sample mixture can be carried out in a continuously moving system, one can directly investigate the end results of these manipulations.

The manipulations which are directly coupled with the standard gas chromatographic system may be called the *ancillary techniques* of gas chromatography.

II. Classification of the Ancillary Techniques

The ancillary techniques of gas chromatography can be classified according to two principles. In the first, these techniques are grouped according to the location of the ancillary system relative to the gas chromatographic separation column, and one distinguishes between systems preceding and following the column. According to the second classification, one distinguishes between ancillary techniques where a chemical reaction takes place and those where an ancillary system is used for additional investigation of the chromatographically separated fractions.

In the present discussion, we are using the second classification; later, when discussing the individual ancillary techniques in more detail, we might also divide them according to the first classification.

The first group of ancillary techniques consists of those manipulations where the chemical composition of the original sample is changed during passage through the whole system. These techniques are usually summarized according to the suggestion of Drawert (1) by the term *reaction gas chromatography*.

The second group of ancillary techniques comprises those systems where another analytical method—usually instrumental—which may also be used independently of the GC system, is applied in addition to, or in place of, the standard gas chromatographic detectors. These methods, which are also used independently of the GC system, facilitate the qualitative identification of the individual sample components.

III. Reaction Gas Chromatography

Reaction gas chromatography is utilized for four reasons:

(*a*) A reaction or a catalyst is studied.

(*b*) A sample which otherwise could not be investigated by gas chromatography is modified in order to permit a GC analysis or a GC characterization of the sample.

(*c*) Sample components, either before or after separation, are modified in order to facilitate their qualitative identification.

(*d*) Fractions in the column effluent are modified in order to enhance detector sensitivity.

For detailed discussion, the methods of reaction gas chromatography can be best divided according to the location of the reaction relative to the separation column. According to this, we divide them into two groups: those which precede and those which follow the separation column. It should be mentioned that sometimes the reaction takes place *in* the column. This is the case, for instance, with packed columns containing silver nitrate and used for the separation of various olefins. However, for our present discussion, we do not consider such systems as part of reaction gas chromatography; their treatment belongs to a discussion on gas chromatographic columns.

A. Reactions Preceding the Separation Column

The first group concerns manipulations which are carried out before the sample proper enters the separation column. In these methods (Fig. 1-1*a*), certain reactions are carried out and the aim is the direct investigation of the reaction products by gas chromatography without any intermediate step such as collection. The reaction products are swept directly into the separation column by the carrier gas. This first group can be divided into three subgroups.

The first subgroup consists of methods where a reaction is performed to study either the change in the reaction products due to modification of the reaction parameters or the behavior of the catalyst used to perform the desired reaction. These methods can generally be termed *microreaction gas chromatographic techniques* because the reaction is performed in a micro scale as compared to the usual pilot-plant type operations.

The second subgroup consists of methods where a sample, usually nonvolatile, is deliberately broken down, generally by thermal means, to smaller, volatile fragments. The fragments are swept into the chromatographic separation column and the original complex sample molecule is then characterized by the relative abundance of these breakdown products. This method is usually called *pyrolysis gas chromatography* and today, it is very important in the investigation of complex, nonvolatile molecules, particularly natural and synthetic polymers.

In the third subgroup, we include those techniques where the vapor of the original sample introduced into the system undergoes certain deliberately performed reactions prior to entering the separation column. The

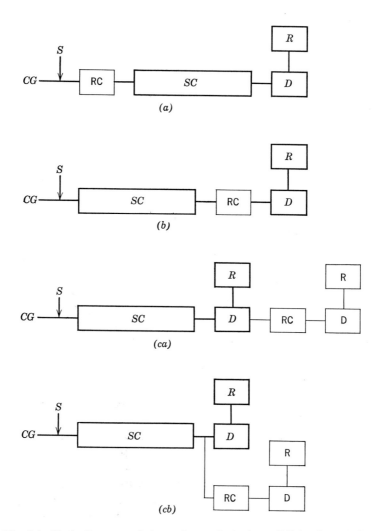

Fig. 1-1. Block diagrams of the various principal possibilities for reaction gas chromatography. The basic parts of the standard gas chromatographic system are indicated with heavy lines and symbols in italics. *CG*, carrier gas; *S*, sample; *SC*, separation column; *D*, detector, *R*, recorder (readout); RC, reaction chamber (reactor); D and R, detector and recorder after the reactor.

aim of these reactions is to modify the original sample components to other simpler substances which can be identified more easily or which permit the identification of the original sample components. Since these manipulations are usually performed in a precolumn preceding the separation column of the gas chromatograph, they may be called *precolumn reactions for structure determination.*

B. Reactions Occurring After the Separation Column

In these systems, the *fractions* in the column effluent undergo chemical reactions with two aims: either to simplify the detection methods or to identify them. In the first case (Fig. 1-1b), the column effluent passes through a catalyst on which one or more fractions undergo a chemical change prior to entering the detector. In the second case, after (Fig. 1-1ca) or parallel to (Fig. 1-1cb) the standard gas chromatographic detector, certain reactions are performed with the fractions in the column effluent, for component identification.

The first method (Fig. 1-1b) is relatively uncommon and its only significant application is the modification of fractions in the effluent either to enhance detector sensitivity or to permit the determination of the radioactivity of individual fractions when working with labeled compounds. The second method (Figs. 1-1ca and 1-1cb) is more important in practical gas chromatography and is often used to identify the fractions in the column effluent via simple and characteristic class reactions.

C. Complex Reaction Gas Chromatographic Systems

In some cases, various basic ancillary techniques are combined and used in more complex systems. For example, one may pyrolyze a polymer in a hydrogen flow and conduct the pyrolysis products through a hydrogenation catalyst prior to the separation column (2,3). In this way, unsaturated pyrolysis products will be reduced and the chromatogram will be simplified. By comparing the chromatogram of the pyrolysis products with and without hydrogenation, the unsaturated reaction products can be more easily pinpointed. Figure 1-2 shows the block diagram of such a system.

Another example for a complex reaction gas chromatographic system is the one described by Fanter, Walker, and Wolf (4, 4a), The principles of this system are shown in Figure 1-3. Here, the sample mixture is first separated in the usual way in a gas chromatograph. The column effluent is split in two; a small portion is conducted into a standard gas chromatographic detector (DI) while the larger part is conducted through a pyrolysis chamber to a second gas chromatographic system. When the detector of

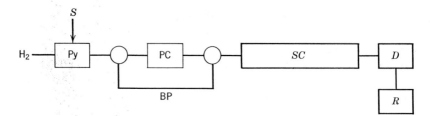

Fig. 1-2. Block diagram of a system incorporating pyrolysis and hydrogenation prior to the standard gas chromatograph. The basic parts of the standard gas chromatographic system are indicated with heavy lines and symbols in italics. H_2, hydrogen (carrier gas); S, sample; Py, pyrolysis chamber; PC, pre-column (containing a hydrogenation catalyst); BP, bypass; SC, separation column; D, detector; R, recorder.

the first system indicates that a peak has emerged, it stops the first carrier gas flow and starts the auxiliary carrier gas flow of the second gas chromatographic system. In this way, the particular fraction is pyrolyzed and the breakdown products analyzed in the second gas chromatographic system. When the second detector (DII) indicates that no more peaks are emerging from the second column (or, if preferred, at a preset time), the first carrier gas flow is resumed until a second peak emerges from the first separation column, at which time the whole process is repeated. According to the authors, the aim of this system is to identify the individual peaks by their

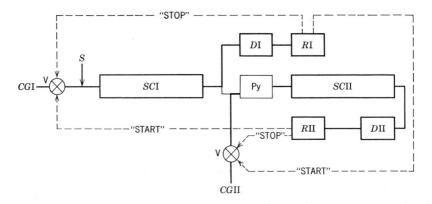

Fig. 1-3. Block diagram of a "stop–start" system coupling gas chromatographs with a pyrolysis chamber, with automatic programming of the individual cycles. The basic parts of the standard gas chromatographic systems are indicated with heavy lines and symbols in italics. CG, carrier gas; SC, separation column; D, detector; R, recorder (readout); Py, pyrolysis chamber; V, solenoid valve.

thermal breakdown products, which are almost as characteristic for any pure substance as the ion fragments in a mass spectrum.

Reaction gas chromatographic systems can also be combined with other ancillary detection methods such as mass spectrometry. Later, the principles of such a system will be outlined.

IV. Utilization of Ancillary Analytical Methods

It was already emphasized that one should consider gas chromatography primarily as a separation method. It is well known that gas chromatographic detectors have remarkable sensitivity, but they generally do not distinguish the individual chemical substances present in the sample. A few specific detectors represent an exception to this general rule, but the application of these detectors is restricted to a few specific groups of chemicals.

There are various methods available which, based on retention data, permit the tentative qualitative identification of the sample components corresponding to the individual peaks (5). However, these methods are never absolute, and if an unequivocal confirmation of the chemical nature of the individual components is needed, one has to utilize additional physical analytical methods such as mass spectroscopy or infrared spectroscopy.

These ancillary analytical techniques can generally be characterized in three ways (Fig. 1-4). In the first two cases, the ancillary technique is directly coupled with the standard gas chromatographic system and the separated fractions are continuously carried to the ancillary instrumentation. The difference between the two variations is only in the location of the ancillary system, that is whether it follows the gas chromatographic detector or is parallel to it.* On the other hand, in the third case, the fractions are first collected from the column effluent and then analyzed with the ancillary system.

Systems of the first two groups may be termed continuous systems while those in the third group are discontinuous systems. The discontinuity is necessary in two cases: if the time of passage of one chromatographic peak is not enough for full investigation with the ancillary system, or if the sample requirement of the latter is such that it would not be satisfied by standard gas chromatographic analysis. Some recently described systems actually combine the two methods so that the gas chromatographic analysis (the carrier gas flow) is discontinued for the time when one fraction is investigated by the ancillary method.

---

* In some cases, the ancillary detector may also serve as a chromatographic detector so that the original detector (D) is eliminated.

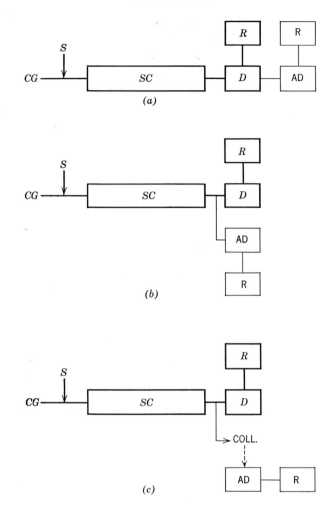

Fig. 1-4. Block diagrams of the various possibilities to combine ancillary analytical methods with the standard gas chromatographic system. The basic parts of the standard gas chromatographic system are indicated with heavy lines and symbols in italics. *CG*, carrier gas; *S*, sample; *SC*, separation column; *D*, detector; *R*, recorder (readout); AD, ancillary analytical method; R, readout; COLL, fraction collection.

As an example, Figure 1-5 illustrates such a system described by Scott et al. (6). Here, the individual fractions are investigated in an IR spectrophotometer and a mass spectrometer. When a peak appears in the GC detector, its main part is condensed in a cooled trap. The normal GC system is then shut off, and the condensed fraction is backflushed with

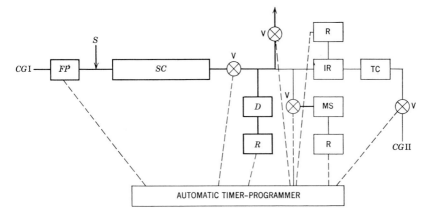

Fig. 1-5. Block diagram of the "stop–start" system of Scott coupling a gas chromatograph with mass spectrometer and IR spectrophotometer, with automatic programming of the individual instrument cycles. The basic parts of the standard gas chromatographic system are indicated with heavy lines and symbols in italics. *CG*, carrier gas; *FP*, flow programmer; *S*, sample; *SC*, separation column; *D*, gas chromatographic detector (FID); *R*, recorder; TC, trap column; IR, infrared cell; MS, mass spectrometer; R, readout; V, solenoid valve.

help of an auxiliary carrier gas line partly into a stainless steel IR cell with 9 cm single path and a silver chloride window, and partly into a mass spectrometer. After the two ancillary instruments finish their work, the operation of the normal GC system is resumed until the next peak arrives at the detector. The carrier gas flow to the GC system is regulated by a flow programmer to prevent high pressures being suddenly applied to the separation column when the chromatographic separation procedure is resumed.

Ancillary analytical methods can also be applied in complex GC systems which already incorporate some other ancillary technique such as pyrolysis. For example, Figure 1–6 shows a system described by Simon and co-workers (7). Here, the original sample is pyrolyzed and the breakdown products separated in the usual way. Subsequently, the column effluent is led through a molecular separator (sample enricher) to a mass spectrometer and the total ion monitor of the latter is used to provide the conventional chromatogram. The mass spectrum is scanned in the usual way.

The most important ancillary analytical methods used in connection with gas chromatography are mass spectrometry, infrared spectroscopy, and proton magnetic resonance spectroscopy. A fairly new and promising complex method combines two chromatographic techniques, gas chromatography and thin-layer chromatography, and the latter is then considered

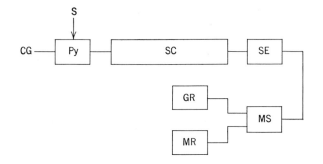

Fig. 1-6. Block diagram of a complex system incorporating pyrolysis and gas chromatographic separation with mass spectrometric detection. CG, carrier gas; S, sample; Py, pyrolysis chamber; SC, separation column; SE, sample enricher (molecular separator); MS, mass spectrometer; GR, gas chromatographic readout (chromatogram); MR, mass spectrometric readout (mass spectrum).

the ancillary technique for fraction identification. Of course, this is similar to the system of using two or more GC columns.

V. General Problems Associated with the Utilization of Ancillary Systems

Although the various ancillary systems differ significantly from one another, there are three common problems associated with their use in connection with a gas chromatographic system.

A. Connecting Lines

The first problem concerns the physical coupling of the individual systems. The ancillary systems, particularly the analytical techniques, are usually individually packaged, which means that there must be some distance between them and the standard gas chromatographic system. At the same time, certain ancillary systems—mainly those used in reaction gas chromatography—are heated to fairly high temperatures and therefore care must be taken that there is no heat exchange between them and the separation column. For both reasons, the ancillary system and the standard gas chromatograph are generally coupled through fairly long tubing and three important rules must be followed:

(1) The connecting lines should have minimum volume, in order to avoid band spreading. For this reason, usually capillary tubing is utilized through which the gas flows with a very high velocity. One should also

avoid any sudden change in tube diameter such as at the reducer fittings because these represent a sizable dead volume which could cause turbulence in the gas flow.

(2) The tubing should be inert with respect to the material passing through it. Metal tubes can be used but one must be sure that they are not active. Generally, this does not represent a problem; however, in certain cases, a passivation with 15% aqueous nitric acid might be useful, as has been shown by Purcell and Ettre (8).

(3) The connecting lines should be kept at a temperature which prevents condensation of the higher boiling compounds. At the same time, care should be exercised that the heating of the connecting lines does not offset the temperature of the thermostated parts of the system. Finally, it is important that the entire connecting line be at a uniform temperature and there are no overheated zones which may cause sample decomposition.

B. Gas Flow Rates

A basic problem of complex GC systems incorporating some ancillary technique is that usually the flow rate necessary for the ancillary technique is either too high or too low for the gas chromatograph. For example, a reactor-chamber type of pyrolysis unit needs flow rates higher than one would use with most GC columns; on the other hand, the flow rate needed for the molecular separator (sample enricher) of a mass spectrometer is usually lower than the optimum flow rates of a packed column. This means that in the complex systems under discussion, flow-splitting devices have to be incorporated in order to enable the application of the optimum conditions at every stage of the system.

Flow-splitting devices are commonly used in gas chromatography with open tubular columns, and problems associated with their construction are well known (9,10). It is important that the same caution be followed in the design of complex GC systems incorporating ancillary techniques.

C. Sample Size

Similar to the wide variation of desired flow rates, the sample size requirements for the various ancillary techniques and associated gas chromatography might greatly differ. This is not a problem if the sample size necessary for a manipulation preceding the separation column is greater than the sample capacity of the column or if an ancillary detection device employed after the column permits only the introduction of samples smaller than used for separation. For these cases, properly selected flow rates and stream splitting permit adjustment to the requirements. However, if the minimum sample requirement of an ancillary detection system is greater than the permissible sample size of a chromatographic column, then the use of special techniques is necessary. Either the gas chromatographic system must be scaled up or the ancillary detection system must be scaled down.

Today, most ancillary detector systems are improved to accept relatively small sample sizes, but in general if gas chromatography is coupled with such an ancillary detection technique, it is necessary to employ sample sizes somewhat higher than is usual in gas chromatography.

A special problem arises in mass spectrometry where sometimes, in continuous sampling, the ratio of the individual fractions to the column effluent must be increased. This is done with the help of special molecular separators (sample enrichers) which are based on the selective diffusion of the carrier gas (11–13).

References

1. F. Drawert, R. Felgenhauer, and G. Kupfer, *Angew. Chem.*, **72**, 555 (1960).
2. B. Kolb and K. H. Kaiser, *J. Gas Chromatog.*, **2**, 233 (1964).
3. B. Kolb, G. Kemmner, K. H. Kaiser, E. W. Cieplinski, and L. S. Ettre, *Z. Anal. Chem.*, **209**, 302 (1965).
4. Anon., *Chem. Eng. News*, **45** (41), 47 (1967).
4a. D. L. Fanter, J. Q. Walker, and J. C. Wolf, *Anal. Chem.*, **40**, 2168 (1968).
5. L. S. Ettre, "Interpretation of Analytical Results," in *The Practice of Gas Chromatography*, L. S. Ettre and A. Zlatkis, Eds., Interscience, New York, 1967, pp. 373–406.
6. R. P. W. Scott, I. A. Fowlis, D. Welti, and T. Wilkins, in *Gas Chromatography 1966*, A. B. Littlewood, Ed., Institute of Petroleum, London, 1967, pp. 318–333.
7. J. Vollmin, P. Kriemler, I. Omura, J. Seibl, and W. Simon, *Microchem. J.*, **11**, 73 (1966).
8. J. E. Purcell and L. S. Ettre, "Steroid Analysis With the Model 900 Gas Chromatograph," in *GC Applications*, No. GC-AP-013, The Perkin-Elmer Corp., Norwalk, Conn., 1968.
9. L. S. Ettre, *Open Tubular Columns in Gas Chromatography*, Plenum Press, New York, 1965, pp. 111–117.
10. R. D. Condon and L. S. Ettre, "Liquid Sample Processing," in *Instrumentation in Gas Chromatography*, J. Krugers, Ed., Centrex, Eindhoven, The Netherlands, 1968, pp. 87–109.
11. J. T. Watson and K. Biemann, *Anal. Chem.*, **36**, 1135 (1964).
12. S. R. Lipsky, C. G. Horvath, and W. J. McMurray, *Anal. Chem.*, **38**, 1585 (1966).
13. R. Ryhage, *Anal. Chem.*, **36**, 759 (1964).

CHAPTER 2

Microreaction Gas Chromatographic Techniques

Paul Steingaszner, *High Pressure Research Institute,*
Budapest, Hungary

Introduction

Not very long ago, investigators wanting to study catalytic reactions had
to choose relatively large laboratory reactors in order to have product

quantities of several hundred milliliters, sufficient for carrying out analytical distillations, because at that time practically no other means was at their disposal for the determination of product composition. This "classical" method of experimentation, besides being time consuming, had the additional disadvantage, especially when working with expensive chemicals, or radioactive compounds that the relatively large amounts of feedstocks necessary made experimentation costly.

The advent of gas chromatography changed the situation considerably; analyses could now be carried out much faster, with a fraction of the material needed previously. But it was not until 1955 that a major breakthrough came when Kokes, Tobin, and Emmett (16) introduced microreaction systems connected to gas chromatographs, enabling researchers to carry out the experiments on a micro scale with milligram quantities of reactants for the determination of reaction data.

The idea of carrying out chemical reactions in a gas chromatographic system was later extended, and methods have been developed where chemical reactions are used for the analytical identification of compounds. It was Drawert (1) who in 1960 first suggested the expression "reaction gas chromatography" to cover all those procedures where the substrate injected is deliberately changed during its passage through a chromatographic system by means of chemical reactions, irrespective of whether this change is brought about to detect and identify components by changing them by pyrolysis, class reactions, etc., or to study the chemical reaction or the catalyst. Beroza and Coad (2) and Ettre and Zlatkis (3) presented excellent reviews on reaction gas chromatographic publications.

The present chapter, however, *will discuss only those chromatographic methods which are used for the study of chemical reactions or catalysts* per se. These manipulations fall into the category defined by the term "microreaction techniques" suggested by Kokes, Tobin, and Emmett, where a gas chromatograph is needed to analyze the stream emerging from a reactor, the size of the latter being determined by the size of the sample necessary to get efficient separation and detectability on the chromatographic column–detector system, irrespective of whether the chemical conversion occurred in a batch, slug, or steady-flow type reactor, and whether the reaction studied is catalytic, thermal, or other.

The basic operating principles, the fields of application, the utilization of the data, and the limitations of the methods will be discussed here, and some examples will be given of the apparatus used. However, it has not been our aim to give a complete literature review; only those applications are considered which, in the opinion of the author, best illustrate the types in question. Also, no attempt has been made to show the methods of calculation of reaction rates, etc.; for this information, the reader is referred to standard textbooks.

With respect to constructional details our aim has been to give as many hints as possible, since inquiries at various leading manufacturers of chromatographic equipment revealed that at the time of writing no commercial microreaction apparatus was available and therefore the "do it yourself" approach was the only possible way to carry out such work. It is hoped that manufacturers will realize the general need for such equipment.

II. Classification of Microreaction Techniques

Microreaction techniques may be divided into the following categories:

(*1*) Batch microreactor operations.

(*2*) Intermittent or periodic microreactor operations (slug or pulse technique).

(*3*) Steady-flow microreactor operations (tail-gas technique).

(*4*) Special microreactor techniques for the determination of catalyst properties other than actual activity tests with the reactant.

Each of the techniques mentioned has its special field of application and limitations.

III. Batch Microreaction Techniques

A. General Characteristics of the Technique

Batch microreaction technique is a method for the study of chemical reactions in an apparatus consisting of a small batch reactor connected to a gas chromatograph by means of a suitable sampling device permitting periodic sampling of the reactor contents. Figure 2-1 gives the block diagram of such a system.

With batch reaction studies the aim is to determine the composition changes occurring as a function of time and reaction parameters, such as

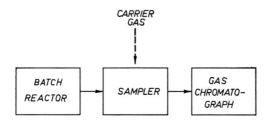

Fig. 2-1. Block diagram of a batch microreaction system.

concentration of individual reactants, temperature of reaction, pressure, and catalyst concentration. Many types of batch reactions may be studied in this setup, such as homogeneous gas-phase and liquid-phase reactions, heterogeneous gas–liquid reactions, with or without a soluble or insoluble catalyst, at atmospheric or elevated pressures; however, practical considerations usually limit the number of cases.

B. Apparatus

The type of reactor chosen depends on the type of reaction to be studied. Homogeneous gas-phase and liquid-phase reactions can be carried out in simple empty vessels with means for temperature and pressure control; usually stirring is needed to give uniform temperature within the reactor. Heterogeneous catalytic gas-phase reactions can be carried out by using a small reactor containing the catalyst, and a gas recycle pump to circulate the contents of the system through the reactor. Heterogeneous catalytic liquid-phase reactions can be carried out by use of a catalyst suspension in a well-stirred reactor.

With the reaction types mentioned, the main requirement is that the composition of sample taken should represent the true measure of the stage reached by the reaction at the moment of drawing the sample from the system. This implies that no further reaction should take place within the sample after it is removed from the vessel. In other words, means must be found to freeze the reaction instantaneously. In some cases, e.g., in photochemical processes, this does not present a problem, but in other cases problems might arise and no general rules can be given since every case needs special consideration.

The size of the batch reactor is dictated by the requirement that sampling should not disturb the operating conditions in the reaction vessel and for that reason the number of analyses needed must be taken into consideration in choosing the reactor size. The frequency of analyses to be carried out depends on reaction velocity, i.e., how fast the concentration changes are taking place in the reaction vessel; there is, of course, another limit on frequency, namely the time required for one gas chromatographic analysis. If fast reactions are the subject of study, more than one chromatograph and a multiport sampling valve may be needed and these must then be operated in a cyclic manner in order to perform the number of analyses required.

Care should be taken in the layout of the sampling system in order to get samples that truly represent the average composition of the contents of the reaction vessel at the moment of sampling. Good mixing in the reactor, short connecting lines, and means for flushing the lines are essential in this respect.

C. Operating Procedure

The reactants are put into the batch microreactor and reaction conditions, such as temperature and pressure, are established to start the reaction. The study of batch reaction kinetics is based on the evaluation of time–concentration patterns obtained under a set of different reaction conditions; therefore, the contents of the reaction vessel have to be submitted to periodic analysis according to a definite time schedule. It is very important to know the exact time elapsed from the start of the reaction and the time the sample was drawn and transferred to the gas chromatograph, in order to get an accurate time–concentration curve.

From the data obtained, if the samples were taken correctly, calculations of rate constants and activation energies can be carried out by known textbook methods and conclusions as to the reaction mechanism can be drawn.

D. Examples of Applications

Some examples of the application of the batch microreaction techniques are given below.

Cher and Hollingsworth (4) developed a repetitive gas chromatographic sampling technique for use with a batch microreactor. They used their technique for the study of gas-phase kinetics of acetone photolysis. The reactor (see Fig. 2-2) was a borosilicate glass vessel, 50 mm in diameter, 176 mm long, with a flat window on one side through which the light source illuminated the contents of the reactor. The reactor was placed in a

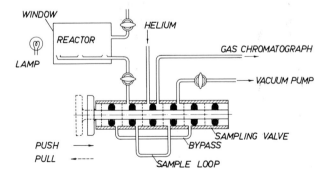

Fig. 2-2. Batch microreaction system of Cher and Hollingsworth for photochemical studies (4).

controlled-temperature muffle furnace and was attached to a manually operated eight-port, five-compartment, push–pull sampling stopcock. In the "pull" position of the stopcock the helium flows to the gas chromatograph, and the vacuum pump evacuates the sampling line, compartment 2, the sample loop, and compartment 4 of the stopcock. Prior to sampling, the sample loop is flushed by opening the stopcock leading to the reactor, and then the sampling stopcock is moved to the "push" position. In this position helium sweeps the contents of the sample loop into the chromatograph. If short connecting capillaries of sufficiently small bore are used, the samples taken represent the momentary composition in the reactor. Samples were taken at predetermined intervals and time–concentration curves were constructed which were treated according to the known rules of kinetic calculations. In this way Cher and Hollingsworth determined rate constants and energies of activation, all in excellent agreement with results measured by conventional methods.

Chambers and Boudart (5) used a batch recirculation system for the study of gold catalyst for the hydrogenation and dehydrogenation of cyclohexene in the vapor phase (see Fig. 2-3). Their microreaction system consisted of a microreactor where a 2-g catalyst sample in a shallow layer was accommodated, a sample stopcock with a sample loop, an all-glass recycle pump, several calibrated-volume containers, a flowmeter, and different connections to gas storage bottles, manometers, the vacuum system, purges, etc. The procedure consisted of steps for evacuating the whole system, filling it with known amounts of reactant, and mixing the contents of the system by prolonged recycling, with the reactor bypassed. After the mixing operation the reactor was opened to the system and periodic analyses were made. Mathematical treatment of the data is deducted and shown. Data obtained were used for the calculation of rate constants, the order of reaction with respect to the components, and the graphical determination of reaction paths or reaction network. The authors were able to show that with this catalyst the two reactions, dehydrogenation to benzene and hydrogenation to cyclohexane, were irreversible, and that addition of minute amounts of oxygen enhanced dehydrogenation selectivity by a factor of 3000.

Kalló and Schay (6) developed a static reactor assembly (see Fig. 2-4) where the solid catalyst sample was put into a microreactor and a gas recycling pump actuated by a square wave generator circulated the reactant gases at a controlled rate up to 1000 ml/min. The technical details of the recycling pump were described.

The same apparatus was used by Fejes and Emmett (7) for their study of the rate of catalytic cracking of propane and butanes over silica-

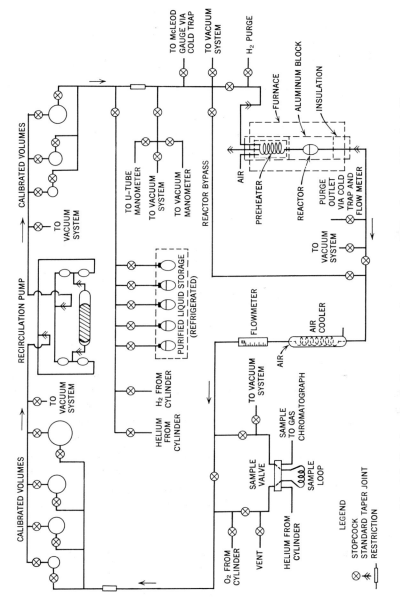

Fig. 2-3. Batch recirculation microreaction system of Chambers and Boudart (5).

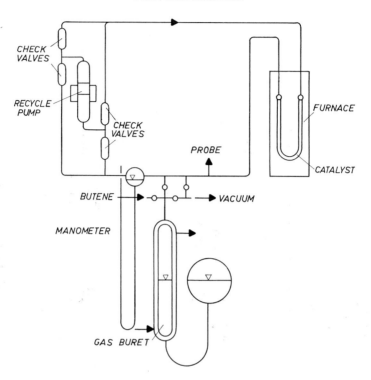

Fig. 2-4. Batch recirculation system of Kalló and Schay (6).

alumina cracking catalyst, by incorporating a constant-volume gas sampling valve which permitted periodic chromatographic analyses of the circulating gas to be made on the chromatograph connected to the reactor system. They were able to show that if the olefins formed were removed from the gas by means of a mercury perchlorate trap, there was a 25-fold increase in reaction rate and that in the absence of olefins the formation of higher molecular weight compounds was completely blocked. Based on these findings a reaction scheme was postulated.

No reference on pressure work with batch-type microreaction techniques has been reported.

IV. Intermittent Microreactor Operations (Slug or Pulse Technique)

A. General Characteristics of the Technique

In intermittent or pulse (slug) microreaction operations, the reactor is kept under a constant-flow carrier gas stream. The reactant sample is injected periodically into the reactor in the form of a "pulse" or "slug,"

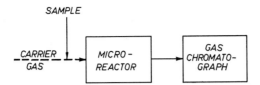

Fig. 2-5. Block diagram of an intermittent microreaction system.

and the whole reaction product resulting from the injected slug is swept from the reactor directly into the chromatograph in order to resolve it into component peaks.

A simplified block diagram of this mode of operation is given in Figure 2-5. Theoretically, any thermal and catalytic reaction can be carried out in this type of apparatus. However, technical and calculational difficulties limit its use to thermal and heterogeneous catalytic vapor-phase operations. Intermittent microreactor experiments are used for catalyst activity determinations, for studying the catalyst activity decline phenomena and for the determination of reaction rates and reaction mechanisms, the two latter types of investigations having some limitations inherent to the method.

B. Differences between the Pulse Technique and Other Techniques

The intermittent type of reactor operation is a unique technique differing basically from true batch or true steady-flow reactor operations. In batch operations, temperature and pressure are constant, and the concentration of the components in the reactor changes as a function of the time elapsed in such a way that at any moment the concentration is uniform throughout the reactor. In steady state operations, temperature and pressure being constant, the reactants enter the reactor at a constant rate and a concentration pattern is developed, the actual concentrations in different sections of the reactor being different but steady in time, whereas in intermittent operations while the temperature and pressure are still held constant, the reactants are added only once and pass like a wave through the reactor, and therefore no steady state conditions can develop. In order to understand the difficulties which may be encountered, let us consider in some detail what happens in an intermittent reactor.

With pure thermal, vapor-phase reactions in an empty reactor, adsorption phenomena being absent, the intermittent reactor behaves as any ordinary coil reactor, and the sample injected is passed through the reactor in the form of a vapor slug, pushed by the carrier gas. If the slug is of appreciable length, effects of longitudinal mixing at the front and back end of the slug are negligible and changes in composition will truly represent the effects of temperature, pressure, and residence time on conversion.

With chemical reaction occurring on solid catalysts, other phenomena will also take place, affecting the results of measurements carried out with intermittent reactors. Three such phenomena are particularly worth mentioning.

First, the reactant being injected in the form of a slug, it usually meets the catalyst that has been stripped of previously injected reactants by the carrier gas stream. In other words, there does not exist a stationary situation with respect to adsorption–desorption equilibria of reactants and products of reaction, since the reactants meet a clean catalyst surface and the partial pressures of the reactants constantly change. Therefore the situation existing in a steady flow reactor usually cannot be reproduced.

Second, the slug of reactants passing through the reactor does not represent a square wave type of pulse, because longitudinal diffusion, especially in pressure experiments, always has a chance to occur, making the partial pressures different in different parts of the slug. This must be taken into consideration, especially if the carrier gas itself is one of the reactants as in the case of hydrogenation or dehydrogenation reactions studied in a stream of hydrogen carrier gas; the partial pressures of hydrogen might be different in the front and in the back end of the slug, creating different and uncontrollable conditions for hydrogenation or dehydrogenation.

Third, the catalyst, usually itself a porous substance, may act as a chromatographic column, as shown by Bassett and Habgood (8) and by exerting a chromatographic separation effect may retard the movement of one or more of the components of the reaction products. In cases of extremely high component adsorbability it may happen that the corresponding reactant or product emerges from the reactor only after a prolonged time. This may have a profound effect on overall conversion, because if one or more of the products are absorbed to a lesser degree than the reactant or reactants, a separation of the components will result, causing a local relative deficiency in the concentration of the corresponding components and creating a concentration pattern differing from the overall concentration conditions. It is evident that in this case a situation exists where local equilibrium mass-balance effects cause a higher overall conversion than predicted by thermodynamics since local deficiency of reaction products will lead to a higher conversion than the overall thermodynamic equilibrium values.

C. Chromatographic Reactors

This observation was the basis of the concept of the chromatographic reactors invented by Dinwiddie and Morgan (9) and Magee (10). This effect was investigated by several authors both experimentally and mathematically. The mathematical treatment describing the phenomena occurring was developed by Roginskii and Rozental (10a, 11); Gaziev (12) and Magee (13) developed a simplified mathematical treatment for the equilibrium reaction

$$A \rightleftarrows B + C,$$

where various cases of infinite and finite reaction rates were taken into consideration. An equation was given for the infinite reaction rates assuming that the molecules of reactant A and product B move down the reactor with identical velocity while the molecules of product C move at a faster rate, giving the time necessary to attain a given conversion as a function of the difference of the rate of movement of the components and of the equilibrium constant K. Numerical solution of the equation resulted in the striking fact that when using a 1-m long column with rates of movement of component A and C differing by 6 cm/sec, even reactions having an equilibrium constant (K) value as low as 2.1×10^{-7} may go to completion. Another example given showed that reactions with a K value of 0.1 and a difference in rate of movement of the components as small as 3.3×10^{-4} cm/sec will give complete conversion.

Gore (14) showed that if the reaction $A \rightleftarrows B + C$ occurs in such a way that component C is negligibly adsorbed on the catalyst as compared to

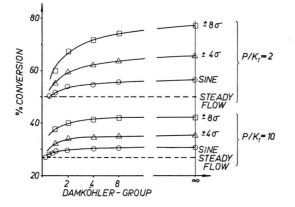

Fig. 2-6. Degree of conversion for various forms of reactant slug as a function of various parameters, according to Gore (14). P = initial partial pressure of the reactant, K_T = the equilibrium constant. The form of the reactant slug is characterized by a sine wave or by a displaced sine wave having the normal error functions truncated at $\pm 4\sigma$ or $\pm 8\sigma$ (σ = the standard deviation), respectively. Damköhler group = $\gamma k(L/u)$ where γ = fractional adsorption of the reactant, k = the forward rate constant, L = the length of the reactant bed, and u = the gas velocity.

component A and B, considerable improvement in overall conversion can be achieved depending on reaction rate, rate of separation, equilibrium conversion, form of the impulse, etc. Computations showed that the improvement in conversion over the normal, steady-flow conversion is higher the higher the reaction rate, the lower the equilibrium concentration, and the "sharper" the feed impulse, i.e., the narrower the injection wave. Improvement might amount to 50% over the thermodynamic conversion in steady-state flow reactors. Figure 2-6 shows some of the results of Gore's computations.

D. Apparatus

Reactors are usually made of narrow diameter tubing, where the catalyst is accommodated. Reactor sizes ranging from 1 to 20 ml are generally used with catalyst charges ranging from 100 mg to about 10 g. No commercial reactors are available and the researcher has to construct his own equipment. For atmospheric pressure work, up to temperatures of about 400°C, glass reactors may be used. For higher temperatures quartz reactors or stainless steel tubing should be utilized. Naturally, in pressure work, metal reactors are preferred. The construction of reactors is quite simple when using special fittings (Swagelok, Ermeto), the only tools needed being a pipe cutter for cutting the tubing to the required length and two wrenches for applying the fittings.

For manual injection by means of a syringe an injection port should be attached to the reactor inlet. In constructing the reactor, dead spaces should be carefully avoided since the material injected might accumulate there and then slowly bleed off, altering the results of subsequent injections. There are some good suggestions in the literature on how to avoid this.

The catalyst, ground to the correct size, is held in the reactor between screens or simply between two glass or quartz wool plugs. The part of the reactor not occupied by the catalyst should be filled with an inert material —quartz chips, glass beads, metal cuttings, etc. In every case periodic checks should be carried out to test whether some material other than the catalyst is present which may cause unwanted chemical reactions. Glass and quartz wool used should be washed with organic solvents prior to use, since most commercial products contain some kind of lubricant used in the manufacturing process. Similarly, the empty reactor tube should be checked for possible catalytic effects.

Reactors are placed in electrically heated furnaces. Induction heating or low voltage current may also be used. In the latter case the reactor walls serve as the resistance.

In order to get realistic results, the temperature distribution of the reactor should be such that maximum temperature occurs only in that part where the catalyst layer is accommodated; all other parts should be at lower temperature in order to avoid unwanted thermal effects. The most accurate temperature measurements can be carried out when using an internal thermo-well with a sliding thermocouple. With this arrangement temperature distribution can be measured most exactly. Less accurate temperature measurements can be obtained with thermocouples welded to the outer wall of the reactor because here radiation effects and heat conductance along thermocouple leads may result in erroneous readings. Usually heater block temperatures are accepted assuming that the difference between the block temperature and the temperature of the catalyst located inside is constant. The validity of this assumption must be checked with every apparatus. Operation with these reactor sizes is essentially isothermal.

Care should be exercised in selecting the correct catalyst pellet size. Large pellets should not be used because in this case only the outer layers of the catalyst pellet will be operating, giving lower conversions relative to the catalyst weight than observed with smaller size catalysts. Reactants by-passing the catalyst in channels between catalyst pellets will further decrease conversion. In order to obtain maximum conversion with a given amount of catalyst, the maximal grain size can be determined empirically by putting always the same weight of catalyst with different mesh size into the reactor and carrying out the investigations at constant conditions. Naturally, if low conversions are needed, e.g., in the study of reaction paths, catalysts with relatively large grain size might be used.

A further question is the selection of correct linear velocity. If the linear velocity is small, back-diffusion of the reactants and products may occur, again obscuring the true picture of the reaction. In such cases, one may determine empirically the correct size of the reactor. Reactors of different shapes, permitting different linear velocities, should be tried and the decision made based on the results obtained.

Injection of the reactants is carried out either manually by means of standard calibrated injection valves, syringes, ampoules, etc., or with semi-automatic or automatic dosing devices which are commercially available.

The reactor is connected to the chromatograph by incorporating it into the carrier gas line leading to the injection port of the chromatograph. When the reaction products also contain low volatility components, heating of the tubing connecting the microreactor and the gas chromatograph is necessary; this can be accomplished by electric heater tapes wrapped around the transfer line.

Standard gas chromatographs which are capable of analyzing the reaction products can be used with microreactors. Due to back-mixing in the reactor, different absorptivity of the reactant and product molecules on the catalysts, and gradual evaporation of the reactants injected, the products emerging from the reactor may assume the form of a broad, elongated peak which is difficult or impossible to resolve by gas chromatography. This shortcoming can be overcome by using a trap somewhere in the transfer line between the reactor and the chromatograph. This trap may be a conventional cold trap; also a part of the chromatographic column may be used as the trap by immersing it in a cooling medium, such as liquid nitrogen or an ice water bath.

For the identification of chromatographic peaks known procedures can be used such as those described by Crippen and Smith (15).

E. Operation of Intermittent (Pulse-Type) Microreactors

The operating procedure with intermittent microreactors is the following: The reactor is filled with the catalyst. It is assembled and the reaction temperature and pressure are adjusted. The carrier gas flow rate is established and the column temperature is set. In the case of catalytic work, the catalyst placed in the reactor usually has to be pretreated (reduction, oxidation, sulfidation, etc.); generally, these operations can be carried out *in situ* before the actual microreaction. With the reaction conditions adjusted, the reactant is injected into the reactor. The amount of liquid injected varies from a few microliters up to about 50 μl; in the case of gaseous substances, volumes of 1–20 ml are used. The injected reactant (if a liquid) evaporates in the carrier gas stream and is swept through the catalyst layer; during its passage the chemical reaction takes place. The reaction products emerging from the reactor pass to the gas chromatograph where they are resolved into peaks from which the composition of the reaction product can be calculated. After the first slug and the termination of the corresponding chromatographic analysis, a second, a third, etc. slug or pulse may be passed over the catalyst. The repetition of pulses at identical temperature and pressure will show whether the activity of the catalyst is changing. If declining conversion on repeated injection is observed, deactivation by poisoning or coke laydown or for some other reasons has taken place. In the case of catalyst deactivation, usually a larger number of slugs is necessary in order to have the catalyst activity "lined out," where on further injections no change or only a small downward trend in activity occurs.

Data obtained by the slug technique must be treated with caution if they are to be used for scale-up, kinetics measurements, etc. because local partial pressure variations occurring during the passage of the slug permit the measurements of kinetics only for first-order reactions and only in the case when chromatographic effects are absent. In order to test for the presence of partial pressure and for chromatographic effects, steady flow experiments should be carried out with the same catalyst charge and catalyst bed geometry using the same flow rates as in the intermittent case. If product composition is found to be the same, the absence of the above-mentioned effects can be assumed.

F. Examples of Applications

Some examples of the applications of the intermittent technique are given below.

Kokes, Tobin, and Emmett (16) were the first who proposed attachment of a microreactor to the inlet of a gas chromatographic column and injection of small amounts of samples into the carrier gas stream passing through the reactor and chromatograph for the investigation of catalytic reactions. The schematic of their microcatalytic unit is shown in Figure 2-7. A glass reaction tube of 8 mm diameter was placed in a vertical electric heater. The catalyst sample, about 1 ml, was held in place by glass wool plugs. Reactor effluents were directly admitted to a gas chromatographic column and the peaks were detected by a thermal conductivity cell. The authors claimed that the apparatus was suitable for exploratory catalyst work as well as for fundamental studies of the mechanism of catalytic reactions, e.g., with radioactive tracer methods.

Bassett and Habgood (8) were the first to point out the further possibilities inherent in the catalytic pulse technique for producing quantitative rate data. According to them, by measuring chromatographically the extent of adsorption occurring under the conditions of the reaction and thus separating the adsorption effects, true surface rates can be calculated. In their paper, they present an analysis of the kinetics of a first-order surface-catalyzed reaction (isomerization of cyclopropane to propylene over Molecular Sieve 13X) where the surface reaction is the rate-controlling step, based on data obtained by the slug technique under the conditions of the chromatographic elution. A piece of 3 mm Pyrex glass U-tubing was used as the reactor, containing about 0.1 g of catalyst held in place by glass wool plugs. The reactor was heated in a small block furnace. A sample volume of 0.20 ml was used for each experiment with helium carrier gas. The products emerging from the reactor were trapped in a similar tube containing the same amount of Molecular Sieve, cooled by Dry Ice; after each run, the trapped products were flashed into the chromatograph by

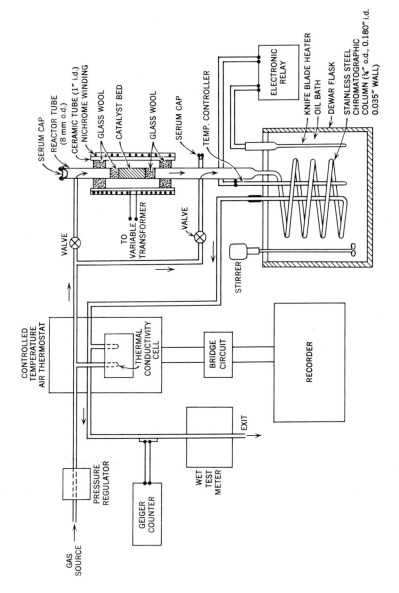

Fig. 2-7. The system of Kokes et al. (16) used for the microcatalytic pulse technique.

rapid heating of the trap. Adsorption constants for the reactant were calculated from retention volume measurements. Equations for the calculation of the surface reaction rate constant and activation energy from product composition data and adsorption constants were given.

The original microcatalytic apparatus, as developed by Kokes et al. involved the injection of reactants through serum caps by means of syringes. This injection method, probably adequate for many studies, suffers, however, from some drawbacks: (1) the reproducibility of data obtained is limited by the precision by which the extremely small volumes of gases and liquids can be injected by means of syringes and (2) in the case of catalysts and reactions sensitive to traces of oxygen or water, diffusion of oxygen or atmospheric moisture through the serum cap and traces of these contaminants adhering to the syringe needle may poison the catalyst. For these reasons, Hall and Emmett (17) developed a simple doser device by which exactly measured amounts of reactant gases can be brought in contact with the catalyst with the total exclusion of air and moisture.

The whole apparatus is shown in Figure 2-8. The doser consists of several glass stopcocks and a calibrated piece of glass tubing, the latter serving as reactant container from which by suitable manipulation of the stopcocks the reactant sample can be flushed into the reactor with the

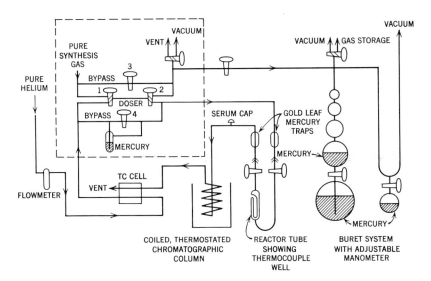

Fig. 2-8. The doser system of Hall and Emmett (17) using calibrated glass capillaries and manually operated stopcocks.

helium carrier gas. The doser capillary between stopcocks *1* and *2* is flushed and filled with the reactant gas. During the filling operation the pure helium carrier gas passes through stopcock *4* into the reactor and chromatograph. The reactant gas trapped in the doser capillary can be admitted to the microreactor by opening stopcocks *1* and *2* and closing stopcock *4*. A vacuum manifold with a manometer and a gas buret attached to the doser can be used for calibration purposes and for the preparation of blends of several reactant gases. For calibration purposes a serum cap was built into the line connecting the microreactor and the chromatographic column. Gold leaf mercury traps were used to prevent mercury vapor from coming into contact with the catalyst.

Using this improved doser device, Hall and Emmett (18) studied the hydrogenation of ethylene over copper-nickel alloys at low temperatures. The microreactor consisting of glass, contained about 1 g of catalyst and was immersed in a special low temperature cryostat with a temperature reproducible to $\pm 0.3°C$. Slugs of 8 ml volume were passed over the catalysts. Catalysts containing nickel and copper in different proportions were investigated. Systematic experiments showed the striking fact that the activity of the catalyst could be profoundly influenced by the way pretreatment of the catalyst was carried out. Based on experimental findings the authors concluded that when the catalysts were treated at high temperature in hydrogen, they took up hydrogen which became a part of the catalyst itself and acted as a promoter. The reproducibility of this effect could be demonstrated. Data obtained at different reaction temperatures were used to calculate apparent activation energies; specific activities were expressed as the ratio of total conversion to catalyst surface area. Catalytic activity at different alloy compositions was found to correlate rather with electronic factors (percent *d* character) than with geometric factors. The doser, when connected to the exit of the microreactor, could be used as an intermittent sampler in steady-flow reactor studies.

Hall, MacIver, and Weber (19) give a detailed description of a semiautomatic microreactor system working with the repetitive slug technique, and show the applicability of their system for the study of propylene polymerization and catalytic cracking of 2,3-dimethylbutane. Their apparatus was a modification of a previous reactor system, the modification consisting of an automatic doser device constructed from solenoid valves operated by a timer (Fig. 2-9). In addition, the operation of the timer was synchronized with the temperature control of the reactor, making possible a programmed sequence of unattended runs, carried out at different temperatures. An automatic integrating system was also used which permitted the printing out of chromatographic analyses.

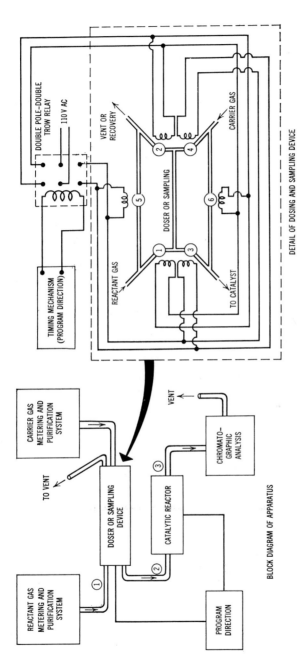

DETAIL OF DOSING AND SAMPLING DEVICE

BLOCK DIAGRAM OF APPARATUS

Fig. 2-9. Improved semiautomatic doser according to Hall et al. (19) with a solenoid-operated valve system.

The doser consisted of six solenoid-operated valves. The reactant gas flows through the central tube (Fig. 2-9), while the carrier gas is flowing through valve *6* directly to the reactor, and from there to the gas chromatograph. In this case valves *1*, *2*, and *6* are open, *3*, *4*, and *5* are closed. In the other cycle, with solenoid valves *1*, *2*, and *6* closed and *3*, *4*, and *5* open, the carrier gas sweeps the doser volume into the reactor, and from here to the gas chromatograph, as a slug. A description is also given of how the timer was coupled to the doser and to the temperature programmer.

The reactor consisted of a piece of glass tubing, having a volume of 8–9 ml. Depending on the type of experiment, 5–6 ml catalyst samples were used.

Products were analyzed as they emerged from the reactor, or in cases where high boiling substances were also formed, these were removed by low temperature trapping. In order to have a measure of conversion in these cases, a marker, usually argon, had to be used.

When secondary products are formed which impair catalyst activity, conversion tends to "line out," i.e., to gradually decrease on subsequent slugs. Extrapolation back to zero slug number gives the intrinsic activity of the catalyst. In certain cases it could be shown that intermittent slug operation gives the same conversion as continuous operation at the same throughput. If no deactivation occurs, data obtained by the slug technique can be used to calculate apparent activation energies from an Arrhenius plot. By plotting conversion data vs. catalyst surface area, the existence of secondary processes could be established from nonzero intercepts of these plots. The automatic doser can also be used as a sampler if connected to the exit of the steady-flow reactor.

Ettre and Brenner (20) described a simple microreactor which could be used with commercially available chromatographic equipments. It consists of stainless steel tubing of $\frac{3}{4}$ in. o.d., $\frac{9}{16}$ in. i.d., and $\frac{1}{2}$ in. length, placed in an electrically heated oven. The reactor is connected to one of two standard six-port sample valves, as shown in Figure 2-10. The gaseous sample enters the second valve, flows through a small three-way valve into the sample volume of the first sample valve. When the sample valve is turned, the carrier gas sweeps the sample from the sample volume tube into the reactor and then to the chromatograph. Direct analysis of reactants can be carried out by bypassing the reactor. With liquid reactants, the microreactor must be installed between the injection port of the chromatograph and the chromatographic column.

Hall and Hassel (21) made further studies of the promoting effect of adsorbed hydrogen on the catalytic activity of metals in ethylene hydrogenation in a microcatalytic equipment (17–19). The volume of the

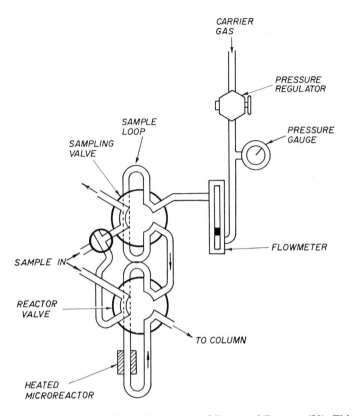

Fig. 2-10. Intermittent operation microreactor of Ettre and Brenner (20). This unit
can be fitted to practically any standard gas chromatograph.

catalyst was about 5 ml; 16 ml slugs were passed over the catalyst by
helium carrier gas, and the exit gases were analyzed chromatographically.
In some experiments hydrogen carrier gas was used. The catalysts were
subjected to different pretreatments in hydrogen and helium streams.
Reversible hydrogen chemisorption measurements were carried out by
attaching the reactors to a conventional BET system, and the amount of
total hydrogen held by the catalyst was determined by deuterium exchange.
It was shown that the activity of copper and copper-nickel alloy catalysts
could be enhanced by pretreatment with hydrogen by causing alteration of
surface properties, whereas iron, cobalt, and nickel catalysts were poisoned
by hydrogen chemisorption, which could result from rather small amounts
of hydrogen.

Gerberich and Hall (22) studied the effect of catalyst hydration state on the isomerization rate of different butenes and on the stereospecificity of reaction over different silica-aluminas and aluminas. The semiautomatic microreactor described by Hall et al. (19) and a combination of pulse and flow techniques was used. Experiments were carried out in three steps. First, small slugs of reactants were passed over the catalyst in a helium stream. Subsequently, flow experiments were carried out with butene–helium mixtures at various temperatures and the exit gases were analyzed in order to get a lined-out catalyst activity. Finally, the "poisoned" catalyst was subjected again to slug experiments. For the slug experiments a trapping technique was used, as desorption of butenes from the catalyst required about 15 min.

From slug data, effective rate constants for a first-order reaction were calculated, assuming no back-reaction, according to Basset and Habgood (8). All data were related to catalyst unit surface area. Since activity decline occurred to a certain extent with all catalysts studied, arbitrarily the sixth slug was always taken as a basis for calculation.

The authors found that in the case of silica-alumina with increasing surface hydroxyl content, the rate of isomerization increased; on alumina, the opposite effect was found, whether data measurements were carried out by the slug technique or by the steady-flow technique. From these observations the authors drew conclusions on the differences in the nature of the catalytic sites responsible for isomerization and postulated that a difference in reaction mechanism should exist.

Gerberich, Lutinski, and Hall (23) used the slug technique to study the effect of the fluorine content of alumina on their catalytic activity for cyclopropane isomerization and 2,3-dimethylbutane cracking, using the semiautomatic microreactor described by Hall et al. (19). Helium carrier gas was used. The first-order rate equations given by Bassett and Habgood were used for the calculation of apparent rate constants and activation energies. Hydrogen content and acidity of the catalyst samples were measured prior to experiments. Based on measured reaction rates and catalyst analyses, conclusions were reached concerning the possible surface configurations responsible.

Keenan and Iyengar (24) studied the structure of chromia-alumina catalysts by investigating the decomposition of nitrous oxide, using the pulse technique. The Pyrex glass reactor of 1 cm internal diameter and 5 cm length, contained 1.4 g of catalyst. The reactor was kept in a multiple-coil furnace for accurate temperature control. One-milliliter slugs of nitrous oxide were introduced into the carrier gas stream and the gases emerging from the reactor were analyzed for undecomposed N_2O.

It was found that when using an inert gas, such as argon, as the carrier gas, the activity of catalysts reduced before the runs declined considerably on subsequent slugs due to oxygen retention on the catalyst. Oxygen content could be reduced and catalyst activity improved either by prolonged flushing in argon or, even better, by the use of hydrogen as carrier gas. If the catalyst was oxidized prior to reaction, it behaved very similarly to oxygen-retaining reduced catalysts. Pure alumina support did not show any oxygen retention. Based on space velocity corrected to reactor temperature and pressure, and on conversion data, the authors calculated the activation energies and first-order rate constants.

From the data obtained, Keenan and Iyensar suggested that the decomposition of nitrous oxide on this catalyst goes through an anion-vacancy mechanism brought about on chromia clusters by flushing with either inert carrier gas or hydrogen, and these vacant sites accept oxygen from the N_2O admitted.

The microcatalytic slug technique has been adapted to the activity testing of commercial cracking catalysts (25) (Fig. 2-11). In the so-called microactivity test a 5-g cracking catalyst sample is put into a glass reactor. A standard charge stock is passed over the catalyst at 510°C at a flow rate of 0.21 ml/min, by means of a motor-driven syringe. The liquid as well as the gaseous products are sampled by means of calibrated stopcocks and

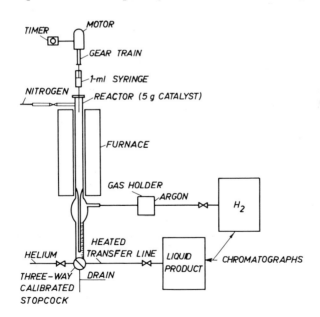

Fig. 2-11. A microreactor system for testing the activity of cracking catalysts.

chromatographed. Activity is defined as conversion over 200°C. The method gives a straight-line correlation when results are plotted against conventional catalyst activity test results.

Dealkylation of alkylbenzenes on cracking catalyst was studied by Mochida and Yoneda (26) by microcatalytic pulse technique in order to show how rate data can be estimated from data representing either reactivity of reactants or activity of catalysts. Rate data were measured on 20–100 mg of catalyst weighed into 4 or 8 mm i.d. glass reactor tubes, and heated to 450°C. Hydrogen carrier gas was used; a sequence of different alkylbenzenes was injected by means of a microsyringe and reaction products were analyzed in an attached chromatograph. It was found that the enthalpy changes of the formation of a carbonium ion by hydride abstraction correlate quite well with the logarithms of apparent rate constants. An equation has been presented giving a correlation between rate and enthalpy change and a characteristic value for each catalyst.

In a further work (27) the same authors using the same technique developed data for the determination of the catalyst characterizing value. The system was extended for the characterization of another acid-catalyzed reaction, isomerization.

Norton (28) investigated polymerization reactions of olefin occurring over synthetic molecular sieves. The microcatalytic apparatus was the same as used by Kokes et al. (16). Three milliliters of molecular sieve pulverized to 100 mesh were placed in the reactor and 5 μl samples of liquid, or 4 ml of gaseous reactant samples, were injected at 3–5 atm pressure into a nitrogen carrier gas flow. The microcatalytic results were used as a qualitative guide for the determination of oligomer distribution. Quantitative data were obtained by conventional bench scale equipment. These microcatalytic experiments showed that the propene polymerization was a reaction with negligible activation energy, whereas with the cracking of 2,3-dimethylbutane there was a definite activation energy.

Sinfelt and Yates (29) studied the kinetics of ethane hydrogenolysis in a microreactor of 1 cm diameter and 8 cm long, that contained 0.20 g noble metal catalyst supported on silica and diluted with ground glass. Ethane and hydrogen were mixed with helium carrier gas, and partial pressure and temperature effects were determined. Reactants were passed over the catalyst for 3 min and then sampled and analyzed. Order of reactions, activation energies, and preexponential factors were calculated and conclusions were drawn about possible reaction mechanisms and the relationship between catalytic activity and electronic properties of the catalyst metals.

While all the manual-injection microreactors mentioned were designed for atmospheric or near atmospheric pressure work, Steingaszner and Pines developed a pulse microreactor for the investigation of high pressure reactions, using manual injection (30). With the pulse technique, as mentioned earlier, one of the difficulties encountered is the fact that the narrow impulse-like injected slug tends to stratify in the reactor, giving rise to a much flatter Gaussian product wave at the outlet, thereby increasing the difficulty of separation of individual components. The situation is worse the higher the pressure in the reactor, as linear velocities are an inverse function of pressure. This was overcome by trapping, i.e., by keeping the first part of the chromatographic column at liquid nitrogen temperature (see Fig. 2-12) until all material emerged from the reactor, and then starting a programmed temperature chromatographic run. The other difficulty, namely, how to inject the reactants manually against pressures of up to

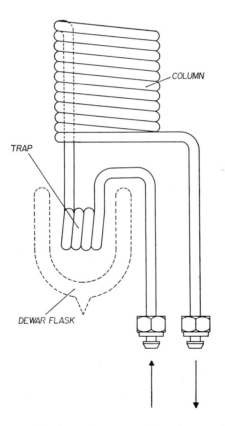

Fig. 2-12. Combined gas chromatographic column and trap (30).

100 atm, was solved by a small septum-retaining disk with a conical inside and a small opening for the syringe needle that was self-sealing even after many punctures.

The reactors were made from Swagelok parts and standard stainless steel tubing of $\frac{1}{4}$ and $\frac{3}{8}$ in. o.d. and 150 mm length. The inlet port, fabricated from special Swagelok parts, contained the septum holder and carrier gas inlet; outlets were standard reducing unions or tee's. A central thermocouple could be placed in the reactors making exact temperature measurement possible. The reactor was enclosed in an electrical heater. The complete system is shown in Figure 2-13. Pressure was controlled at the inlet, and gas flow rate was set by means of a precision valve inserted into the line connecting the reactor and the chromatograph.

The trapping technique not only made the analytical separation of high pressure reactor effluents possible, but at the same time, increased the separating efficiency and capacity of the analytical chromatographic columns. It was possible to separate low boiling hydrocarbons on a column which was not able to separate the same components at room temperature, and also to successively collect volumes up to 500 μl in the trap and carry out consequent preparative chromatographic separation on a $\frac{1}{4}$-in. analytical column.

This pressure-pulse microreaction equipment has been used by the authors to investigate the reactions of mono- and dihydric alcohols over reduced metal catalyst under pressure at elevated temperatures (30a). A new and interesting reaction was discovered, namely that when primary alcohols were injected into the microreactor containing a reduced nickel-kieselguhr catalyst, and using hydrogen as the carrier gas, then, besides the known reactions of reductive dehydroxylation and dehydroxydemethylation, ether formation also occurred, the latter reaction being highly sensitive to the method of catalyst preparation. It could be shown that minute amounts of alkalies in the catalyst profoundly depressed ether formation and by using alkali-free catalyst preparations the selectivity for ether formation could be increased appreciably. By this new reaction aliphatic ethers from aliphatic normal alcohols and cyclic ethers from 1,4-butanediol and 1,5-pentanediol could be prepared with high yields. A new compound, dineopentyl ether, could be made in this way. By the repeated pulse technique several hundred milligrams of this compound could be separated on a $\frac{1}{4}$-in. analytical column in one run, a quantity sufficient for the determination of physical data, NMR analysis, etc.

Catalytic oxidation of hydrocarbons is a problem interesting in many fields such as in monitoring the methane concentration in coal mines, in automobile exhaust control, and in hydrocarbon fuel cells. Anderson et al.

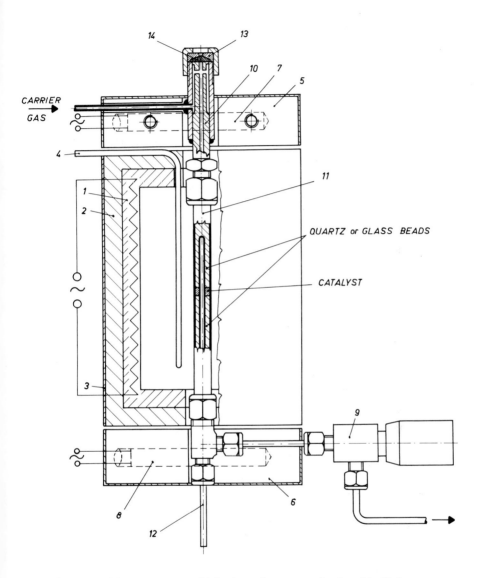

Fig. 2-13. High pressure manual injection pulse reactor, developed by Steingaszner and Pines. *1*, ceramic heater block; *2*, insulation; *3*, stainless steel sheet mantle; *4*, thermowell; *5*, upper heater block; *6*, lower heater block; *7*, *8*, cartridge heaters; *9*, precision microvalve; *10*, sample inlet block; *11*, microreactor; *12*, internal thermocouple; *13*, septum retaining cup; *14*, rubber septum.

(31) made a microcatalytic study using the atmospheric pressure impulse technique of Kokes et al. (16) for screening catalysts and to provide kinetic data on methane oxidation. The authors used oxygen as carrier gas which was passed at a controlled rate or 40 ml/min over 5 ml catalyst in a quartz reactor heated by electric resistance winding. Gases leaving the reactor passed through dryers and then through a thermal conductivity cell. At appropriate time intervals, measured volumes (0.66 ml) of methane were introduced to the catalyst bed. The technique consisted in setting the temperature of the catalyst, injecting methane, analyzing the effluent, and then repeating the same sequence at another temperature. Methane conversion rate constants were expressed by an empirical first-order rate equation containing terms of conversion and space velocity. Apparent activation energies and preexponential factors were calculated from Arrhenius plots in the usual way. From the results the order of catalytic activity of different metals could be determined.

Barron et al. (32) made an investigation on the mechanism of different reactions of hexanes and methylcyclopentane over several platinum catalysts. The microreaction technique described earlier (33) was used with supported platinum catalysts; the chromatograph, attached to the microreactor, analyzed the products emerging from the latter in a stream of hydrogen carrier gas. The reactions studied were hydrogenolysis, isomerization, and dehydrocyclization. From their results conclusions as to the mechanism of the reactions mentioned were drawn.

Chapman and Hair (34) investigated the role of surface hydroxyl groups in catalytic cracking of cumene on porous glass catalysts. The surface hydroxyl groups were partially replaced by fluoriding and the cumene cracking activity was determined in a pulse-type microreactor similar to that described by Ettre and Brenner (20). No details of the experimental technique were given. Based on activity measurements and other analyses they suggested that only Broensted-type acid sites are present at the surface; they could not detect Lewis-type acidity neither by chemical analysis or by ESR measurements.

Ogasawara and Cvetanovic (35) investigated the isomerization of n-butenes occurring over aluminas by a method which consisted of injecting sample pulses into a "reactor" formed from a long coil filled with catalyst installed in a chromatograph in place of the column, trapping the products at the outlet at liquid nitrogen temperature, and analyzing the product condensate separately by gas chromatography. This method does not belong exactly to the same class of operations discussed in the present chapter; it is mentioned, however, since it shows an additional possibility of carrying out microcatalytic reactions in a gas chromatograph without a special microreactor. The authors calculated adsorption equilibrium constants for the compounds in question from measured retention data and found that in their case no difference in adsorption among the components could be detected and therefore their results represented true reaction rate data.

A paper by Norton and Moss (36) gives an excellent example of how the pulse technique can be used for the fast screening of catalysts and for establishing the range of best operating conditions. The particular reaction studied was the oxidative dealkylation of alkylaromatic hydrocarbons. The microreactor tube was constructed of a 40 cm long, $\frac{1}{4}$-in. stainless steel tubing, with a wall thickness of 0.035 in. Catalysts, ground to 30–60 mesh were used in 6.5 ml quantities and vibrated to ensure a uniform packing. The microreactor was placed in a heater and attached to a chromatograph. Air and different oxygen–nitrogen mixtures were used as the carrier gas. Injection of feed samples was carried out by means of calibrated 1–10 μl microdippers. Product composition, conversion, and yield data were calculated from the chromatograms. Experimental design and statistical methods for investigation of chemical reactions for optimum yields were used to correlate effects of reaction parameters upon yields and conversions with the best catalyst found in the screening tests.

The heterogeneous catalytic decomposition of methanol over oxide catalysts was the subject of study by Wichterlova and Jiru (37). Nitrogen carrier gas is split into two streams, the first being led through a sampler into a microreactor containing 1 g catalyst, then into the chromatographic column and to the detector cell, while the second stream led through the reference side of the detector cell and from here into a variable-temperature methanol saturator; the gas leaving the saturator was conducted back to the sample loop connected to the sample valve, and served as feed for the next pulse.

The intermittent microcatalytic technique can be very effectively used to follow reactions occurring between gases and solids. Some such reactions are reduction, oxidation, or sulfidation of catalysts. By injecting measured slugs of the reacting gas or liquid into a carrier gas stream passing over the catalyst bed and measuring either the amount of unconverted reactant emerging or the amount of reaction product formed (e.g., water in oxide reduction), the degree of conversion of the solid can be calculated, and kinetic measurements can also be carried out.

Giordano, Bossi, and Paratella (38) studied the reoxidation of reduced oxide catalysts in a microreactor consisting of a U-shaped glass column, 5 mm i.d., with a variable length of 20–90 mm, filled with spherical catalyst particles of 0.063–0.480 mm diameter (see Fig. 2-14). Helium was

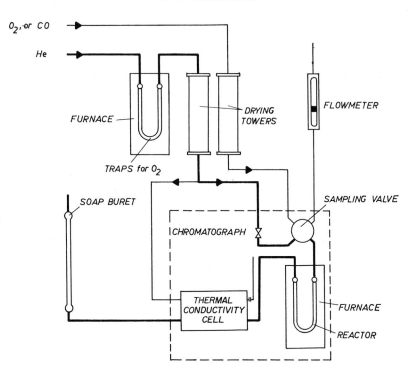

Fig. 2-14. Intermittent microreactor for the study of gas–solid reactions (38).

used as the carrier gas. The column containing the catalyst was installed in place of the chromatographic column. Slugs of reactant gas were evenly pulsed into the carrier gas stream and effluent composition was measured by the detector response. The reduction was carried out by carbon monoxide, the oxidation by oxygen. In their paper, the authors developed equations expressing the conversion for different assumed rate-controlling steps in terms of a dimensionless time. By plotting the assumed conversion curves and comparing these with actual curves, conclusions about the rate-controlling step(s) could be drawn.

Cabicar and Pospisil (39) investigated the process of reduction of oxide catalysts by using hydrogen as the reducing agent, the hydrogen being at the same time the carrier gas. Milligram quantities of the catalysts investigated were placed in a reactor column and the hydrogen stream emerging from the column was led through the detector cell. With water being the only reaction product, the deflection obtained at the recorder chart showed the time pattern of the reduction that took place.

V. Steady-Flow Microreactor Techniques (Tail Gas Technique)

A. General Characteristics of the Technique

The combination of a microreactor and a gas chromatograph where the reactant is fed continuously into the microreactor and the continuously emerging reaction products are sampled periodically by means of a sampler built into the exit line of the reactor and diverted to the gas chromatograph falls into this category. The block diagram of such a unit is shown in Figure 2-15. The equipment consists of a device for the continuous feeding of liquid and/or gaseous reactants, a microreactor containing the catalyst, pressure controls, a sampling valve, and a gas chromatograph. Liquid and/or gaseous reactants coming from the corresponding feeders sweep through the microreactor, and the reactor effluent passes through the sampling valve to the vent. By actuating the sampling valve, controlled amounts of reactor effluent can be diverted to the chromatograph in order to analyze the reaction products. This technique is sometimes called the "tail gas technique." By reversing the order of the microreactor and the sampling valve and omitting the liquid feed pump, a steady-flow microreactor can always be converted into an intermittent-operation pulse-type unit and *vice versa*.

Since during operation, the reaction parameters—pressure, temperature, feed rate, etc.—are kept constant, a steady concentration and temperature profile is developed in the catalyst bed, shifting only if the activity of the catalyst undergoes a change. This technique is based on the same operating principles as large-scale continuous industrial reactors, therefore it can be expected that results obtained from steady state microreaction experiments will give the maximum information on the effects of reaction parameters, etc. on the yield. This setup is also the best for the determination of kinetic data, since all chromatographic and partial pressure effects are absent, in contrast to the pulse-technique measurements where these can represent a

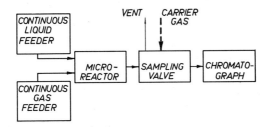

Fig. 2-15. Block diagram of a steady-flow microreactor.

serious drawback. Using thin catalyst layers, differential reactor measurements can be carried out giving information on the true reaction kinetics.

Heterogeneous catalytic and thermal reactions with substrates which are gaseous under the reaction conditions can be studied with this type of equipment.

B. Apparatus

Liquid reactant injection rates are usually in the range of 1–50 ml/hr. It is very important to have a truly steady flow, therefore special feeders must be used. For low pressure work a carrier gas stream may be saturated with the reactant in a thermostated saturator. For higher pressure work special feed pumps must be used because the usual plunger or diaphragm-type reciprocating pumps do not give reproducible flow rates in this range. Large diameter, hydraulically or mechanically driven positive displacement plunger pumps can be used. Such a pump is described by Webb, Dallas, and Campbell (40) for liquid feed rates of 2–30 ml/hr. The pump is operated by a synchronous motor through a gear train providing ten fixed feed rates. A limit switch stops the pump automatically when the plunger reaches the end of its travel. Recharging is done manually by cranking the piston down to the initial position.

Plunger pumps with hydraulic drive are also known. However, designs in which the hydraulic fluid is pushing the piston directly should be avoided because of the possibility of contaminating the reactants with the hydraulic fluid leaking through the piston packing rings. In order to prevent packing losses which might seriously influence the actual feed rates, the pump itself may be filled with mercury and the reactant can be displaced from a vessel placed in the pressure line. Precision pumps for the feed rates mentioned are also available commercially.

The reactor volume varies between about 1 and 50 ml; and catalyst samples ranging from about 100 mg to 10–20 g are used. Reactors have to be built according to the same principles as outlined earlier, for intermittent microreactor operations. Again, care should be exercised in selecting the proper reactor geometry and catalyst pellet size. Reactors for pressure work have pressure control microvalves fitted to the outlet, expanding the gaseous reactor effluent to the pressure of the chromatograph unit.

Effluent sampling is done by means of sampling valves connected to calibrated volume sample loops, built into the reactor exit line. In order to prevent condensation, the transfer lines, the sampling valve, and the sample loop have to be heated either in a heated box or by use of heater tapes.

For unattended operation a programming unit may be used, which carries out a preset sequence of runs.

C. Examples of Applications

Some examples of steady-flow microreaction equipment and its application are given below.

The most detailed description of a highly automated steady-flow microreaction apparatus was given by Harrison, Hall, and Rase (41) (see Fig. 2-16). The reactor system was designed for a maximum temperature of 800°C at 1500 psig (105 atm) pressure. Reactors of different size were made of standard stainless steel tubing and fittings, permitting fixed bed or fluidized bed catalytic experiments. The reactor was immersed in a sand bath fluidized by means of preheated air in a larger diameter stainless steel pipe; this arrangement gave excellent heat transfer characteristics and uniform temperature distribution through the sand bath. The feed system consisted of the variable gear-ratio mechanical pump already mentioned (40) and a gas cylinder from which the gases passed to a Deoxo unit, dryer, thermal conductivity cell, mass-flow meter, and a metering valve before entering the reactor. The thermal conductivity cell served to give an indication on stationary conditions. Liquid reactants were preheated in an electrical evaporator before being mixed with the gas. The reactants passed to the reactor in the fluidized sand bath, and from here, through the loop of a sampling valve and the reference side of the thermal conductivity cell, and then vented. Periodic sampling was carried out through the sampling valve and the sample was swept into the chromatograph. The quoted paper discussed in detail the various instrumental aspects and also described an automated sampler actuating system.

Hall and Rase (42) carried out a microcatalytic study in order to show the effect of treatments resulting in increase in dislocation density on catalyst activity with the apparatus previously described. Extremely pure lithium fluoride was used as the catalyst and ethanol as the substrate which was dehydrogenated to acetaldehyde and hydrogen. Reactors having an 0.180-in. i.d. and lengths varying from $\frac{13}{32}$ to 4 in. were used. Dislocation densities of catalysts were changed by different mechanical treatments and the catalytic activities were determined. The authors found that when dislocation densities were plotted against the reaction rates observed, smooth curves were obtained. It was also suggested how to extend these findings to other, crystalline and noncrystalline, catalysts.

Schwab and Knözinger (43) studied the catalytic decomposition of methyl formate on different metals. Their apparatus consisted of an

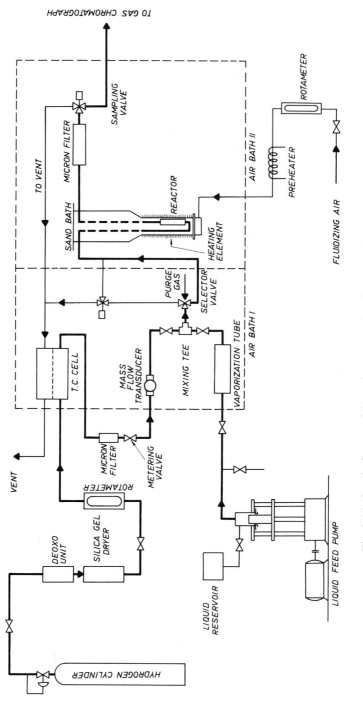

Fig. 2-16. Steady-flow precision microreactor described by Harrison et al. (41).

evaporator, a glass microreactor, a sampling stopcock, and a gas chromatograph. The carrier gas was saturated in the evaporator at constant temperature with the reactant. By turning the sampling stopcock a well-defined amount of reactor effluent could be directed to the chromatograph. The microreactor consisted of an empty horizontal tube where the catalyst was placed in a small container. In their article the authors discuss in detail how the results of chromatographic analyses can be applied to kinetic calculations. Starting from the Damköhler equation, they developed an equation for small concentration changes in order to express the dependence of peak height of a component on the rate constant, concentration, order of reaction, length of the reaction zone, and flow rate. Further expressions were also given for the calculation of the activation energy from peak heights, and for the determination of the influence of diffusion. The data obtained were used to suggest a possible mechanism of methyl formate decomposition.

The same apparatus was used by Knözinger and Köhne (44) for the study of the catalytic dehydration of aliphatic alcohols over γ-alumina. Based on the results obtained, a reaction mechanism was suggested to explain the formation of ethers and olefins.

MacIver, Wilmot, and Bridges (45) conducted a study to determine the differences between η- and γ-alumina by carrying out steady-flow, atmospheric pressure runs in an all-glass flow microreactor with 1-pentene, 2,4-dimethylbutane, and p-xylene to determine the catalytic activity of these samples in typical acid-catalyzed reactions such as isomerization and cracking. Pentene isomerization runs were carried out over 0.5 g samples of 50–100 mesh catalyst with helium–argon carrier gas saturated at $-15°C$ with pentene corresponding to 1–4.8×10^{-4} moles/min (about 0.4–2.1 g/hr). For the p-xylene isomerization runs, 1 g catalyst samples were used, and the argon carrier gas was saturated at $22°C$ with the reactant to give a feed rate of 5×10^{-5} mole/min (about 0.03 g/hr) of p-xylene.

Hartwig (46) studied the isomerization and hydrocracking reactions of n-hexane in a steady-flow microcatalytic reactor, connected to a gas chromatograph through a sampling valve. Palladium deposited on different supports was used as the catalyst. The reactor contained 2 ml of catalyst and had a bed depth of 5 cm. Hydrogen was used as the carrier gas and also as one of the reactants; the gas was conducted through a thermostated saturator containing n-hexane where its concentration in the gas stream was controlled. By carrying out runs at different temperatures, Hartwig was able to show how reaction product composition and isomer distribution depend on the temperature and the type of catalyst. No rate data were given.

48 PAUL STEINGASZNER

Stauffer and Kranich (47) investigated the kinetics of the catalytic dehydration of primary alcohols to olefins and ethers over γ-alumina in order to derive a set of equations to predict kinetic behavior of homologous compounds. A differential type microreactor was used with gas recycling and continuous alcohol feed, the exit gases being analyzed by means of a chromatograph attached to the reactor outlet through a sampling valve (Fig. 2-17). The reactor consisted of a ⅜-in. standard stainless steel pipe-

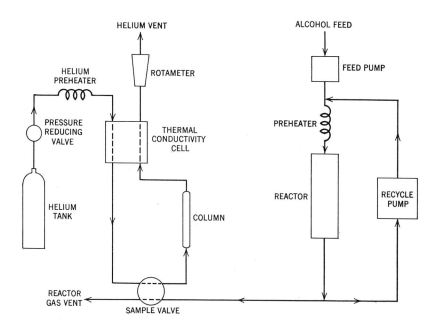

Fig. 2-17. Differential steady-flow microreactor of Stauffer and Kranich (47).

nipple. The alcohol feed and the recycling gases were preheated in a coil placed in a hot air bath together with the reactor. A home-made bellows-type gas recycling pump was used, in order to have low conversion level and at the same time a gas flow rate sufficiently high to avoid diffusion effects. Reaction rates expressed as moles of product formed per hour per unit weight of catalyst were plotted against temperature. Rate data for olefin formation could be represented by parallel lines showing the same reaction mechanism; however, ether formation showed different slopes. Based on their results and comparative literature data, authors gave some suggestions with respect to the possible mechanism involved.

VI. Special Applications of Microreactor Techniques

Gas chromatographic methods have been developed for the measurement of physicochemical properties of catalysts, such as the surface area and specific surface area of catalyst metals deposited on carriers, acidity and acidity distribution, adsorption equilibrium constants of reactants, etc. A short review on some of these methods will be given here.

Such measurements can be carried out with the standard gas chromatographic equipment with some modification and addition of necessary items such as sampling valves and furnaces. The methods involve measurements of gas–solid interactions between the catalyst filled in the column of the gas chromatograph and an adsorbate introduced into the system.

A. Adsorption Equilibria

Tamaru (48) proposed a method by which adsorption equilibrium on a catalyst can be studied in a gas chromatographic apparatus based on the fact that the more strongly the catalyst adsorbs a given species the longer the time needed for the same component to appear on the recorder chart, i.e., the longer the retention time. By carrying out measurements at different temperatures the heat of adsorption can be calculated.

Ozaki, Shigehara, and Ogasawara (49) determined the reversible adsorption of hydrogen on a nickel catalyst, the latter being held in a chromatographic column. Their procedure consisted of measuring the retention volume of a deuterium pulse relative to helium at different temperatures and pressures. Measurements have been carried out on a nickel–kieselguhr catalyst in different reduction states. The effect of water vapor on the retention volume was also investigated. From their data the extent and type of reversible adsorption occurring could be determined as a function of the parameters investigated and heats of adsorption were calculated. The catalytic activity for deuterium exchange and for the hydrogenation of ethylene was examined by the pulse flow technique; based on these data, a relationship between hydrogen adsorption and catalyst activity was suggested.

The type and relative magnitude of forces on catalyst surfaces can be estimated gas chromatographically at temperatures approaching those of catalytic interest, by measuring differences in retention time or retention volume with compounds having different acidities or basicities. An example of this type of work was given in a publication of Moseley and Archibald (50). The catalysts to be investigated were filled in a column made of a 30 cm long ¼-in. o.d. U-tube placed in the heater furnace of a chromatograph. Adsorbates were injected with a microsyringe into helium carrier gas. Retention volumes relative to the air peak were determined and corrected for pressure, temperature, and component volatility. Oxide-type cracking catalysts of different acidity were used with *m*-xylene and *n*-octane adsorbates; some measurements in the presence of water vapor and pyridine have also been carried out. Heat of adsorption values were calculated from the retention volumes obtained at different temperatures. Results obtained could be used to elucidate the nature of binding of the adsorbed species.

B. Number of Active Sites

Turkevich, Nozaki, and Stamirez (51), when studying the nature of active centers of synthetic zeolite catalysts, developed a method for the determination of the number of catalytically active sites. The catalyst was placed in a microreactor, and pulses of reactants were passed over it. Between the pulses, controlled amounts of quinoline, a catalyst poison irreversibly adsorbed, were sent over the catalyst and the decline in catalyst activity, as measured by the product composition resulting from the next reactant slug, was determined. The particular reactions chosen for testing were cumene cracking, ethylene polymerization, and butene double bond isomerization, all these reactions needing an acid-type catalyst. By this so-called "titration" method, the number of active sites could be determined. Carrying out the titrations with zeolites decationized to different degrees, the relative strength distribution of the sites could be established.

C. Catalytic Metal Surface Area

Gas chromatographic methods for the determination of crystallite size of reduced metal catalysts have been developed.

Cremer and Roselius (52) and Cremer (53) described a chromatographic method for the determination of crystallite size of metals dispersed on catalyst supports. Hydrogen admitted as a slug undergoes a change in retention time when passing a bed containing sites adsorbing hydrogen: due to adsorption, the retention time increases and the changes in retention time can be related to the size of the crystallites. The method permits relative crystallite size measurements.

Weidenbach and Fürst (54) developed a straightforward chromatographic method for the determination of crystallite size of noble metal catalysts such as used for gasoline reforming. The catalyst sample to be investigated is placed into a microreactor, the latter replacing the column of the chromatograph. The measurement of crystallite size is based on the fact that if a reduced platinum crystallite is exposed to oxygen at room temperature, each surface platinum atom adsorbs one atom of oxygen. If the oxidized sample is placed in nitrogen carrier gas and known amounts of hydrogen are injected, the surface oxygens will consume hydrogen and form water, while the unused hydrogen will appear as a peak on the recorder chart. From the difference between the volumes of the injected and unused hydrogen, the amount consumed can be established; from this value, the surface area of the platinum crystallites can be calculated.

D. Surface Area

Nelsen and Eggertsen (55) developed a gas chromatographic method for surface area determination which involves passing a stream of nitrogen–helium mixture of known composition over the sample placed in a piece of tubing replacing the chromatographic column, the effluent being monitored by the thermal conductivity cell. When the sample tube is immersed in liquid nitrogen, adsorption of nitrogen occurs, giving a negative peak on the recorder chart. After equilibrium is obtained, the recorder pen returns to its original position. Upon removal of the cooling bath desorption will occur resulting in a peak of the same size but on the opposite side of the baseline. By repeating runs with different carrier gas composition, a complete adsorption–desorption isotherm can be obtained from which, with help of the standard Brunauer-Emmett-Teller calculations, the specific surface area of the sample investigated can be established. The theoretical and practical aspects of this method and its various modifications are well documented (56–58). The method is not restricted to studies of nitrogen adsorption.

E. Diffusivity

Leffler (59) used gas chromatography for the determination of effective diffusivities of catalysts. Nitrogen was pulsed into a stream of helium carrier gas which was passing over a catalyst column kept at $-78°C$, and the peak broadening at different gas velocities was measured. From the data obtained, the constant of the mass transfer term in the van Deemter equation (60) was calculated. From this constant, by using the equation of Habgood and Hanlan (61) the values of effective diffusivity were established and found to agree reasonably with literature data.

VII. Summary

An attempt has been made to introduce the reader to the field of micro-reaction techniques, and to show some examples of how microreactors coupled to gas chromatographs can be used for the investigation of chemical (mainly catalytic) reactions.

Microreaction techniques were categorized as batch, intermittent, and steady-flow operations. All these methods are excellent for catalyst screening. However, for exact kinetic measurements, the batch and steady-flow types should be preferred because the inherent limitations of the intermittent reaction techniques restrict their use to first-order reactions and even there, to the special case when chromatographic effects are absent.

Some examples were also cited for the application of gas chromatography for the determination of catalyst properties other than activity.

All these methods possess the common advantage that time, number of operators, and amount of chemicals needed for experimentation can be reduced considerably making research faster and less expensive compared to classical methods.

References

1. F. Drawert, R. Felgenhauer, and G. Kupfer, *Angew. Chem.*, **72**, 555 (1960).
2. M. Beroza, and R. A. Coad, *J. Chromatog.*, **4**, 199 (1966).
3. L. S. Ettre and A. Zlatkis, Eds., *The Practice of Gas Chromatography*, Interscience, New York, 1967.
4. M. Cher and C. S. Hollingsworth, *Anal. Chem.*, **38**, 353 (1966).
5. R. F. Chambers and M. Boudart, *J. Catalysis*, **5**, 517 (1966).
6. D. Kalló and G. Schay, *Acta Chim. Acad. Sci. Hung.*, **39**, 183 (1963).
7. P. Fejes and P. H. Emmett, *J. Catalysis*, **5**, 193 (1966).
8. D. W. Bassett and H. W. Habgood, *J. Phys. Chem.*, **64**, 769 (1960).
9. J. A. Dinwiddie and W. A. Morgan, U.S. Pat. 2,976,132 (1961).
10. E. M. Magee, Can. Pat. 631,882 (1961).
10a. S. Z. Roginskii, M. I. Yanovskii, and G. A. Gaziev, *Dokl. Akad. Nauk SSSR*, **140**, 1125 (1961).
11. S. Z. Roginskii and A. L. Rozental, *Dokl. Akad. Nauk SSSR*, **146**, 152 (1962).
12. G. A. Gaziev, V. Yu. Filinovskii, and M. I. Yanovskii, *Kinetika i Kataliz*, **4**, 688 (1963).
13. E. M. Magee, *Ind. Eng. Chem., Fundamentals*, **2**, 32 (1963).
14. F. E. Gore, *Ind. Eng. Chem., Process Design Develop.*, **6**, 10 (1967).
15. R. C. Crippen and C. E. Smith, *J. Gas Chromatog.*, **3**, 37 (1965).
16. R. J. Kokes, H. Tobin, and P. H. Emmett, *J. Am. Chem. Soc.*, **77**, 5860 (1955).
17. W. K. Hall and P. H. Emmett, *J. Am. Chem. Soc.*, **79**, 2091 (1957).
18. W. K. Hall and P. H. Emmett, *J. Phys. Chem.*, **63**, 1102 (1959).
19. W. K. Hall, D. S. MacIver, and H. P. Weber, *Ind. Eng. Chem.*, **52**, 421 (1960).
20. L. S. Ettre and N. Brenner, *J. Chromatog.*, **3**, 524 (1960).
21. W. K. Hall and J. A. Hassel, *J. Phys. Chem.*, **67**, 636 (1963).
22. H. R. Gerberich and W. K. Hall, *J. Catalysis*, **5**, 99 (1966).
23. H. R. Gerberich, F. E. Lutinski, and W. K. Hall, *J. Catalysis*, **6**, 209 (1966).
24. A. G. Keenan and R. D. Iyengar, *J. Catalysis*, **5**, 301 (1966).

25. Anon., *Oil Gas J.*, **64**, 84 (Sept. 26, 1966).
26. J. Mochida and Y. Yoneda, *J. Catalysis*, **7**, 386 (1967).
27. J. Mochida and Y. Yoneda, *J. Catalysis*, **7**, 393 (1967).
28. C. J. Norton, *Ind. Eng. Chem., Process Design Develop.*, **3**, 230 (1964).
29. J. H. Sinfelt and D. J. C. Yates, *J. Catalysis*, **8**, 82 (1967).
30. P. Steingaszner and H. Pines, *J. Catalysis*, **5**, 356 (1966).
30a. H. Pines and P. Steingaszner, *J. Catalysis*, **10**, 60 (1968).
31. R. B. Anderson, K. C. Stein, J. J. Feenan, and L. J. E. Hofer, *Ind. Eng. Chem.*, **53**, 809 (1966).
32. Y. Barron, G. Maire, J. M. Muller, and F. G. Gault, *J. Catalysis*, **5**, 428 (1966).
33. G. Maire, G. Plouidy, J. C. Prudhomme, and F. G. Gault, *J. Catalysis*, **4**, 556 (1965).
34. I. D. Chapman and M. L. Hair, *J. Catalysis*, **2**, 145 (1963).
35. S. Ogasaware and R. J. Cvetanovic, *J. Catalysis*, **2**, 45 (1963).
36. C. J. Norton and T. E. Moss, *Ind. Eng., Chem. Process Design Develop.*, **3**, 23 (1964).
37. B. Wichterlova and P. Jiru, *Chem. Listy*, **59**, 1451 (1965).
38. N. Giordano, A. Bossi, and A. Paratella, *Chem. Eng. Sci.*, **21**, 621 (1966).
39. J. Cabicar and M. Pospisil, *Chem. Listy*, **60**, 528 (1966).
40. C. R. Webb, C. G. Dallas, and W. H. Campbell, *Ind. Eng. Chem.*, **54**, 28 (1962).
41. D. P. Harrison, J. W. Hall, and H. F. Rase, *Ind. Eng. Chem.*, **57**, No. 1., 25 (1965).
42. J. W. Hall and H. F. Rase, *Ind. Eng. Chem., Fundamentals*, **3**, 158 (1964).
43. G. M. Schwab and H. Knözinger, *Z. Physik. Chem. (Frankfurt)*, **37**, 230 (1963).
44. H. Knözinger and R. Köhne, *J. Catalysis*, **3**, 559 (1964).
45. D. S. MacIver, W. H. Wilmot, and J. M. Bridges, *J. Catalysis*, **3**, 502 (1964).
46. M. Hartwig, *Brennstoff-Chem.*, **45**, 234 (1964).
47. J. E. Stauffer and W. L. Kranich, *Ind. Eng. Chem., Fundamentals*, **1**, 107 (1962).
48. K. Tamaru, *Nature*, **183**, 319 (1959).
49. A. Ozaki, Y. Shigehara, and S. Ogasawara, *J. Catalysis*, **8**, 22 (1967).
50. R. B. Moseley and R. C. Archibald, *J. Catalysis*, **2**, 131 (1963).
51. J. Turkevich, F. Nozaki, and D. Stamirez, *Proceedings of the Third International Congress on Catalysis, Amsterdam*, North-Holland Publ. Co., Amsterdam, 1965, p. 587.
52. E. Cremer and L. Roselius, *Angew. Chem.*, **70**, 42 (1958).
53. E. Cremer, *Angew. Chem.*, **71**, 512 (1959).
54. G. Weidenbach and H. Fürst, *Chem. Techn. (Berlin)*, **15**, 589 (1963).
55. F. M. Nelsen and F. T. Eggertsen, *Anal. Chem.*, **30**, 1387 (1958).
56. H. W. Daeschner and F. H. Stross, *Anal. Chem.*, **34**, 1150 (1962).
57. L. S. Ettre, N. Brenner, and E. W. Cieplinski, *Z. Physik. Chem. (Leipzig)*, **219**, 17 (1962).
58. L. S. Ettre, "Application of the Continuous Flow Method for Surface Studies," *Application Brochure* No. SO-AP-002, The Perkin-Elmer Corp., Norwalk, Conn., 1966.
59. A. J. Leffler, *J. Catalysis*, **5**, 22 (1966).
60. J. J. van Deemter, F. J. Zuiderweg, and A. Klinkenberg, *Chem. Eng. Sci.*, **5**, 271 (1956).
61. H. W. Habgood and J. F. Hanlan, *Can. J. Chem.*, **37**, 843 (1959).

CHAPTER 3

Pyrolysis Gas Chromatography

Robert W. McKinney, *W. R. Grace & Company,*
Washington Research Center, Clarksville, Maryland

I. Introduction

Since 1960 well over 400 papers have been published which describe the pyrolysis (thermal decomposition) of materials combined with the use of gas chromatography for the examination of the pyrolysis products. An extensive bibliography published in 1964 illustrates the wide variety of materials to which the technique has been applied (1). A comprehensive review has been published by Levy (2).

Many applications of pyrolysis gas chromatography (PGC) involve materials which are too nonvolatile for gas chromatographic analysis and these materials are pyrolyzed to produce volatile fragments which, hopefully, will provide useful information about the original sample. The chromatogram of pyrolysis products is commonly called a *pyrogram*. PGC is also useful in obtaining information on submicrogram size samples which are too small for analysis by other techniques and also for the analysis of insoluble materials. PGC is relatively inexpensive compared to nuclear magnetic resonance, mass spectrometry, or infrared spectrometry, particularly if the gas chromatographic equipment is already available.

Many commercial pyrolysis units are available, but examination of the literature indicates the use of custom-built pyrolysis devices is still most common. This results in a major difficulty in standardizing PGC in that there are, at present, no commonly used designs for pyrolysis units. Nearly every article describes a pyrolysis system which differs to some degree from all others. The gas chromatography systems used also include numerous combinations of instruments, columns, detectors, temperature conditions, etc., and the result is that it is not possible to reproduce exactly work done in different laboratories. Kingston and Kirk (3) have demonstrated, in a statistical study of PGC of alkaloids, that with a given system the variability of PGC is within practical limits.

II. Technique

To date the most commonly used pyrolysis devices are filament units and microreactors. Filament pyrolysis involves placing a sample either directly on a wire filament or in a small container held within the filament which is heated by passage of electrical current through it. Microreactors are preheated to the desired pyrolysis temperature before the sample enters or is moved into the hot zone. In both cases a gas flows past the filament or through the microreactor during pyrolysis. With some designs the pyrolysis is carried out in the carrier gas stream of the gas chromatograph and the pyrolysis products formed are swept onto the column. Other designs provide for the gas flowing through the pyrolysis unit to pass into a gas sampling valve for injection into the gas chromatograph. Trapping pyrolysis products for subsequent injection into the gas chromatograph is done infrequently.

Although these two pyrolysis systems are most commonly used at present, other promising systems being developed and evaluated will also be described.

A. Filament Units

1. Electrical Types

A typical filament pyrolysis unit is shown in Figure 3-1. The current passing through the filament is regulated by a powerstat and indicated by a dc ammeter. The power supply is an 8-V transformer whose output is rectified by two silicon diodes. The instrument carrier gas passes over the filament and carries the volatile fragments onto the column. Generally the sample is dissolved in a suitable solvent and the solution applied to the coil. The solvent may then be removed by warming the coil by passage of a small current.

One criticism of pyrolysis of materials on metal filaments is that the metal might have a catalytic affect on the pyrolysis. This does not appear to be a significant problem since no difference in pyrograms is obtained when nichrome or gold-plated nichrome filaments are used (5,6) or with platinum or glass-coated platinum filaments (7). Stanford (8) did find that the pyrolysis of atropine deposited directly on a filament produced

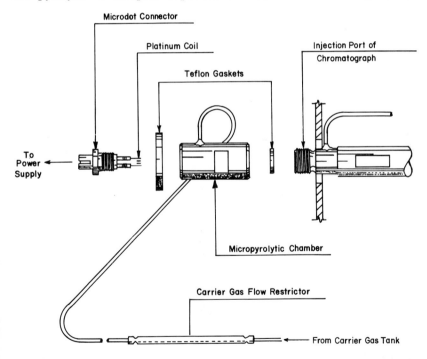

Fig. 3-1. Exploded view of pyrolytic accessory (4).

qualitatively reproducible pyrograms but the ratio of peak heights or areas to sample weight varied widely. This difficulty was overcome by placing the sample inside a 0.5 mm o.d. glass capillary which was inserted in the filament coil.

Figure 3-2 shows a nichrome ribbon filament unit which, when used with an open tubular gas chromatography column and argon triode detector, permitted use of samples as small as 10^{-8} g (9). Sample solutions were deposited on the filament by a micropipet. Using dilute polymer solutions, the pyrolysis of films as thin as 100 Å could be studied.

One problem associated with the use of filaments is that a significant period of time is required from the time the voltage is applied until the filament reaches the final pyrolysis temperature. Since milligram or microgram quantities of sample are used, appreciable pyrolysis may occur at lower unknown temperatures. Figure 3-3 shows a series of time–temperature curves obtained for a nichrome filament arranged in a bridge circuit so that the filament temperature could be measured by following the change

Fig. 3-2. Ribbon filament unit (5). A, Carrier gas inlet; B F, connections to filament C; D, outer glass tube; E, inlet end of open tubular column.

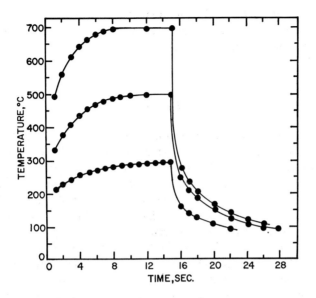

Fig. 3-3. Effects observed for a filament nominally at 300°, 500°, and 700°C for 15 sec (5).

in resistance during heating. Times of the order of 8–12 sec are required to reach the final maximum temperature. Dimbat and Eggertsen (7) produced large differences in pyrograms using the same platinum filament heated at different rates. By considerably increasing the initial applied voltage for 1 sec (boosting), then decreasing to a lower value, Lehrle and Robb (5) were able to obtain time–temperature curves more nearly like the ideal square wave with the final temperature reached in about 1 sec.

Some qualitative applications of PGC make use of "fingerprints" which are pyrograms of materials for which peaks are not necessarily identified but which can be reproduced with suitable precision. Comparison of pyrogram fingerprints of samples with the fingerprints of known compounds or composites of pyrograms of known compounds is often used, particularly with polymeric materials. For such applications, knowledge of the exact pyrolysis conditions is not necessary. It is necessary, however, to be able to reproduce the exact pyrolysis conditions.

Jennings and Dimick (4) found that the platinum filaments they used had a finite life. This was attributed to carbon from pyrolyzed compounds dissolved in the hot platinum filament resulting in a change in the emissivity. As the coil aged, higher currents were required to obtain the same filament temperature.

Determination of the surface temperature of a hot filament is also a problem. The usual techniques employed involve the use of a thermocouple close to the filament surface, optical pyrometers, or calibration with compounds of known melting point.

2. RF-Heated Types

A PGC system described by Simon et al. (9,10) is free of the above-mentioned disadvantages (Fig. 3-4). Reproducible pyrolysis conditions are obtained by coating a thin layer (micrograms) of the sample on a ferromagnetic conductor which is heated to a constant known temperature in 20–30 msec by induction heating. The energy uptake of cylindrical ferromagnetic conductors in high frequency induction fields decreases drastically at the Curie temperature of the material used. Under proper conditions the final temperature is determined by the Curie temperature of the particular material used and is independent of variations in experimental conditions. A wide range of pyrolysis temperatures is made available by choice of a conductor of the proper composition. Typical heating curves are shown in Figure 3-5.

A relatively simple system was described by Giacobbo and Simon (11) for the pyrolysis of quantities of material weighing less than 1 μg. A ferromagnetic wire (1 mm diameter × 30 mm long) was dipped in a dilute sample solution. The wire was then introduced through two stopcocks, using a magnet, into the vertical pyrolysis unit (Fig. 3-6). The wire was heated to pyrolysis temperature (800°C) in less than 0.1 sec. Sample weights as low as 10^{-7} g were pyrolyzed.

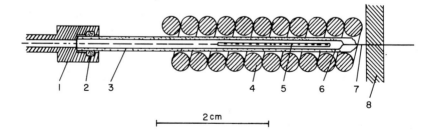

2 cm

Fig. 3-4. Schematic of a pyrolysis unit (9). *1*, Carrier gas connector; *2*, O-ring; *3*, glass capillary; *4*, high frequency induction coil (copper) connected to an oscillator working at 0.45 MHz; *5*, ferromagnetic conductor (wire: 0.5 × 20 mm); *6*, sintered glass beads; *7*, platinum iridium capillary (0.3 mm i.d.); *8*, serum cap inlet system for a gas chromatograph.

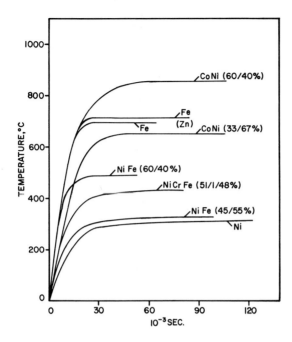

Fig. 3-5. Warmup curves for different ferromagnetic conductors (10). Diameter: 0.5–0.6 mm. Oscillator frequency: 0.45 MHz.

B. Microreactor Units

Microreactors may either be empty or filled with materials such as glass or quartz wool or glass beads. One of the simplest of all pyrolysis units to construct employs a metal pyrolysis tube attached to a gas sampling valve in place of the gas sample loop (12). The sample is placed in the tube, and the valve is opened to flush the tube with helium and then closed. A container of Wood's alloy at 500°C is placed around the tube for 30 sec after which the gas sampling valve is opened to carry the pyrolysis products into the gas chromatograph.

Samples are introduced into a microreactor by pushing a sample container (boat) into the hot zone, by dropping the sample into a vertical chamber, or by direct injection with a syringe or solid sample injection system.

Figure 3-7 shows a microreactor unit capable of pyrolyzing a number of samples without disturbing the PGC system (13). One section of the quartz pyrolysis tube is enclosed in a furnace and maintained at the desired

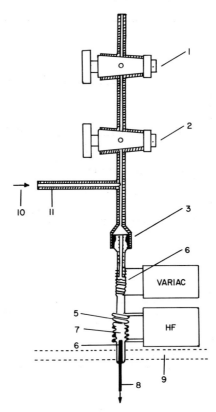

Fig. 3-6. Schematic of a pyrolysis apparatus (11). *1, 2*, Polytetrafluoroethylene stopcock; *3*, O-ring seal (used only with a serum cap injection port); *4*, heating coil for preheating the carrier gas or evaporation of solvent from injection probe; *5*, HF induction coil; *6*, glass wool; *7*, ferromagnetic conductor and sample; *8*, steel capillary, 0.5 mm i.d. (open tubular column or inlet to analytical column); *9*, insulation; *10*, carrier gas inlet; *11*, glass capillary (1.7 mm i.d., 3.5 mm o.d.).

pyrolysis temperature. The weighed samples, in ceramic boats, are moved one at a time into the hot zone. After pyrolysis the boat is withdrawn. The pyrolysis products are carried from the pyrolysis unit to the gas chromatograph by means of the open gas sampling valve. Any residue left in a boat can be recovered for weighing or analysis.

Figure 3-8 is a design used by Honaker and Horton (14). The pyrolyzer is essentially a quartz tube heated by a small tube furnace, a design similar to that shown in Figure 3-7. In this design the unit is attached directly to the sample port of the gas chromatograph.

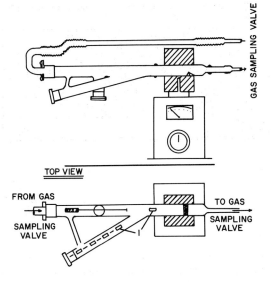

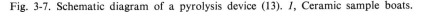

Fig. 3-7. Schematic diagram of a pyrolysis device (13). *1*, Ceramic sample boats.

A different microreactor design is shown in Figure 3-9 (15). The quartz tube packed with Pyrex glass beads is placed inside a heated metal block. The carrier gas passes through the tube into the gas chromatograph port. Samples, either liquid or solids dissolved in a suitable solvent, are injected by a microsyringe through a septum inlet into the pyrolysis tube.

Burke (16) obtained reproducible pyrograms of detergents (alkylbenzene sulfonates) in dilute aqueous solution by direct injection of the solutions into the hot zone of a microreactor by means of a microsyringe. Injection of small quantities of dilute solutions results in very small concentrations

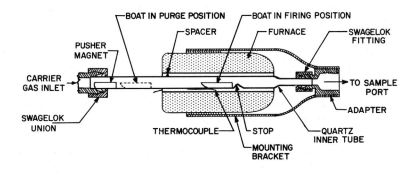

Fig. 3-8. Schematic of a pyrolyzer (14).

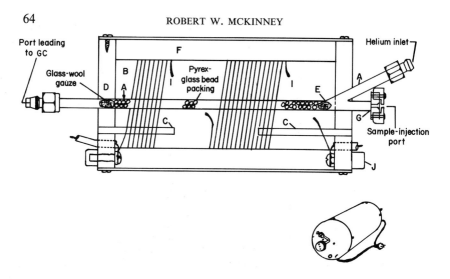

Fig. 3-9. Schematic of a pyrolysis tube and thermostat (15). *A*, Stainless steel tube; *B*, aluminum block; *C*, thermocouple well; *D*, transite; *E*, silver soldered joint; *F*, asbestos insulation; *G*, weld; *H*, silicone rubber septum strip; *I*, nichrome wire heating coil; *J*, lavite cap.

of material at any point in the hot zone and therefore a minimum of secondary reactions occur. An indentation in the quartz pyrolysis tube blocked by several folds of platinum screen eliminated the possibility of any material being injected beyond the hot zone.

C. Comparison of the Filament and Microreactor Types of Pyrolysis Systems

Both filament and microreactor systems have some advantages which should be considered before the selection of a pyrolysis unit. Here, we list them briefly; a more thorough evaluation of the relative merits of the two systems can be found in the summary (Section VI).

(*1*) Filament Systems

    (*a*) Primary pyrolysis products are rapidly swept away from the hot surfaces, minimizing secondary reactions.

    (*b*) The hot portion of the system is smaller in area than that of microreactors, providing less chance for pyrolysis of initially formed fragments.

    (*c*) Heat transfer to a sample coated on a filament is faster than for a sample in a boat moved into a microreactor hot zone.

    (*d*) Smaller samples can be pyrolyzed.

(2) Microreactors
  (a) Quantitative analyses are easier since sample weights can be determined.
  (b) The exact pyrolysis temperature can simply and accurately be measured and held constant.
  (c) Solids, liquids, and gases can be pyrolyzed.
  (d) Residues can easily be recovered for weighing and analysis.
  (e) A change in characteristics of the system due to aging is less likely.

D. Other Pyrolysis Systems

Several pyrolysis techniques which do not involve direct thermal heating have been described. Martin and Ramstad (17) have pyrolyzed materials with intense radiation from either a carbon arc source or a xenon flash tube. The reaction times ranged from 1 sec to 1 msec. Significantly different pyrograms were obtained for cellulose pyrolysis products with "flash" pyrolysis induced by intense radiant energy for a millisecond duration compared to products obtained for slower, lower temperature pyrolysis resulting from less intense radiation for several seconds.

Sternberg and Little (18) employed low current, high voltage discharge for the flash pyrolysis of solid samples. A high degree of reproducibility was obtained by this technique. The pyrogram obtained for a sample of polyethylene is quite similar to the polyethylene pyrogram obtained by Kolb et al. (24,25) using a conventional filament pyrolyzer. Johns and Morris (19) also used high voltage discharge for the pyrolysis of solids and liquid samples.

Barlow, Lehrle, and Robb (20) placed samples between two brass electrodes through which the energy from a large condenser was discharged. Temperatures of many thousand degrees were believed to have been attained but unfortunately, the principal products are of low molecular weight and provide little information about the sample.

An important advantage to the use of very short duration (flash) pyrolysis techniques is the possibility of greatly reducing the extent of secondary reactions of pyrolysis products.

E. The Complete Pyrolysis Gas Chromatography System

Much published PGC work stresses the design of the pyrolysis system with much less emphasis placed on the gas chromatographic system used. This can result in poor resolution of pyrolysis products with the accompanying loss of the potential offered by a well-designed pyrolysis unit. An insensitive detector results in the need to pyrolyze large samples with the resultant increase in secondary reactions. An example of a well-balanced system is described by Simon et al. (9). Microgram samples are pyrolyzed using the unit shown in Figure 3-4. A temperature-programmed open tubular (capillary) column is used for separation of the pyrolysis products. The exit of the column is connected to a mass spectrometer via a molecular separator. The total ion current monitor of the mass spectrometer is used to provide the pyrogram, and mass spectra of pyrolysis products as small as 0.1 μg can be obtained.

III. The Problem of Secondary Reactions

Minimizing secondary reactions is one of the most important factors in the design of pyrolysis experiments. With an increasing extent of secondary reactions it is more likely that the pyrogram will consist of peaks from low molecular weight fragments which provide little useful information. Furthermore, the reproducibility of pyrolyses under a given set of conditions descreases as secondary reactions increase.

The most important factor in reducing secondary reactions is the quantity of material pyrolyzed. Ideally the sample should be sufficiently small to undergo instant complete pyrolysis and the products should be swept away from any hot surfaces by the gas flow. If the pyrolysis is not rapid, fragments may react with each other or undergo further decomposition. One limitation on the reduction of sample size, of course, is the type of gas chromatography detector being used.

Simon et al. (10) found that biphenyl was produced as a secondary reaction product when sodium benzoate was pyrolyzed. Reduction of the quantity of sodium benzoate from 20 to 2 μg resulted in a reduction of the relative amount of biphenyl by a factor of about 2. On further reduction of the sample size to 0.2 μg, no biphenyl was observed and the authors concluded a thousandfold decrease in the relative biphenyl concentration had occurred. Burke (16) pyrolyzed samples of sodium dodecylbenzenesulfonate ranging in weight from 0.2 to 2 mg in a boat–microreactor system. Both the pyrograms and the amount of residue remaining after pyrolysis varied with sample weight. With an increase in sample weight the higher molecular weight peaks on the pyrograms increased and the residue increased. However, an aqueous solution placed in a boat and evaporated to a thin deposit of approximately 0.1 mg gave reproducible pyrolyses with minimum residue.

Pyrolysis in a static system leads to a high degree of secondary reaction since the initial products remain near a hot filament or in the hot zone of a microreactor and continue to undergo reactions. Burke (16) investigated the effect of flow rate on the pyrograms of a detergent. The flow was varied from 0 to 80 ml/min. With the carrier gas off (zero flow rate), the residue after pyrolysis was abnormally large and peaks appeared on pyrograms which were not observed with a flowing system. Increasing the flow rate to 30 ml/min caused the high molecular weight peaks to disappear. Further increase in flow rate from 30 to 80 ml/min did not affect the pyrograms. No differences in pyrograms were observed with different carrier gases (hydrogen, helium, nitrogen, and argon).

Barrall et al. (21) found that isobutylene production from pyrolysis of polyisobutylene increased 15% when the carrier gas flow was decreased from 35 to 20 ml/min. Production of the dimer 2,4,4-trimethyl-2-pentene decreased under the same conditions as did the peaks on the pyrogram due to higher molecular weight products. These results indicate that relatively high molecular weight fragments can be swept out of the pyrolysis area before pyrolysis is complete by too great a gas flow.

The pyrolysis temperature will also have some effect on the extent of secondary reactions. As the temperature increases, collisions of initially formed fragments or radicals will increase and lead to further (secondary) reactions. It is generally necessary to vary the pyrolysis temperature to find the optimum for each material, that is, the temperature which is high enough to cause rapid pyrolysis with the formation of useful products but not so high that low molecular weight products of little interest (methane, acetylene, carbon monoxide, etc.) are formed. The temperature for maximum production of isobutylene, for example, was found to be different for each of four different polybutenes (21).

IV. Identification of the Pyrolysis Products

Identification of volatile pyrolysis products is accomplished by the usual techniques used in gas chromatography. Some techniques not commonly used are worth mentioning.

A technique for determination of the molecular weight of pyrolysis products was described by Parsons (22). Using a gas density detector and carrier gases with a range of molecular weights, the molecular weights of polymer pyrolysis products were bracketed. Positive or negative peaks are obtained depending on whether the unknown has a higher or lower molecular weight than the carrier gas. Fluorinated and chlorinated hydrocarbon gases with molecular weights up to 200 can be used as carrier gases. A standard is added to the pyrolysate and two carrier gases are used. The molecular weight of a given pyrolysis product can then be calculated by the following equation:

$$\frac{A_x\,(M_s\,-\,M_1)}{A_s\,(M_x\,-\,M_1)} = \frac{A'_x\,(M_s\,-\,M_2)}{A'_s\,(M_x\,-\,M_2)}$$

where M_x is the molecular weight of the unknown and M_s that of the standard; M_1 and M_2 are the respective molecular weights of carrier gases 1 and 2; A_x and A'_x are the peak areas of the unknown, while A_s and A'_s are the peak areas of the standard.

It is frequently a possibility that relatively high molecular weight pyrolysis products may be sufficiently volatile to be carried out of the pyrolysis unit yet not be sufficiently volatile to appear on a pyrogram. Thus, pyrolysis products, particularly from polymer samples, may not be detected either as peaks on a pyrogram or as a residue left after pyrolysis. Rogers (23) used thin-layer chromatography for separation and analysis of such high molecular weight compounds. An activated thin-layer chromatography plate was held within 1 mm of the end of the pyrolysis tube and drawn past the orifice as the pyrolysis temperature increased.

Pyrolysis of materials containing alkyl groups generally results in the formation of olefins. Because of the number of olefin isomers which may be produced and the difficulty in obtaining samples of the various olefins, the pyrolysis products may be hydrogenated by inserting a suitable catalyst in line before the analytical column and using hydrogen as the carrier gas (7,24–26). Figure 3-10 shows the schematic of a system used for pyrolysis

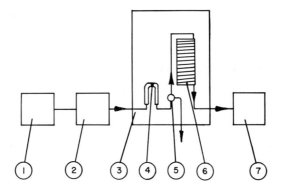

Fig. 3-10. Schematic of the instrumental setup used for hydrogenating pyrolysis gas chromatography, employing open tubular columns (25). *1*, Carrier gas supply; *2*, pyrolysis unit; *3*, oven; *4*, hydrogenation column; *5*, split; *6*, chromatographic separation column; *7*, flame ionization detector.

of polyethylene and hydrogenation of the unsaturated products using a 5% palladium-on-kieselguhr catalyst. For ease of handling, the catalyst was mixed with Celite and then coated with SE-30 in order to avoid adsorption of the higher boiling compounds.

Determination of the weight of the total pyrolysis products and the residue, if any, from a weighed sample can be accomplished by pyrolysis of a sample in a boat in a microreactor. The residue left in the boat can be weighed and the total pyrolysis product can be collected for weighing in a weighed Molecular Sieve 5A trap cooled in liquid nitrogen (21,27).

V. Applications of Pyrolysis Gas Chromatography

A. Miscellaneous Simple Molecules

It is often possible to obtain markedly different pyrograms from molecules whose structures differ only slightly or even from different isomers. Figure 3-11 shows the pyrograms obtained by Janák (28) for two similar barbituric acid derivatives. Jennings and Dimick (4) obtained unique

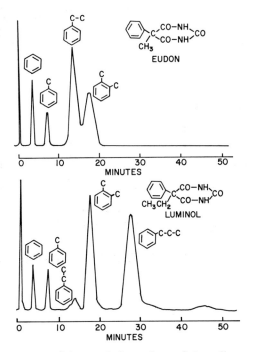

Fig. 3-11. Chromatogram of the pyrolytic products of the sodium salts of barbituric acid phenyl derivatives in the area of aromatic hydrocarbon fragments (28).

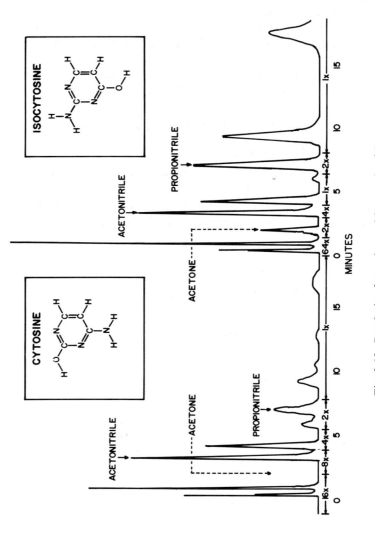

Fig. 3-12. Pyrolysis of cytosine and isocytosine (4).

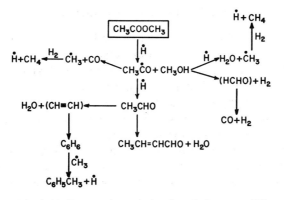

Fig. 3-13. Proposed pyrolysis of methyl acetate (29).

pyrograms from pyrolysis of cytosine and isocytosine (Fig. 3-12). Wolf and Rosie (29) undertook the exploration of the relationship between temperature and thermal decomposition for typical organic groupings by investigating the pyrolysis of a selection of 20 simple organic compounds. Simple molecules were pyrolyzed at a series of temperatures and identification of all pyrolysis products achieved by use of a number of gas chromatography columns. A computer was used to calculate molar percentages of each product and to present these as a function of temperature. With this information for methyl acetate, mechanisms were proposed to account for the pyrolysis products as indicated in Figure 3-13.

In their investigation of the 20 simple molecules it was found that at temperatures above 800°C the products consisted mainly of hydrogen, methane, carbon monoxide, and water, except in the case of benzene, which decomposed only slightly. The greatest number of peaks was produced by pyrolysis at 650–850°C. The decomposition ranges of the different functional groups were in most cases overlapping; thus, in a multifunctional organic molecule, decomposition of only one type of functionality by suitable selection of temperature did not appear to be a practical analytical method.

Even though unique pyrograms may be obtained for similar structures, it is usually necessary to have some knowledge of the type samples being pyrolyzed before determination of structural characteristics or identification are possible. For the simplest technique of comparison of pyrogram fingerprints, the need for additional information regarding the type of sample is obvious. Even if all the pyrolysis products are identified it will still generally be necessary to have some additional knowledge of the pyrolyzed material.

B. Pharmaceutical Compounds

Janák (28) showed that pyrograms of the hydrocarbons from pyrolysis of 14 barbiturates were qualitatively and quantitatively different and could be related to the particular barbiturate structure. Kirk and Nelson (30,31) have carried out a detailed study of a number of barbituric acids. They have shown that the pyrogram fingerprints could be used to differentiate 27 barbituric acids. Figure 3-14 shows the retention times (log scale) and relative peak heights of the pyrolysis products. The identification of the pyrolysis products disclosed the formation of nitriles characteristic of the particular barbiturate structure. For six of seven 5,5-substituted barbiturates, the characteristic nitriles were major pyrolysis products (Table 3-1).

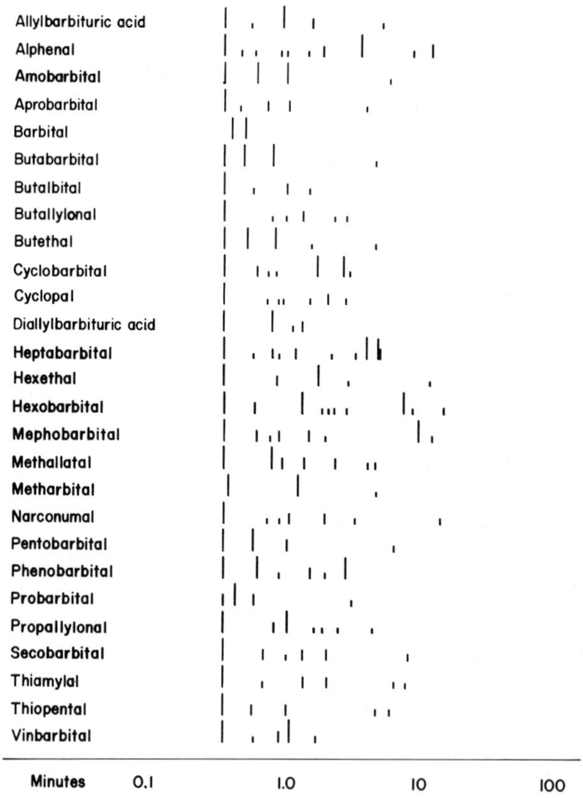

Fig. 3-14. Results of chromatographing at 150°C pyrolysis products of sodium salts of barbituric acids (30).

TABLE 3-1

Barbiturates Studied and Their Major Pyrolysis Products (31)

Barbiturate pyrolyzed	5,5-Substituent	Major pyrolyzate	Identification
Na barbital	Ethyl, ethyl	2-Ethylbutanenitrile	Firm
Na butethal	Ethyl, n-butyl	2-Ethylhexanenitrile	Firm
Na hexethal	Ethyl, n-hexyl	2-Ethyloctanenitrile	Virtually certain
Na probarbital	Ethyl, isopropyl	5-methyl-2-ethyl-butanenitrile	Reasonably certain
Na amobarbital	Ethyl, 3-methyl-butyl	5-Methyl-2-ethyl-hexanenitrile	Probable
Na pentobarbital	Ethyl, 1-methyl-butyl	3-Methyl-2-ethyl-hexanenitrile	Probable
Na metharbital	1-Methyl-5,5-diethyl	(1) Methyl isocyanate (2) Still unknown	Probable

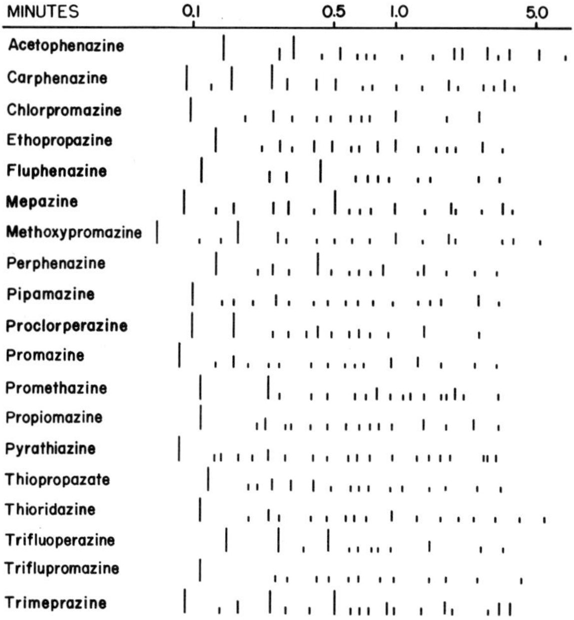

Fig. 3-15. Retention times of the pyrolysis products of 19 phenothiazines relative to the retention time of the major pyrolyzate of sodium pentobarbital (32).

The following reaction shows the relationship of the nitriles to the barbiturates:

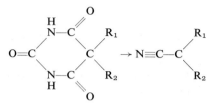

PGC of phenothizines also provided pyrogram fingerprints which could be used for identification (32). Figure 3-15 shows the logarithmic representation of the relative retention values of the pyrolysis products of 19 phenothiazines. Gudzinowicz et al. (33) identified the pyrolysis products of chlorpromazine, chlorpromazine-*S*-oxide and chlorpromazine-*N*-oxide and, based on this information, proposed thermal decomposition mechanisms. Brodasky (34) used PGC as a means of differentiating a variety of antibiotics and for characterization of lincomycins (structure **3-1**). The R

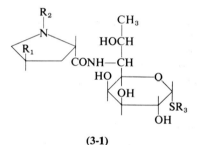

(3-1)

groups can be identified by PGC since the pyrolysis products include HSR_3 (mercaptans) and substituted pyrrols (structure **3-2**) which can be identified.

(3-2)

Acceptable quantitative values were obtained for determination of the co-produced 4-ethyl analog of lincomycin in lincomycin by measurement of the 3-ethylpyrrol formed on pyrolysis:

	R₁	R₂	R₃
Lincomycin	propyl	methyl	methyl
4-Ethyl analog	ethyl	methyl	methyl

Stanford (35) pyrolyzed 5–30 µg quantities of atropine and bilirubin. Straight lines were obtained by plotting sample weight pyrolyzed vs. heights or areas of selected peaks.

C. Organic Phosphates and Thiophosphates

Zinc dialkyldithiophosphates are commonly used additives in lubricants. These compounds have the general structure shown by structure **3-3**. They

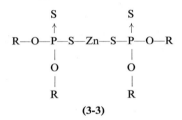

(3-3)

are resistant to hydrolysis so that the alcohols from which they were made cannot readily be regenerated for identification. Poor thermal stability prevents determination of parent peaks in the mass spectrometer. It has been shown that PGC can provide considerable information related to the alkyl groups (36–38). On pyrolysis the C—O bond is broken with the formation of olefins characteristic of the alkyl groups. For example, in the case of zinc di(4-methyl-2-pentyl)dithiophosphate, the olefins 4-methyl-pentene-1 and 4-methylpentene-2 are formed by the following reaction:

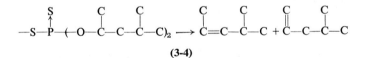

(3-4)

The olefins result from the initial formation of a carbonium ion followed by elimination of a proton from a beta carbon atom with no skeletal isomerization. When protons on two beta carbon atoms are available, appreciable quantities of more than one olefin are produced. However, if no proton is available on a beta carbon atom, olefins are formed by skeletal rearrangement as in the case of neopentyl groups. Results obtained by Legate and Burnham (36) for some model phosphate and thiophosphate compounds are shown in Table 3-2.

TABLE 3-2
Pyrolysis of Model Phosphate and Thiophosphate Compounds (36)

Compound	Principal volatile products
Pb salt of *O,O*-di-*n*-amyl thionothiophosphate	1-Pentene
Zn salt of *O,O*-di-*n*-dodecyl thionothiophosphate	1-Dodecene
Zn salt of *O,O*-dineopentyl thionothiophosphate	2-Methyl-1-butene; 2-methyl-2-butene; C$_4$ and lighter olefins
K salt of *O,O*-di(4-methyl-2-pentyl) thionothiophosphate	*cis* and *trans* 4-Methyl-2-pentene; 4-methyl-1-pentene; 2-methyl-2-pentene
Zn salt of *O,O*-dicyclohexyl thionothiophosphate	cyclohexene
2-Ethylhexyl diphenyl phosphate	2-Ethyl-1-hexene; phenol

D. Total Oxygenates in Organic Materials

Meade et al. (39) have perfected a PGC procedure for determination of total oxygenates in organic materials. Total oxygenate values ranging from 5 ppm (in kerosene) to 26% (benzoic acid) were reported. Samples of gases, solids, and volatile or nonvolatile liquids were pyrolyzed over carbon at 1050°C. Under these conditions oxygenated organic compounds form methane, hydrogen, and carbon monoxide. This can be measured using a Molecular Sieve 5A column and thermal conductivity detector. Organic compounds containing halogens, boron, nitrogen, sulfur, and phosphorus were analyzed without interference.

E. Polymers

By far the greatest application of PGC has been the study of polymers including homopolymers, copolymers, and polymer mixtures. A number of polymers can be pyrolyzed to give predominantly one product. Poly-(methyl methacrylate), poly(ethyl methacrylate), and polystyrene "unzip" to give monomer as the major pyrolysis product. Polyvinyl chloride produces hydrogen chloride as a major product. In all cases it is necessary to determine experimentally the pyrolysis temperature at which the maximum yield of the major product is obtained.

Lehrle and Robb (5) have done quantitative analyses of copolymers and polymer mixtures by PGC. Homopolymer samples of known weight were

TABLE 3-3

Principal Degradation Products as a Function of Temperature (5)

No.	Polymer	Principal Pyrolysis products	Pyrolysis temperature, °C								
			150	250	350	450	550	650	750	850	950
1	PMMA	MMA	—	4	9	62	100	5	—	—	—
		Acetylene	—	—	—	—	—	1	2	4	1
2	PVC	HCl	—	—	—	15	100	—	—	—	—
		Acetylene	—	—	—	—	6	9	14	15	25
3	PMMA–PVC, 20%–80% mixture	MMA	—	—	1	3	8	—	—	—	—
		HCl	—	—	—	82	100	—	—	—	—
4	MMA–VC, 30%–70% random copolymer	MMA	—	—	—	—	6	—	—	—	—
		HCl	—	29	97	100	69	1	—	—	—
5	MMA–VC, 20%–80% block copolymer	MMA	—	—	—	1	1	—	—	—	—
		HCl	—	—	—	50	100	—	—	—	—

For each sample, the largest peak obtained is taken as 100% and the peaks for the principal pyrolysis products are given as percent of the largest peak.

MMA = methyl methacrylate; VC = vinyl chloride; PMMA = poly(methyl methacrylate); PVC = poly(vinyl chloride).

77

TABLE 3-4A

Percentage of Vinyl Chloride in Vinyl Chloride–Vinyl Acetate Copolymers (5)

Copolymer	By chlorine estimation (mean of 2 results)	By IR analysis ($\pm 1\%$)	By degradation ($\pm 2\%$)
049	60.8 ± 0.6	54.7	55.8
047	69.4 ± 0.1	64.4	65.2
075	74.1 ± 0.3	72.3	72.2
076	69.1 ± 0.9	66.7	67.8
R46/82	81.8 ± 4.4	84.8	83.9
R51/83	87.9 ± 1.0	89.0	87.7

pyrolyzed at the selected temperature and a calibration curve obtained by plotting the area of the major peak vs. sample weight. The calibration curves were used for analysis of copolymers and polymer mixtures. Table 3-3 shows the principal products as a function of temperature for five polymer samples. Generally there is little, if any, difference between the pyrograms of homopolymer mixtures and copolymers. There are differences, however, for the poly(methyl methacrylate)–poly(vinyl chloride) system, as can be seen in Table 3-3. In such cases calibration must be done with copolymers of known composition. Tables 3-4A–3-4C show the results obtained for copolymers and polymer mixtures by PGC.

Most polymers do not yield monomer or a single major product upon pyrolysis. However, useful information on pyrolysis mechanisms and polymer structure can be obtained even from complex pyrograms and for complex samples.

TABLE 3-4B

Percentage of Methyl Methacrylate in Styrene–Methyl Methacrylate Copolymers

Analysis of polymer by radioactive assay[a]	By degradation
7.2	7.6 ± 0.2
14.5	13.1 ± 0.3
48.4	46.8 ± 0.9

[a] Polymer synthesized with labeled methyl methacrylate units.

TABLE 3-4C

Analyses of Polymer Mixtures

Polymer A	Polymer B	Percentage of A from original weight taken	Percentage of A by degradation
Vinyl chloride	Methyl methacrylate	80.0	81.7 ± 1.4
Methyl methacrylate	Styrene	50.0	50.8 ± 1.8
Methyl methacrylate	Ethyl methacrylate	75.0	74.3 ± 1.2
Methyl methacrylate	Ethyl methacrylate	51.0	52.2 ± 1.2
Methyl methacrylate	Ethyl methacrylate	30.0	29.7 ± 0.6
Methyl methacrylate	Ethyl methacrylate	25.0	24.5 ± 1.4

Figure 3-16 is a pyrogram of polyethylene which consists of a series of groups of three peaks with smaller peaks between the groups of three (25, 40). Addition of a precolumn containing a hydrogenation catalyst (5% palladium on kieselguhr mixed with Celite) and use of hydrogen as carrier gas resulted in the pyrogram shown in Figure 3-17. Two peaks in each triplet disappeared, leaving only the first peak in each original triplet.

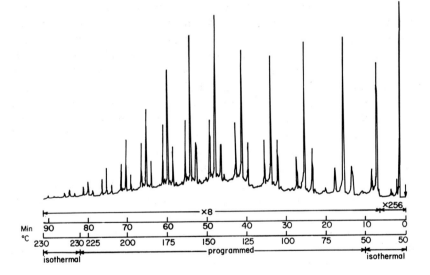

Fig. 3-16. Pyrogram obtained from the pyrolysis of polyethylene (density: 0.960) (25). Column: 25 m × 0.25 mm i.d. open tubular, coated with OS-138 poly(phenyl ether) liquid phase. Column temperature: programmed as given. Carrier gas (He) flow rate at outlet: 1.5 ml/min. Split ratio: 1/50. The peaks in one triplet are: *1*, *n*-paraffin. *2*, α-olefin. *3*, α,ω-diolefin.

ROBERT W. MCKINNEY

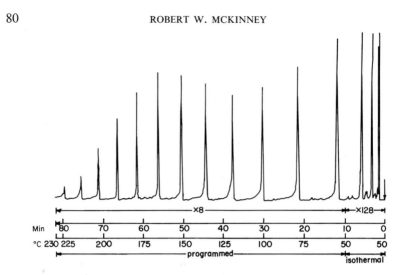

Fig. 3-17. Pyrogram obtained from the pyrolysis of polyethylene (density: 0.960) with subsequent hydrogenation of the thermal breakdown products (25). Conditions same as for Figure 3.16 except hydrogen used as carrier gas.

Therefore the hydrogenation products of the other two peaks are identical to the first peak. These results indicate each triplet consists of an n-paraffin plus two n-olefins with the same carbon number as the n-paraffin. Since the olefins were produced by pyrolysis, it was assumed the double bonds are located at the ends of the molecules and the peaks would be due to α-olefins and α,ω-olefins. These assumptions were verified by addition of n-paraffins and α-olefins to a polyethylene sample before pyrolysis and observation of the relative increase in peaks. Also, the C_{12} triplet was collected from repeated pyrolyses in a solution of potassium permanganate in glacial acetic acid. Esterification produced methyl undecanoate and dimethyl sebacate from oxidized n-dodecene-1 and n-dodecadiene-1,11. This information was used to propose a mechanism for the thermal decomposition of polyethylene. The small peaks between the triplets were assumed to be from the branching in the polyethylene. Evidence for this was the lack of these peaks in the pyrogram of polymethylene.

A PGC polyethylene characterization system has been developed based on indices derived from the C_1–C_6 portions of polyethylene pyrograms (41). The two indices, BI_2 and BI_2, are defined as

$$BI_1 = \frac{\text{butadiene peak height}}{\text{butenes peak height}}$$

$$BI_2 = \frac{\text{propylene + pentenes peak heights}}{\text{butenes peak height}}$$

TABLE 3-5

Analysis of Blends with Various Content of
High and Low Density Polyethylene (41)

| Blend No. | Composition, % | |
	Prepared ($\pm 1\%$)	Found
1 XHB–48	20	19.8
Marlex	80	80.2
2 XHB–48	30	28.6
Marlex	70	71.4
3 XHB–48	50	50.5
Marlex	50	49.5
4 XHB–48	70	68.1
Marlex	30	31.9

An equation was derived for determination of the percentage of low density polyethylene in a blend of low and high density polyethylenes using the BI_1 values for each component of the blend. An empirical equation was also derived for determination of the degree of crystallinity using BI_2 values. Tables 3-5 and 3-6 show results for analysis of high and low density polyethylene mixtures and determination of percent crystallinity.

The analysis of phenol–formaldehyde polycondensates is difficult by any conventional analytical techniques. Recent work on these complex materials indicates that a PGC analytical system might eventually be

TABLE 3-6

Comparison of Crystallinity Degree Found on the Basis
of Gas Chromatography, NMR, and Density Data (41)

| Sample | Crystallinity degree in % determined by | |
	GC	Other methods
Hostalen	84.5	89.4 (NMR)
Marlex	83.0	87.0 (NMR)
Okiten F–12	52.6	52.5 (density)
Alkathene XHB–48	51.9	51.5 (density)
Okiten WNC–18	48.4	48.5 (density)
Polythene G–03	49.1	49.1 (density)

developed (42,43). Most of the bridges between aromatic rings are —CH_2—
groups which are bonded in *ortho* and *para* positions with respect to the
phenolic —OH group. Thus pyrolysis of a resin prepared from formaldehyde and phenol should produce phenols with methyl groups in *ortho* and
para positions. The phenolic pyrolysis products found experimentally were
phenol and, in decreasing amounts, 2- and 4-methylphenols, 2,4- and 2,6-
dimethylphenols, and 2,4,6-trimethylphenol. The presence of phenols with
methyl groups in the 3 or 5 positions would indicate such phenols were
present in the original synthesis materials.

Resins were prepared using phenol, 3-methylphenol, 3,5-dimethyl-
phenol, and mixtures of the three. The phenols observed on the pyrograms
were divided into three groups: those with no methyl in position 3 or 5,
those with one methyl in position 3 or 5, and those with two methyl groups
in positions 3 and 5. The total phenols corresponding to each group was
used to calculate the mole percent composition of the phenols used to
prepare the resins. Table 3-7 shows the results obtained for the analysis of
seven resins.

Vinyl coating resins modified with other resins and plasticizers have been
quantitatively analyzed by PGC using an internal standard (44). A solution
of the polymer to be used as the internal standard was weighed with a known
amount of sample and diluted with dioxane. The ethyl methacrylate polymer used as the internal standard when pyrolyzed alone yielded 98% ethyl
methacrylate and 2% methyl methacrylate. Results of the analysis of a
number of coatings are shown in Table 3-8. The results obtained using an
internal standard are compared to those calculated as area percent.

TABLE 3-7

Relation Between the Composition of the Phenolic Mixture Used to Prepare
the Resins and the Composition of Their Pyrolysis Products (42)

Compound	A	B	C	D	E	F	G
Mole % in the synthesis products:							
Phenol	100	—	—	33.3	50.0	33.3	16.7
3-Methylphenol	—	100	—	33.3	33.3	16.7	50.0
3,5-Dimethylphenol	—	—	100	33.3	16.7	50.0	33.3
Mole % of the pyrolysis products, characteristic of:							
Phenol	100	4.7	3.6	32.4	48.4	29.6	16.9
3-Methylphenol	trace	93.8	4.6	34.4	34.4	21.1	47.0
3,5-Dimethylphenol	—	1.8	91.8	33.2	17.2	49.3	36.1

TABLE 3-8

Semiquantitative Analysis of Methacrylate and Styrene
Polymers in Organic Coatings (44)

| System | Present, % | Found, % | |
		Internal standard	Total area
Poly(butyl methacrylate)–poly(vinyl chloride)–dioctyl phthalate	45.8	43.0; 44.8; 43.3	76.9
Poly(butyl methacrylate)–nitro-cellulose–butyl phthalate	64.3	62.7; 59.7; 62.9	74.1
Poly(methyl methacrylate)–butyl benzyl phthalate	68.8	69.6; 69.2	95.7
Polystyrene–rosin ester–tricresyl phosphate	33.8	36.0; 36.3	85.7
Styrenated alkyd	44.0	44.8; 43.2; 43.7	61.3
Styrenated alkyd	35.1	33.1; 31.0; 33.9	59.7

F. Microorganisms

Identification of microorganisms by morphological, biochemical, and serological techniques is time consuming. PGC is being evaluated as a possible rapid means of classifying and possibly identifying microorganisms (45–48). The reproducibility of pyrograms of dried microorganism

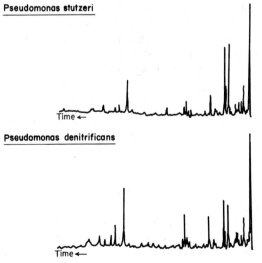

Fig. 3-18. Pyrograms of microorganisms (48).

samples has been demonstrated. Although the pyrograms are complex, those of different microorganisms differ primarily in peak height or area ratios, not in the presence or absence of peaks (Fig. 3-18). Valid comparisons may require not only carefully controlled PGC conditions but also careful control of the media in which microorganisms are grown (48). Development of a reliable system for classification and identification of microorganisms will probably require only the evaluation of the relative differences in the major peaks by computerized data-handling techniques and comparison of PGC patterns of standards with unknown samples.

G. Pyrolysis of Gas Chromatograph Eluants

Levy and Paul (49) used PGC as a means of identification of effluent compounds from a gas chromatograph. Peaks of interest eluted from a gas chromatograph with a nondestructive (thermal conductivity) detector were passed into a flow-through quartz tube pyrolyzer and the pyrolysis products analyzed in a second gas chromatograph. A standard set of conditions was used to obtain reproducible "cracking patterns."

To obtain the relationship between the cracking pattern and molecular structure, several families of compounds were run. These included normal alkanes, straight-chain alpha olefins, mercaptans, alcohols, and saturated and unsaturated methyl esters. Considerable information relating molecular structure and pyrolysis mechanisms and products was obtained. The pyrolysis mechanism for paraffins was found to be explained best by a modified free radical theory of Rice (50). Good agreement was obtained between the theoretically calculated pyrolysis products and the experimental results obtained for hexadecane.

Fanter, Walker, and Wolf (51, 52) used the same technique with a 36-in. gold coil between chromatographs as the pyrolysis unit. They also employed stop–start (interrupted) elution of peaks from the first gas chromatograph to allow time for analysis of pyrolysis products in the second instrument. Flow interruption for up to 3 hr was possible without effecting separation.

VI. Summary

Pyrolysis gas chromatography is growing rapidly in both technique and application. A wide variety of pyrolysis unit designs have been described in the literature, most of which are either microreactor or filament (electrical or RF-heated) types. Considerable development in unit design can be expected in the future to provide instantaneous pyrolysis with a minimum of secondary reactions. Standardization of designs to a few types for uniform reproduction of PGC experiments and data between laboratories does not appear likely to occur during this period of rapid growth.

PGC applications at present are primarily qualitative. The use of computerized techniques for handling PGC data could greatly increase the speed and applicability for identification by comparison of pyrogram fingerprints with a library of standard fingerprints. Quantitative analytical data can also be obtained and the use of PGC for quantitative analysis should increase rapidly.

The factors to consider in construction or purchase of a pyrolysis unit are cost, the types of materials to be examined, the sample sizes available, the type of data required, and the gas chromatography equipment available for use with the pyrolysis unit.

Simple electrically heated filament units can be constructed at very low cost. The filament can be placed directly in the carrier gas stream of a gas chromatograph. RF-heated filament systems and microreactors require more expensive components. The internal volume of a microreactor may be such that the necessary gas flow through the system is too great to be allowed to flow directly into a gas chromatograph. In such cases, a splitting system is necessary.

The types of materials which may be pyrolyzed with filament systems is limited primarily to materials which can be dissolved or suspended and coated on the filament or solids which are placed within a filament coil or spiral or in a container held in a filament coil. Solids, liquids, solutions, and gases can all be pyrolyzed in microreactors. Furthermore, samples such as coated glass filaments which require that a large volume of sample be pyrolyzed require the use of a microreactor.

Very small samples are more readily handled with filament systems where small volumes of dilute solutions can be coated on a wire or ribbon and pyrolyzed directly in the carrier gas stream of the gas chromatograph.

If only fingerprint data are required, a simple filament system is adequate. However, if accurate knowledge of the pyrolysis temperature is required or if pyrolyses are to be done at a series of known temperatures, a microreactor is more suitable. Filaments appear to be subject to aging which results in a change in temperature with use—a problem not encountered with microreactors. The pyrolysis of known sample weights and recovery of residues for weighing and/or analysis can conveniently be done only with microreactors.

The gas chromatography equipment available for examination of pyrolysis products may impose a limitation on the type of pyrolysis system used. The products from pyrolysis of microgram size samples would best be analyzed using a gas chromatograph equipped with an ionization detector. Samples which produce very complex pyrograms may require open tubular columns for adequate resolution of pyrolysis products.

References

1. R. W. McKinney, *J. Gas Chromatog.*, **2**, 432 (1964).
2. R. L. Levy, *Chromatog. Rev.*, **8**, 49 (1966).
3. C. R. Kingston and P. L. Kirk, *Bull. Narcotics*, **17**, 19 (1965).
4. E. C. Jennings, Jr. and K. P. Dimick, *Anal. Chem.*, **34**, 1543 (1962).
5. R. S. Lehrle and J. C. Robb, *J. Gas Chromatog.*, **5**, 89 (1967).
6. C. E. R. Jones and A. F. Moyles, *Nature*, **191**, 663 (1961).
7. M. Dimbat and F. T. Eggertsen, *Microchem. J.*, **9**, 500 (1965).
8. F. G. Stanford, *Analyst*, **90**, 266 (1965).
9. J. Vollmin, P. Kriemler, I. Omura, J. Seibl, and W. Simon, *Microchem. J.*, **11**, 73 (1966).
10. W. Simon, P. Kreimler, J. A. Vollmin, and H. Steiner, *J. Gas Chromatog.*, **5**, 53 (1967).
11. H. Giacobbo and W. Simon, *Pharm. Acta Helv.*, **39**, 162 (1964).
12. E. A. Raddell and H. C. Strutz, *Anal. Chem.*, **31**, 1890 (1959).
13. K. Ettre and P. F. Váradi, *Anal. Chem.*, **35**, 69 (1963).
14. C. B. Honaker and A. D. Horton, *J. Gas Chromatog.*, **3**, 396 (1965).
15. G. G. Smith, W. H. Wetzel, and B. Kosters, *Analyst*, **86**, 480 (1961).
16. M. F. Burke, Ph.D. Thesis, Virginia Polytechnic Institute, 1966, University Microfilms Inc., Ann Arbor, Michigan.
17. S. B. Martin and R. W. Ramstad, *Anal. Chem.*, **33**, 982 (1961).
18. J. C. Sternberg and R. L. Little, *Anal. Chem.*, **38**, 321 (1966).
19. T. Johns and R. A. Morris, *Develop. Appl. Spectry.*, **4**, 361 (1965).
20. A. Barlow, R. S. Lehrle, and J. C. Robb, *Polymer*, **2**, 27 (1961).
21. E. M. Barrall II, R. S. Porter, and J. F. Johnson, *J. Chromatog.*, **11**, 177 (1963).
22. J. S. Parsons, *Anal. Chem.*, **36**, 1849 (1964).
23. R. N. Rogers, *Anal. Chem.*, **39**, 731 (1967).
24. B. Kolb and K. H. Kaiser, *J. Gas Chromatog.*, **2**, 233 (1964).
25. B. Kolb, G. Kemmner, K. H. Kaiser, E. W. Cieplinski, and L. S. Ettre, *Z. Anal. Chem.*, **209**, 302 (1965).
26. P. W. O. Wijga, *Z. Anal. Chem.*, **205**, 342 (1964).
27. R. S. Porter, A. S. Hoffman, and J. F. Johnson, *Anal. Chem.*, **34**, 1179 (1962).
28. J. Janák, *Nature*, **185**, 684 (1960).
29. T. Wolf and D. M. Rosie, *Anal. Chem.*, **39**, 725 (1967).
30. D. F. Nelson and P. L. Kirk, *Anal. Chem.*, **34**, 899 (1962).
31. D. F. Nelson and P. L. Kirk, *Anal. Chem.*, **36**, 875 (1964).
32. C. R. Fontan, N. C. Jain, and P. L. Kirk, *Mikrochim. Acta*, **2–4**, 326 (1964).
33. B. J. Gudzinowicz, H. F. Martin, and J. L. Driscoll, *J. Gas Chromatog.*, **2**, 265 (1964).
34. T. F. Brodasky, *J. Gas Chromatog.*, **5**, 311 (1967).
35. F. G. Stanford, *Analyst*, **90**, 266 (1965).
36. C. E. Legate and H. D. Burnham, *Anal. Chem.*, **32**, 1042 (1960).

37. W. W. Hanneman and R. S. Porter, *J. Org. Chem.*, **29**, 2996 (1964).
38. S. G. Perry, *J. Gas Chromatog.*, **2**, 54 (1964).
39. C. F. Meade, D. A. Keyworth, V. T. Brand, and J. R. Deering, *Anal. Chem.*, **39**, 512 (1967).
40. E. W. Cieplinski, L. S. Ettre, B. Kolb, and G. Kemmer, *Z. Anal. Chem.*, **205**, 357 (1964).
41. D. Deur-Siftar, *J. Gas Chromatog.*, **5**, 72 (1967).
42. J. Martinez and G. Guiochon, *J. Gas Chromatog.*, **5**, 146 (1967).
43. C. B. Honaker and A. D. Horton, *J. Gas Chromatog.*, **3**, 396 (1965).
44. G. G. Esposito, *Anal. Chem.*, **36**, 2183 (1964).
45. E. Reiner, *Nature*, **206**, 1272 (1965).
46. Anon., *Chem. Eng. News*, **43**(39), 69 (1965).
47. E. Reiner, *J. Gas Chromatog.*, **5**, 65 (1967).
48. V. I. Oyama and G. C. Carle, *J. Gas Chromatog.*, **5**, 151 (1967).
49. E. J. Levy and D. G. Paul, *J. Gas Chromatog.*, **5**, 136 (1967).
50. F. O. Rice, *Free Radical (Collected Papers of F. O. Rice)*, The Catholic University Press, Washington, D.C., 1958.
51. C. J. Wolf and J. Q. Walker, *Chem. Eng. News*, **45**(41), 47 (1967).
52. D. L. Fanter, J. Q. Walker, and C. J. Wolf, *Anal. Chem.*, **40**, 2168 (1968).

CHAPTER 4

Precolumn Reactions for Structure Determination

Morton Beroza and May N. Inscoe, *Agricultural Research Service, U. S. Department of Agriculture, Beltsville, Maryland*

I. Introduction

The widespread use of gas chromatography and its remarkable ability to separate small quantities of complex mixtures of compounds have directed research toward identification of the resulting infinitesimal amounts of compounds isolated in pure form. All too often, a chemist may process large quantities of material at great expense only to find the active ingredient appearing as one small peak on the complex chromatogram of the concentrate with no real clue as to its nature or structure. Yet, thanks to recent innovations, solutions to difficult problems of this kind are being reported almost daily, and identifications have been made of compounds that have never been seen. Spectrometric (IR, MS, NMR) analysis, prominently featured in this book, has been invaluable, both in terms of minimum amount of sample needed and of quantity of structural information derived. Retention time parameters (e.g., the retention index), detectors that indicate the presence of certain elements or groupings in a molecule, physical properties (e.g., partition coefficients), derivatization and other chemical reactions, either before, during, or after gas chromatography have been employed with much success, especially when these techniques are used together, along with other forms of chromatography.

Though much progress has been made in devising methodology and instrumentation to identify substances available in these minute amounts, this field must still be considered one of the relatively underdeveloped areas of chemistry. The rapid progress in spectrometric analysis may be ascribed mainly to the application of electronics and sophisticated means of manufacture to provide the necessary instrumentation. The use of classical chemical reactions as a means of identification, such as reactions to establish the presence of functional groups, has not kept pace with these spectrometric developments. Such reactions can provide valuable information to supplement spectrometric and other probes for structure analysis, and can frequently allow identifications to be made more rapidly and with greater certainty than can the use of more standardized techniques.

A. Scope

This chapter does not attempt to give an exhaustive array of available methodology but concentrates instead on presenting a comprehensive discussion of *selected* precolumn methodology and on citing different types of reactions useful for structure analysis. Concepts rather than details are emphasized, and currently used reactions, rather than those of historical interest, are given. Methods discussed are those designed specifically for use with gas chromatography. Thus, the normal preliminary separations for fractionating a crude extract or the preparation of a derivative to facilitate the chromatography of a compound are not included. (Pyrolysis, discussed in Chapter 3, is also excluded.) In most cases the analyst starts with a pure compound (single peak on a gas chromatogram), though some of the methods may also be used with mixtures to gain information on the nature of the compounds present.

Precolumn reactions are those that occur before the compounds or their reaction products enter the GC column. As will be noted, many of these reactions may also be used to advantage following the GC separation column. Reactions may be carried out prior to injecting the reaction mixture, or they may be conducted within the gas chromatographic pathway after injection, a process known as reaction gas chromatography. This process, the integrated version of a reaction + GC, has been the subject of several reviews (1–3); various chapters of this book—including the present one—discuss techniques which fall within the broad scope of reaction gas chromatography.

Precolumn reactions may be categorized as general—those applicable to a wide variety of compounds—or specific—those limited to determining a single functional group, and sometimes further limited to one type of compound. General reactions will be discussed first and emphasized because of their wider applicability and greater importance in structure determination.

The techniques are of greatest value when the amount of material available is too small for all but the most sensitive spectrometric probes. Reactions useful at the microgram rather than at higher levels are therefore emphasized here. Even when ample material is available, the more sensitive methodology can be used profitably because smaller amounts can be separated from a mixture more rapidly and need for preparative GC may be eliminated. Furthermore, although the methods usually work better with somewhat larger amounts of material, this is true only up to a certain point, because better separations and more accurate retention times are usually obtained with small quantities. Thus, in the use of the retention index system, gas chromatography of small amounts is recommended (4).

Techniques discussed must of necessity be weighted heavily in favor of those we have devised or used in our efforts to isolate and identify compounds found in insects or natural products. However, the field is still very much in the developmental stage, and the problems encountered in our experience are likely to be similar to those encountered in other areas.

B. General Comments on Technique

The use of minute amounts of material imposes on the analyst the need to exercise special care and cleanliness in his operations. Solvents and other reagents usually must be purer than those normally available even though they are used in smaller amounts. Technique and manipulative skill are most important, although this need for dexterity may often be minimized by the use of specially built equipment. It is unfortunate that many valuable techniques are not being fully exploited for the lack of such specialized equipment. Some proficiency in the basic skills needed for the construction of apparatus, such as silver soldering and glass blowing, can be of great value to the analyst.

C. Sample Collection

The sample to be studied by precolumn reactions is often a solute emerging from a GC column that must be collected before it can be analyzed further. Efficiency of the collection depends on quantity of substance, flow rate of carrier gas, vapor pressure of substance, speed of cooling, temperature of exit port, type of collection device, and many other parameters. Accordingly this subject can be discussed only briefly here. For more information, the reader should consult other monographs on the subject (5,6).

If adequate material (usually milligrams) is available, losses in collection can usually be tolerated and few precautions need be taken. The most commonly used collection device is a thin-wall capillary tube, which may be inserted into a GC exit port having a snugly fitting hole or a silicone rubber orifice, or it may be attached with Teflon tubing. The heated gases and exit port quickly establish a temperature gradient down the length of the tube because of the tube's low mass. Recoveries are good because gradual cooling of the effluent minimizes fog formation. If the diameter of the tube is too small, the solute may condense in the tube and be blown out; if too large, the solute may not condense adequately. Should high efficiency be required, long tubes are used with cooling (Dry Ice in a cup piled around the tube) at the distal end. Alternatively, Teflon tubing is attached at the distal end of the capillary tube and the effluent bubbled into a volatile solvent; the solvent can then be rinsed through the capillary and tubing with a hypodermic syringe and needle to recover the compound. We have used Teflon tubing (0.06 in. i.d.) with a hypodermic needle (Luer fitting, Becton-Dickinson Co.,* Rutherford, N.J.) at the inlet and a test

* Mention of a proprietary product or company does not necessarily imply endorsement of the product or company by the U. S. Department of Agriculture.

tube containing some solvent at the outlet; as each successive peak emerges from a nondestructive detector or splitter, another needle assembly is attached to the Luer fitting on the exit port.

Trapping on packing can be very efficient, even with volatile solutes at submicrogram levels. A device for trapping a solute on packing and reinjecting it into a gas chromatograph or other analytical instrument is shown in Figure 4-1 (7). It utilizes Luer fittings on the exit port H and on 2- or 4-in. stainless steel collection tubes B containing GC packing. As the solutes emerge, they are trapped individually in collection tubes (usually cooled). Each solute is then reinjected into the gas chromatograph by attaching a hypodermic needle A and syringe F as shown in Figure 4-1 and allowing the tube to be heated in an injection port heater (not shown) for 30 sec before forcing a syringeful of carrier gas (loaded from a side needle C) through the collection tube into the carrier gas stream. The device is assembled from commercially available materials by silver soldering and some machining of parts. With this device good recoveries were obtained with compounds ranging from methyl formate (b.p. 32°) to methyl eicosanoate. With a simpler similar device, employing 0.030-in. i.d. copper tubing without packing, compounds boiling as low as ethane ($-88°$C) have been trapped and transferred with good efficiency (8).

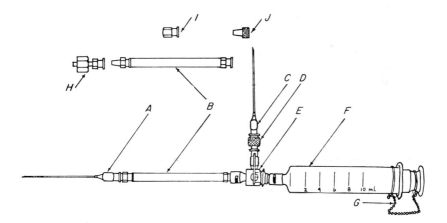

Fig. 4-1. Collection and transfer device (7). A, Hypodermic needle, 25 gauge; B, collection tube with Luer fittings containing GC packing; C, hypodermic needle, 25 gauge; D, adapter; E, three-way stopcock; F, 10-ml syringe with Luer-Lok fitting; G, chain to prevent syringe opening beyond 10-ml mark; H, Luer fitting that attaches to exit port; I and J, caps to close collection tube for storage. See reference 7 for details of construction and use.

II. Carbon-Skeleton Chromatography

Carbon-skeleton chromatography has been described in a series of papers, which have progressively expanded the utility of the method (9–14). By stripping off all functional groups, it helps determine the carbon skeleton of compounds, which may be considered the simplest starting point of any structure determination. The addition of retention indices to the retention time of the carbon skeleton, discussed in Section II-F, is potentially a valuable approach to the elucidation of chemical structure.

A. Apparatus

The general arrangement of the equipment used is shown in Figure 4-2. A precolumn containing a hot catalyst is attached to a gas chromatograph, usually one with a flame ionization detector, and the hydrogen carrier gas passes through the precolumn. Nitrogen or helium is introduced at the usual hydrogen inlet to achieve optimum response of the flame detector. When a compound is injected, the hydrogen carrier gas (20–60 ml/min) sweeps it over the catalyst (usually at 280–300°C) and in the process saturates multiple bonds and removes functional groups containing oxygen, nitrogen, sulfur, and halogens. The hydrocarbons produced flow from the precolumn into the GC column, where they are separated and may be identified by their retention times or other means. A wide variety of compounds—acids, alcohols, aldehydes, amides, amines, esters, ethers, halides, sulfides, unsaturates—have been analyzed.

The apparatus for carbon-skeleton (CS) chromatography shown in Figure 4-3 is available from the National Instruments Laboratory, Rockville, Md. The catalyst assembly, shown schematically in Figure 4-4, contains an aluminum tube with a catalyst chamber 22 cm long by 4.5 mm

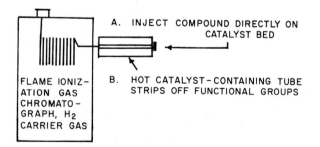

Fig. 4-2. Carbon-skeleton chromatography.

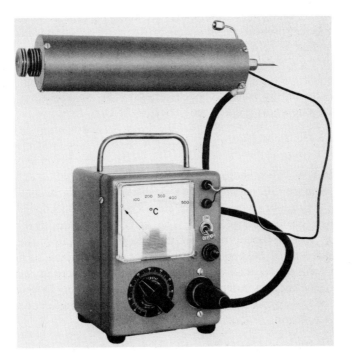

Fig. 4-3. Carbon-skeleton determinator (12).

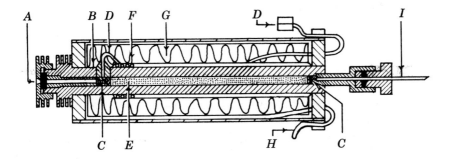

Fig. 4-4. Cross section of catalyst assembly for carbon-skeleton chromatography (2). *A*, Injection port (fins keep septum cool); *B*, aluminum tube; *C*, glass wool; *D*, hydrogen inlet (wrapped around F); *E*, catalyst; *F*, heating jacket; *G*, insulation; *H*, heating jacket electric cord. Needle stock, *I*, attaches assembly to gas chromatographic injection port.

in diameter. A heater surrounding the tube keeps the catalyst at the desired temperature and preheats the incoming carrier gas. The heater is surrounded with insulation to help maintain a constant catalyst temperature. A larger version of the CS determinator, meant for use with thermal conductivity detectors for micropreparative purposes, has a bore diameter twice as large and holds four times as much catalyst (13). An apparatus that fits into the injection port is described in Section II-D.

B. Catalyst

The catalyst, 1% palladium (Pd) on a support, is prepared in the same way as ordinary GC packing. $PdCl_2$ is dissolved in 5% acetic acid and added to the support, usually 60/80 mesh GasChrom P (Applied Science Laboratory, State College, Pa.), with sufficient sodium carbonate to neutralize the HCl that forms when the catalyst is activated; the mixture is evaporated to dryness in a rotating evaporator. (Failure to neutralize the HCl will cause cracking of carbon–carbon bonds in subsequent analyses.) The catalyst is activated in the catalyst assembly by heating it in a stream of hydrogen ($\frac{1}{2}$ hr each at 125°C, 200°C, and reaction temperature). The following reaction occurs:

$$PdCl_2 + H_2 \longrightarrow Pd + HCl \text{ (neutralized with } Na_2CO_3) \tag{1}$$

For analysis of carboxylic acids or their derivatives (esters, amides), Chromosorb P (Johns-Manville, New York, N.Y.) is the preferred catalyst support; precautions necessary with it will be discussed later. In the analysis of amines, the support should be made slightly alkaline with additional sodium carbonate.

C. Reactions

Three reactions may occur in carbon-skeleton chromatography:
1. Hydrogenation: the addition of hydrogen to multiple bonds
2. Hydrogenolysis: the breaking of a bond and the addition of hydrogen to the points of cleavage
3. Dehydrogenation: hydrogen abstraction, usually from cyclohexane rings to give aromatic structures

1. Typical Analyses

With the most favorable conditions (catalyst temperature 300°C, flow rate 20 ml/min) the following reactions are typical.

Carbon–halogen bonds are cleaved to give the parent compound.

$$CH_3CH_2CH_2CH_2CH_2 \vdots Br \longrightarrow CH_3CH_2CH_2CH_2CH_3 \tag{2}$$

(Cleavage shown by dotted line.)

When either pentane or pentyl bromide is injected, the retention time of the peak is the same, showing that the reaction is practically instantaneous. HBr, which is formed in the reaction, does not register on a flame ionization detector, nor do other inorganic products such as H_2O, CO_2, CO, H_2S, HCl, HI, NH_3. If there were more halogen substituents on the compound, the product would still be pentane, i.e., the carbon skeleton.

Carbon-sulfide bonds are readily cleaved to give the parent hydrocarbon:

$$CH_3(CH_2)_{10}CH_2 \vdots SH \longrightarrow CH_3(CH_2)_{10}CH_3 \tag{3}$$

With thiophene two reactions occur: cleavage of the carbon–sulfur bonds, and saturation of the double bonds:

$$\begin{array}{c} HC\!\!-\!\!-\!\!CH \\ \parallel \qquad \parallel \\ HC \diagdown \diagup CH \\ S \end{array} \longrightarrow CH_3CH_2CH_2CH_3 \tag{4}$$

When an oxygen-containing group is present, results are different. Should the oxygen function be on an end carbon (primary), this end carbon may be removed in part or entirely with the oxygen function to give the next lower homolog of the carbon skeleton:

$$CH_3(CH_2)_7CH_2 \vdots CHO \longrightarrow CH_3(CH_2)_7CH_3 \tag{5}$$

$$(CH_3)_2CHCH_2 \vdots COOH \longrightarrow (CH_3)_2CHCH_3 \tag{6}$$

With the oxygen function on an internal carbon (*sec* or *tert*), the parent hydrocarbon is obtained:

$$CH_3CH_2CHCH_2CH_2CH_2CH_3 \longrightarrow CH_3CH_2CH_2CH_2CH_2CH_2CH_3 \tag{7}$$
$$\qquad \vdots$$
$$\qquad OH$$

Results with amine functions are similar to those with oxygen functions.

Paraffin hydrocarbons are not affected in passing the hot catalyst, and multiple bonds are saturated.

A summary of these reactions is shown in Figure 4-5. Some compounds produce several fragments. An ester yields a fragment or fragments from

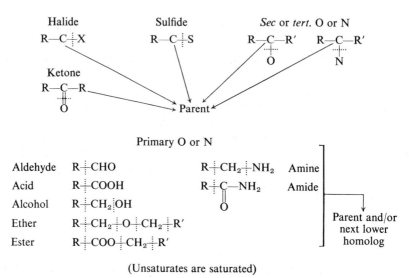

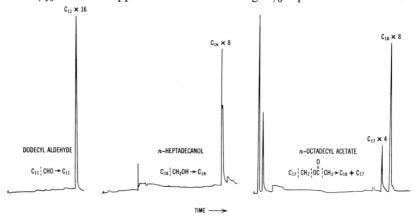

Fig. 4-5. Reactions observed with carbon-skeleton chromatography.

both the alcohol and acid moieties of the compound without preliminary saponification; all fragments will be observed if they are separable on the GC column. Aldehydes, alcohols, acids, and their derivatives give very little of the parent compound; the major product is the next lower homolog.

The hydrocarbons produced separate well on nonpolar columns. Six-foot $3/16$-in. o.d. copper columns containing 5% squalane or SE-30 on

Fig. 4-6. Typical temperature-programmed (11°C/min) carbon-skeleton chromatograms (12). Attenuations shown above or next to peak.

60/80 mesh Chromosorb W are useful with compounds in the C_1–C_8 and C_5–C_{20} range, respectively.

Some typical chromatograms are shown in Figure 4-6. Dodecyl aldehyde gives a peak with a retention time identical with that of undecane. Heptadecanol, because it is a primary alcohol, gives hexadecane, the next lower homolog of the carbon skeleton. With octadecyl acetate, both the parent C_{18} and some C_{17} hydrocarbon are obtained from the octadecyl group. Thus the products and their relative amounts may give an indication of the functional group in the original compound.

2. Carboxylic Acids and Derivatives

Yields of hydrocarbons from acids and acid derivatives, such as esters and amides, are low with the Pd on GasChrom P catalyst. Good yields from these compounds may be obtained by utilizing Chromosorb P in place of GasChrom P as the catalyst support (14). However, the Chromosorb P support, being much more adsorptive than GasChrom P, delays passage of compounds through the reactor, and the resulting retention times are not reliable. This defect is overcome with compounds larger than C_{12} by temperature programming the GC column. The products of the injected compound emerging from the catalyst bed over a period of a few minutes accumulate at the head of the GC column and do not move until the column temperature rises sufficiently; the peaks then appear on schedule with practically no tailing. Chromatograms illustrating these effects are shown in Figure 4-7. With temperature programming, the products from methyl stearate give sharp peaks, but isothermally products move through the GC column as they emerge from the catalyst bed and consequently tail badly.

With low-molecular-weight acids and acid derivatives, the reaction products may be held up in a cold U-shaped trap (e.g., 12-in. length of 0.030-in. i.d. copper tubing) interposed between the CS determinator and the injection port of the gas chromatograph; after allowing a few minutes for the products to elute completely, they are released into the chromatograph by removing the coolant and rapidly immersing the trap in a hot mineral oil bath.

This trap-and-release technique may be used for micro preparative purposes by cold-trapping the products of repeated injections and then, by sudden heating of the trap, releasing the collected material into a gas chromatograph equipped with a thermal conductivity detector for collection at the exit port.

Products obtained with the Chromosorb P catalyst are the same as those

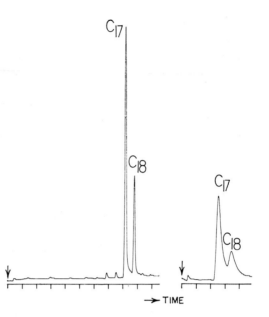

Fig. 4-7. Chromatograms of methyl stearate with 1% Pd on Chromosorb P catalyst. Left, column temperature-programmed 9°C/min starting at 65°C. Right, column isothermal at 195°C. Each division under chromatogram is 2 min (14).

obtained with the GasChrom P catalyst, except that more of the products, especially the parent hydrocarbon, are obtained from acids and their derivatives.

3. Long-Chain or High-Molecular-Weight Compounds

It is not possible to use a single set of conditions to analyze all compounds by CS chromatography. The commercially available CS determinator with its 9-in. length of catalyst gives no product response for compounds with chain lengths greater than 20 carbon atoms. This defect can be overcome by shortening the catalyst bed. A 4-in. length of catalyst performs well with compounds having carbon chains between C_{12} and C_{30}. The catalyst length in the CS determinator is easily reduced to this length by filling the rear 5 in. of the catalyst tube with glass beads, inert support, or a spacer and then adding 4 in. of catalyst to complete the filling. In this way sharp peaks have been obtained from cholesterol acetate and squalane (C_{30}). With compounds having more carbon atoms, the catalyst length should be decreased still further.

4. Dehydrogenation — Cyclohexane and Aromatic Structures

One reaction encountered in CS chromatography that was unexpected at first was the dehydrogenation of cyclohexane rings to benzenoid structures, because it occurred in a hydrogen atmosphere with a hydrogenation catalyst. It was found that as the catalyst temperature was increased from 200 to 360°C, the conversion of benzene to cyclohexane decreased while the conversion of cyclohexane to benzene increased. In other words, higher temperatures favored aromatization while lower temperatures favored hydrogenation of cyclic structures. These data are consistent with the exothermic heat of hydrogenation of benzene (actually 49.8 kcal/mole). Most of the substituted cyclohexanes that we checked were completely dehydrogenated at a catalyst temperature of 350°C.

This dehydrogenation of cyclohexane ring structures to aromatic ones can be used in determining the carbon skeleton of a polycyclic structure. The analyst not only obtains the retention times of the product(s), but he can trap the product (by passing the effluent into ethanol or hexane) and determine its ultraviolet or fluorescence spectrum, or secure a mass spectrum of the trapped product. Advantages of this technique over zinc or selenium dehydrogenation are the comparatively high yield of product, the ease and speed of conducting the reaction, and, most important, the small amount of sample required. Analyses at different catalyst temperatures should be useful in demonstrating the presence of cyclohexane or cyclohexene rings by the increasing amounts of aromatic structures formed as the temperature is raised.

Polycyclic structures have been found to give distinctive patterns in CS chromatography. As shown in Figure 4-8, almost identical chromatograms were obtained when five compounds having the same carbon skeleton were analyzed with the catalyst at 300°C. When fused ring structures occur, both cis and trans isomers are formed, e.g., both fully saturated structures, cis and trans decalin, formed from naphthalene.

Keulemans and Voge (15) studied dehydrogenation of C_5–C_8 naphthenes to aromatics in a precolumn (16) with a platinum–alumina–halogen catalyst at 350°C in a hydrogen carrier gas. Rowan found this catalyst best in his dehydrogenation work but used helium as the carrier gas (17). He suggested using dehydrogenation in the analysis of hydrocarbons. Aromatization, with helium as the carrier gas, has been advanced as a valuable technique for the identification and structure elucidation of terpenoids (18,19). Hydrogenation and dehydrogenation "spectra" were obtained from terpenes and alkylbenzenes with noble metal catalysts at 360–460°C (20).

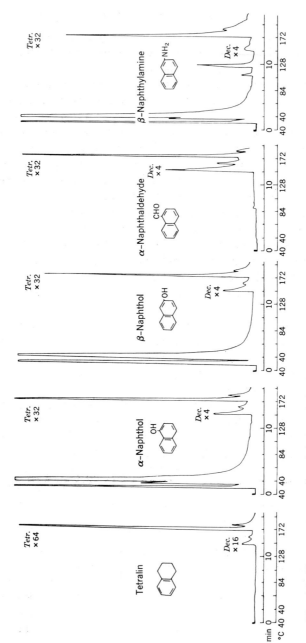

Fig. 4-8. Chromatograms of naphthalene derivatives subjected to CS chromatography. *Dec* = decalins; *Tetr* = tetralin. Attenuations shown above or next to peak. Analytical column held isothermal for 2 min at 40° and then programmed at 11°C/min (12).

102

D. Notes on Operation

Because CS chromatography is applicable to a wide variety of compounds, variations in its procedure are necessary to obtain correct results. The following notations on the use of the technique and suggestions for securing proper operation should be helpful in its application to specific problems.

Carbon-skeleton chromatography is not advanced as a quantitative technique because some cracking of carbon–carbon bonds occurs and stoichiometric yields are not obtained. Decreasing catalyst temperature (e.g., 260°C) lessens cracking but may delay retention times. Results are based on the principal products. In some cases quantitative analysis should be possible, depending on the type of compound being analyzed and the degree of accuracy required. Prospects for quantitative analysis of hydrocarbons, sulfides, ethers, aldehydes, and primary alcohols are probably very good because they react very smoothly with the catalyst. Conditions (catalyst length and temperature) could be made optimum for a given analysis.

The system must contain no leaks, and it should be tested when hot. The commercial apparatus, being made to fit all gas chromatographs, is equipped with needle stock tubing to fit into the injection port septum of the gas chromatograph. This connection is one source of leakage, and the injection port septum on the CS determinator is another. The injection port nut may not be tight enough or the septum may spring a leak. The simplest means of checking for leaks is by a pressure-drop test. After the GC column is disconnected from the injection port and the exposed port is closed with a plug (e.g., Swagelok 400-P for $\frac{1}{4}$ in.), the pressure is increased to about 40 lb with the adjustment screw handle of the pressure regulator, and the handle is then backed off. After a minute or two to allow for pressure equilibration, there should be no drop in pressure for several minutes. If there is, a leak is present. Periodic checking for leaks is essential, especially since the septa are exposed to high temperatures in this technique and their useful life is shortened considerably. Septa should be conditioned in an oven at 250°C overnight before use to minimize bleeding.

With the usual CS determinator (4.5 mm bore), the injection of 0.01–0.03 μl (10–30 μg) gives good results, although satisfactory data have been obtained from 1 μg or less. With the larger CS determinator (9 mm bore) as much as 1.6 μl (ca. 1.5 mg of an ester) has been injected at one time without loss of efficiency. However, with compounds other than esters, good conversion efficiency is obtained with much smaller samples (0.1–0.4 μl depending on the type of compound); usually, the conversion efficiency improves as the amount of sample is decreased. It is best not to overload the catalyst.

The catalyst is the least controlled variable of the analysis, and periodic checking of its performance with a known compound is suggested. The first few injections on a new catalyst may not give good results. With a new CS determinator the catalyst may last a day or two. As the apparatus is used, catalyst charges last longer. After the apparatus has been in operation for several months the useful life of the catalyst may be 2 or 3 weeks, and 50–100 analyses are normal with one charge. Loss of activity can be caused by injecting nonvolatile material (polymers, oils) on the catalyst. Catalyst activity deteriorates more rapidly at higher temperatures; at 350°C, the catalyst may suffer severe deterioration in a day. Catalyst deterioration also accompanies the analysis of halogenated compounds because they produce hydrogen halides; a few analyses of small amounts of these compounds can be tolerated, but—as noted—excessive amounts of these acids on the catalyst can induce undesirable cracking of carbon–carbon bonds.

It is usually necessary to inject the compound into the catalyst bed if retention times are to be valid. Once the functional groups are stripped off, the more volatile hydrocarbon products travel through the catalyst at the same speed as the standards. The time lag for compounds introduced ahead of the catalyst to reach the catalyst may cause the retention times of the products to be greater than they should be.

The injection of a compound in small amounts, e.g., 1 µg or less, is most conveniently done in a solvent. Hydrocarbon solvents that do not produce peaks in the region of interest may be used. The introduction of the compound neat is, of course, best. Brownlee and Silverstein (21) have recently described an apparatus and an integrated procedure for trapping a compound from the gas chromatograph in a glass tube, sealing it, and then releasing the compound neat into the CS unit by breaking the sealed tube in the carrier gas stream just ahead of the CS unit. They used as little as 2 µg per determination.

In many instances, a catalyst bed only a few inches long may be used. Such a short bed may be useful not only with high-molecular-weight compounds as noted, but also with polar compounds. Compounds with several functional groups, especially phenolic hydroxyls, secondary alcohols, ketones, and acids are difficult to bring through the catalyst bed unless it is shortened to a few inches. [Alternatively, derivatives (e.g., acetyl or trimethylsilyl) may be used.] A several-inch length of a catalyst may be placed in the injection port liner as shown schematically in the upper diagram of Figure 4-9, thus eliminating the need for CS apparatus.

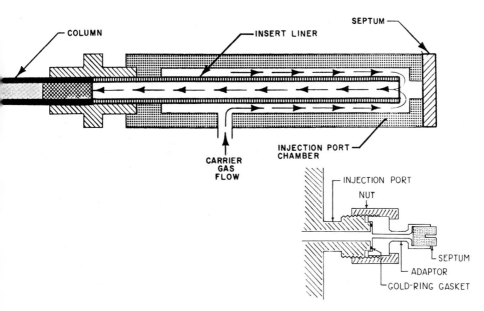

Fig. 4-9. Cross sections. Upper, insert liner that holds CS catalyst in the injection port (schematic); lower, injection port modification to keep septum cooler (22).

Many gas chromatographs are equipped with such liners or may easily be modified to place a catalyst bed several inches long within the injection port. The catalyst temperature is then set by adjusting the injection port temperature, which may usually be read from a meter on the GC control board. Septum bleed at the high temperatures needed for carbon-skeleton chromatography can be minimized considerably by using the modified injection port shown in the lower diagram of Figure 4-9.

There are several advantages in using temperature programming of the analytical column with CS chromatography (catalyst temperature is kept constant) aside from its use in the analysis of carboxylic acids and their derivatives, as already mentioned. It makes it possible for both large and small hydrocarbon products to be identified in a single analysis, and it makes such identifications possible in a reasonable period of time. Peaks are sharper than those of isothermal runs. The GC column may be held isothermally at a low temperature to allow small hydrocarbons to be eluted, and the column then temperature programmed to identify the larger hydrocarbons. Temperature programming also helps eliminate a difficulty encountered with isothermal analyses by clearing the GC column for the next analysis. Under certain conditions, some compounds, e.g., secondary alcohols, may not be completely converted to the hydrocarbon, and the unconverted compounds may appear in a subsequent analysis where they are easily recognized because their peaks are much broader than those of the hydrocarbons. By temperature programming or by raising the temperature of the GC column for a few minutes at the end of the analysis, such compounds are eluted, and interference is eliminated.

The reliability of identification by retention time decreases as the molecular weights of the products increase because the number of possible isomers becomes very much greater, even though saturation of multiple bonds greatly reduces the number of possibilities. The determination of the mass spectrum of high-molecular-weight products is therefore recommended. Mass spectral patterns of hydrocarbons are well understood, and chances of identification by this route are therefore excellent.

The selection of palladium as the catalyst of choice was dictated by the higher temperatures that could be attained without cracking the carbon skeleton and by the fact that it is not poisoned by sulfur compounds at the temperatures used. Reactions with platinum gave products in somewhat different yield, but because of cracking, its useful temperature range was limited to 200–240°C. Cracking with other metals used in hydrogenations —rhodium, ruthenium, and nickel—was too extensive for analytical CS applications.

The possible occurrence of rearrangements has been considered but thus far none has been encountered.

E. Applications

The following interesting applications of CS chromatography have been reported:

The structures of two dibasic acid fragments from an insecticidal alkaloid of the Chinese thundergod vine were only partially determined by conventional methods. Formation of nonane in the CS chromatography of the two fragments established their identities, which were then verified by NMR (23). Cleavage of the two compounds took place as shown in structures **1** and **2**.

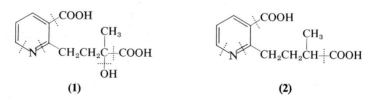

(1)   (2)

Silverstein and co-workers (24) determined the structure of the insect sex attractant, brevicomin (3), with the aid of CS chromatography; the product again was nonane:

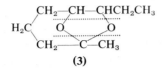

(3)

A phthalate isolated from thermally oxidized corn oil (25) was identified as the di-2-ethylhexyl ester (26) as shown by equation 8. The 3-methyl-heptane was identified by retention time and mass spectrometry. In a

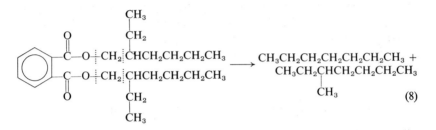

(8)

similar manner CS chromatography helped determine the structure of the diastereoisomers of 2-ethyl-1,3-hexanediol (27).

Isoprene reacted with crotonic acid to give a single product although two were theoretically possible (eqs. 9). The CS catalyst temperature was

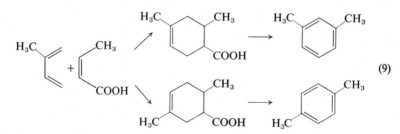

(9)

set high (330°C) to favor aromatization, and *m*-xylene was identified by its retention time and ultraviolet spectrum (28). The upper middle structure of equation 9 is therefore the correct one.

Asai, Gunther, and Westlake (29) have shown that CS chromatography can be useful in characterizing insecticide residues. Broderick (30) has suggested its use for identifying flavor components such as alcohols and terpenes.

Schomburg (31) mentioned the utility of CS chromatography for qualitative identifications, and Crippen and Smith (32) cite its use for the identification of alcohols. The technique has been used with steroids (33), sesquiterpene hydrocarbons (34,35), carbonyl compounds isolated from cotton buds (36), and volatile compounds formed during the autoxidation of sunflower oil (37). A survey paper by Thompson and his co-workers (38) describes their microhydrogenolysis technique and discusses its application in the identification of sulfur compounds from petroleum and other compounds.

A related post-column reactor containing 5% platinum on kieselguhr has been advanced by Klesment (39) for identifying hydrocarbons. When they pass the catalyst, olefins are hydrogenated and cyclohexanes and naphthenes are dehydrogenated in a 5% hydrogen in argon carrier gas. A tube containing activated charcoal following the reactor adsorbs the hydrocarbon products, and changes in the hydrogen concentration of the carrier gas, which register on a thermal conductivity detector, help in identifying the hydrocarbons.

F. Use with the Retention Index System

Carbon-skeleton chromatography frequently offers a good starting point for use with retention indices, which are a most valuable tool for structure determination (4,40–42). (Temperature effects on indices will be ignored for purposes of this discussion.)

The system is based on indices (I) derived from the *adjusted* (gas holdup subtracted) retention time of a compound expressed relative to the adjusted retention times of the straight-chain alkanes. The indices of the hydrocarbons are obtained by multiplying the number of carbon atoms in the alkane by 100; e.g., $I_{decane} = 1000$, $I_{undecane} = 1100$. A compound with an index of 1050 thus has an elution time half-way between that of decane and undecane on a logarithmic scale. A major feature of the system is that the indices of the carbon skeleton and the increments for the functional groups of a compound are additive. For example, the index of 2-chloro-5-bromodecane, found to be 1280 on a given column, could be derived from the sum of the index of the alkane, I_{decane} (1000), and the individual index increments, ∂I_{Cl} (120), and ∂I_{Br} (160) (∂I = differences of indices *on the same column*). A further feature is that the difference in indices (ΔI) of a given functional group *on two given columns*, usually polar (P) and nonpolar (NP), ($\partial I^{P} - \partial I^{NP} = \Delta I$) has a characteristic value, and these differences are likewise additive for several groups; e.g., when $\Delta I_{Cl} = 20$ and $\Delta I_{Br} = 40$, the ΔI of the compound having the two func-

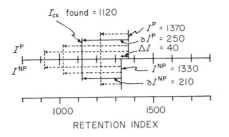

Fig. 4-10. Number of possible structures based on retention indices (I) is reduced when carbon skeleton is known (see text). P = polar stationary phase. NP = nonpolar stationary phase.

tional groups is 60. The index of the carbon skeleton remains essentially constant regardless of stationary phase.

Knowledge of the carbon skeleton from CS chromatography makes the retention index system more effective because it reduces the number of possible structures that may be derived from the adjusted retention times on a polar and nonpolar column. A *hypothetical* example is illustrated in Figure 4-10: $I^P = 1370$ and $I^{NP} = 1330$. Therefore $\Delta I = 40$. Without other information, the value of 40 may represent many possibilities and could include more than one functional group. If the functional group were an ether, the following might be possibilities for $\Delta I = 40$:

Possible functional groups	I of functional groups		I of CS (calculated by difference)
	I^P	I^{NP}	
—$OCH_2CH_2CH_3$	450	410	920
—OCH_2CH_3	350	310	1020
—OCH_3	250	210	1120
—O— (epoxy)	150	110	1220

If the index of the carbon skeleton were determined by CS chromatography to be 1120, then $\partial I^P = 250$ and $\partial I^{NP} = 210$ for the functional group(s) (solid arrows on Figure 4-10; dotted lines are rejected possibilities). Even without knowing what the carbon skeleton is, the analyst thus has sufficient information to determine the functional group present, which must be the methoxy group, rather than just a difference value. What is more, retention index values on additional columns can be obtained to supplement these data. The presence of the oxygen-containing functional group on an end carbon will frequently give the next lower homolog in addition to the parent hydrocarbon and thereby signal its position on the molecule.

With sufficient material, the presence of certain elements and functional groups can be determined from infrared, ultraviolet, NMR, and mass spectral data and from other available sources. When several functional groups are known to be present, the known functional group and CS indices may be subtracted from the index of the unknown, and the problem may then be simplified to the identification of only a single remaining group.

In using CS chromatography with retention indices the analyst may find it more convenient to consider the eliminated group as the functional group in those compounds giving solely the next lower homolog; e.g., —CH_2OH rather than —OH could be considered the functional group of a primary alcohol. This procedure would also automatically distinguish between primary and secondary groups, which are likely to have different retention indices depending on their position in the molecule.

III. Hydrogenation

Hydrogenation is widely used to determine the structure of unsaturated compounds and to determine the olefin content of a wide variety of mixtures, generally by comparing chromatograms of materials before and after hydrogenation. Because unsaturates may contain *cis* or *trans* double bonds, conjugated structures, branched or cyclic structures, epoxides, and other groupings, reliance on GC retention times for identifying such compounds is sometimes risky.

It is a simple matter to hydrogenate many unsaturated compounds instantaneously and *quantitatively* in the GC pathway by using hydrogen as the carrier gas and including a precolumn containing a hydrogenation catalyst. This technique was used first for the analysis of hydrocarbons (17), then for methyl esters of fatty acids (43–45), and recently for a wide variety of compounds (46). Olefin content of mixtures of hydrocarbons and methyl esters of fatty acids is now being determined routinely from chromatograms made with and without the hydrogenation step. The catalyst is usually palladium, platinum, or nickel deposited on a gas chromatographic support, such as GasChrom P or Chromosorb P.

Mounts and Dutton (44) showed that methyl esters of fatty acids could be hydrogenated without the hydrogenolysis encountered in carbon-skeleton chromatography by employing milder conditions—catalyst temperature 200°C, H_2 flow rate 60 ml/min, 43 × 6.3 mm catalyst bed. Their catalyst, usually 1–3% nickel on kieselguhr, was held in a small precolumn that screwed onto the injection port; 0.1–2μl of a sample was injected for analysis. They were able to study the kinetics of the hydrogenation process by injecting the sample at different depths in the catalyst bed, thereby varying the effective catalyst length.

Further investigation of the procedure led Beroza and Sarmiento (46) to devise simple, easily constructed oven and injection-port hydrogenators. A wide variety of compounds may be hydrogenated instantaneously with these hydrogenators, though some hydrogenolysis is encountered with certain types of compounds.

They found that the hydrogenation of methyl esters of fatty acids is complete and quantitative with a 6-mm length (25 mg) of the highly efficient catalyst ["neutral" 1% Pd on GasChrom P (11)] used in CS chromatography. Because the gas chromatography of the fatty acid methyl esters is conducted at 175–200°C, the hydrogenator may be kept in the oven; the hydrogenator of their unit (¼-in. fittings) was simply a Swagelok fitting (No. 400-6-316) fastened to a 1-in. length of ¼-in. o.d. stainless steel tubing (Fig. 4-11A) inserted between the injection port and the GC column. The catalyst was placed in the stainless steel tube between glass wool plugs previously rinsed with solvent and dried. Because only 25 mg of catalyst is required, it can also fit easily into the injection port. The catalyst may then be maintained at one temperature and the analytical column at another; this arrangement permits analyses of low-molecular-weight compounds since the temperature of the analytical column may be selected to provide a convenient retention time. The injection-port hydrogenator, shown in Figure 4-11B, is identical with the oven hydrogenator except that it has a length of 11-gauge stainless steel needle stock tubing silver soldered into it; the tubing holds the catalyst within the injection port. (Dimensions will have to be varied for different chromatographs.) Hydrogenation may also be accomplished by placing the catalyst in the first ½ in. of the GC column (in or out of the injection port); however,

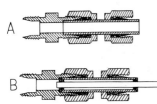

Fig. 4-11. Hydrogenator designs. *A*, Oven hydrogenator; *B*, injection-port hydrogenator (46).

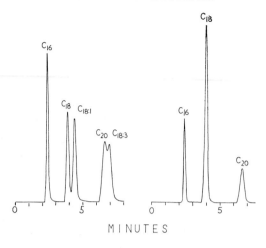

Fig. 4-12. Chromatograms of a mixture containing methyl palmitate (C_{16}), stearate (C_{18}), oleate ($C_{18:1}$), linolenate ($C_{18:3}$), and arachidate (C_{20}) made without hydrogenator (left) and with injection-port hydrogenator (right) (46).

this procedure is not advisable because the catalyst cannot be completely removed, and even minute amounts can cause partial hydrogenation in subsequent analyses.

A typical chromatogram of a mixture of methyl esters of fatty acids, made with and without the injection-port hydrogenator, is shown in Figure 4-12. The analysis is quantitative, and the chromatogram shows no

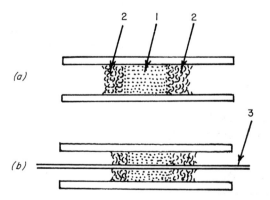

Fig. 4-13. Packing of catalyst (46). (a) Conventional. (b) With tubular bypass. 1, catalyst; 2, glass wool; 3, 30-mm length of 22-gauge stainless steel tube (50 mm length of 25-gauge tube used with injection-port hydrogenator).

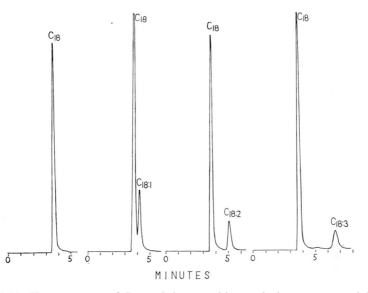

Fig. 4-14. Chromatograms of C_{18} methyl esters with oven hydrogenator containing tubular bypass: (left to right) methyl stearate, oleate, linoleate, linolenate (46).

evidence of any unsaturated compound in the mixture after passing through the hydrogenator. Although the usual sample size is 10–500 μg, as little as 1 μg can be analyzed qualitatively.

The insertion of a piece of needle stock tubing protruding from each end of the catalyst in either the oven or injection-port hydrogenator, as shown in Figure 4-13b (conventional packing shown in Figure 4-13a), allows a small amount of injected compound to pass through the tube unaltered while most of the compound is hydrogenated. Great care must be exercised to keep catalyst out of the needle stock bore. The resulting chromatograms of methyl stearate, oleate, linoleate, and linolenate are shown in Figure 4-14. Separation factors (ratio of the retention time of the original to that of the fully hydrogenated analog), which may be determined in this manner, are characteristic of each compound. [Ackman (47) has shown that on polar columns the separation factor is least for the most highly centralized arrangement of double bonds.]

Separation factors and other data obtained with a variety of compounds are given in Table 4-1.

The catalyst temperature did not appear to be at all critical when used between 140 and 250°C, and separation factors were not altered by the rapid passage through 25 mg of catalyst. Hydrogenolysis to hydrocarbons

TABLE 4-1

Compounds Analyzed with Hydrogenator (46)

Type	Compound[a]	No. of multiple bonds	Separation factor[b]	Catalyst temperature, °C
Alcohol	2,6-Octadien-1-ol, 3,7-			
	dimethyl-, *cis*- (nerol)	2	1.92	150
	dimethyl-, *trans*- (geraniol)	2	2.22	150
	6-Octen-1-ol, 3,7-dimethyl-			
	(citronellol)	1	1.53	150
	1-Octanol, 3,7-dimethyl-	0	1.00	150
	6-Octen-2-ol, 3,7-dimethyl-	1	1.47	150
	Dodecyl alcohol	0	1.00	150
	1-Nonyn-3-ol, 3-methyl-	1	1.62	150
	5-Decyn-4,7-diol, 2,4,7,9-			
	tetramethyl-	1	0.79	150
	Cinnamyl alcohol	1	1.88	200
	1-Propanol, 3-phenyl-	0	1.00	200
Amide	Crotonamide, *N,N*-dibutyl-			
	3-methyl-	1	1.68	200
	Chrysanthemumamide,			
	N,N-diethyl-[c]	1	1.67	200
	Tetradecanamide,			
	N,N-dimethyl-	0	1.00	150
Amine	Hexadecylamine, *N,N*-			
	dimethyl-	0	1.00	200
	Pyridine, 2-hexyl-	0	1.00	150
Carbonyl				
compounds	Undecenal[d]	1	1.29	150
	Undecanal[d]	0	1.00	150
	Lauraldehyde[d]	0	1.00	150
	5-Nonene-2-one, 9-cyclo-			
	hexylidene-6-methyl-	2	1.50	200
	Cinnamaldehyde, *trans*-[d]	1	1.91	200
	Hydrocinnamaldehyde[d]	0	1.00	200
Ester	Sorbic acid, heptyl ester	2	3.22	150
	Hexanoic acid, heptyl ester	0	1.00	150
	Butyric acid, 2,4-hexadienyl			
	ester	2	2.17	150
	Hexanoic acid, 2-ethyl-,			
	allyl ester	1	1.14	150
	propyl ester	0	1.00	150

(continued)

TABLE 4-1 (*Continued*)
Compounds Analyzed with Hydrogenator (46)

Type	Compound[a]	No. of multiple bonds	Separation factor[b]	Catalyst temperature, °C
Ester	Chrysanthemumic acid, methyl ester[c]	1	1.39	150
	Fumaric acid, diethyl ester	1	0.93	150
	Maleic acid, diethyl ester	1	1.39	150
	Succinic acid, diethyl ester	0	1.00	150
	Oleic acid, methyl ester	1	1.12	200
	Linoleic acid, methyl ester	2	1.38	200
	Linolenic acid, methyl ester	3	1.77	200
	Stearic acid, methyl ester	0	1.00	200
Ether	Ether, allyl tetradecyl	1	1.40	150
	Ether, isobutyl phenyl	0	1.00	150
Halide	Decane, 1-bromo-[d]	0	1.00	150
	Octadecane, 1-chloro-[d]	0	1.00	200
Others	Dodecane, 1,2-epoxy-	0	1.00	150
	1-Dodecanethiol[d]	0	1.00	150
	Tridecanenitrile	0	1.00	150

[a] Retention times of saturated products of the olefins were compared in many instances with authentic compounds to establish their identity. A vertical line connecting the olefin and its saturated analog indicates that such a comparison has been made and that it showed agreement. A 6-ft, $\frac{1}{4}$-in. o.d. aluminum column containing 15% diethylene glycol succinate on 60/80 mesh GasChrom P was used except with the halides; for them a 4-ft, $\frac{1}{4}$-in. o.d. stainless steel column containing 10% SE-30 on 60/80 mesh GasChrom P was employed.

[b] No correction made for gas holdup.

[c] Chrysanthemumic acid is 2,2-dimethyl-3-(2-methylpropenyl) cyclopropane-carboxylic acid.

[d] Some breakdown to hydrocarbon (hydrogenolysis) at both 150 and 200°C.

occurred to some extent with aldehydes, mercaptans, and halogen compounds. This breakdown probably prevents quantification of the data but causes no difficulty in identifications because a qualitative result is sufficient. With other compound types, conditions were sufficient to completely saturate multiple bonds, and yet mild enough to allow functional groups to survive.

Catalyst activity should be periodically checked with known compounds to determine whether hydrogenation is complete. Sufficient catalyst can be activated once a week by heating it for 30 min at 200°C in the hydrogen carrier gas; the catalyst in the hydrogenator is then replaced as needed. The recharged hydrogenator is ready for use 5 or 10 min after it reaches the desired temperature. Samples with polymers or other non-volatile material can weaken the catalyst, probably by covering active sites, and thus make it necessary to change the catalyst more frequently.

Although the hydrogenators were placed ahead of the GC column for these studies, they can be used in any part of the chromatographic pathway. For example, a hydrogenator placed after the GC column may have special advantages. A compound separated by a GC column can then be passed into a mass spectrometer with and without the hydrogenator. The number of double bonds can be determined from the difference in molecular weights of the parent ions. This arrangement would be particularly handy in the analysis of mixtures.

Littlewood and Wiseman used the CS hydrogenation catalyst in a hydrogenation coulometer devised for use as a GC detector (48). A passing unsaturate absorbed hydrogen, and the coulometer sensed the difference in hydrogen concentration between the entering and leaving gas stream, causing hydrogen to be generated electrolytically to make up the difference; the amount generated was registered on a strip-chart recorder. Littlewood and Wiseman estimated the speed of hydrogenation to be about a milli-second; this is in line with the rapid hydrogenation observed with 25 mg of catalyst.

The use of hydrogenation technique to solve a difficult problem is illustrated in Figure 4-15. The reaction of chloroprene with crotonic acid produces one of two possible compounds. The structure of this product, which could not be resolved by NMR or other means, was determined by

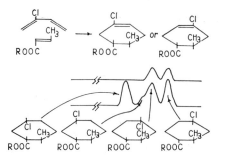

Fig. 4-15. Hydrogenation of Diels-Alder reaction product to determine structure (28).

hydrogenation to be the one having the chlorine atom *para* to the ester group; the two resulting peaks had retention times coinciding with those of two known compounds which were among the four possible products expected.

Precolumn hydrogenation has been used in determining the structure of heptenes (49), unsaturated oxygen-containing compounds formed during olefin oxidations (50), and esters of fatty acids (51,52). Lecerf and Bezard (53) determined chain lengths of the individual unsaturated fatty acid esters in a mixture using a hydrogenator attached to the GC column outlet; the resulting saturated esters were trapped on glass wool, eluted with pentane, and rechromatographed.

IV. Location of Double Bonds by Ozonolysis

When unsaturation is found, the question invariably arises as to its position in the molecule. Ozonolysis when used with gas chromatography is a reliable choice for these determinations, even at levels as low as 1 μg. The compound is ozonized, the ozonide cleaved by oxidative, reductive, or pyrolytic treatment, and the products subjected to analysis by gas chromatography. Double-bond position is deduced from the chain lengths of the fragments. Advances in the field of analytical ozonolysis were pioneered by workers interested in fatty acid analysis (54,55); only the most recent work will be mentioned.

Davison and Dutton (56) fitted a soldering gun with a U-tube having an injection port and carrier gas inlet at one end, and at the other, needle stock tubing sharpened to penetrate a septum. The sample is injected into the unheated U-tube and ozone then passed over it to form the ozonide. Ozone in the U-tube is flushed out with carrier gas and the products are then injected into the gas chromatograph (with the carrier gas flowing) via the needle stock tubing while the U-tube is rapidly heated to 250°C by pressing the trigger of the soldering gun. A short plug of zinc oxide is included at the head of the GC column to retain the carboxylic acids that form in the pyrolytic cleavage. Analyses are said to be quantitative.

Pyrolytic cleavage is said to form an acid and an aldehyde (eq. 10).

$$
\underset{\substack{| \qquad | \\ RCH{-\!-\!-\!-}CHR'}}{O{-}O{-}O} \xrightarrow[\text{or}]{} \begin{array}{l} RCHO + R'COOH \\ R'CHO + RCOOH \end{array} \qquad (10)
$$

At the moment, it is not clear whether the two possible reactions occur to an equal degree or whether neighboring groups or other compounds present influence the proportion of products.

A compact microreactor for ozonolysis–pyrolysis, applicable also to esterification, saponification, hydrogenation, or halogenation reactions, has been described by Dutton, Davison, and Bitner (57,58).

Nickell and Privett (59) reported the use of what they call "controlled pyrolysis" of ozonides. They ozonize their compound at −65°C in pentane containing ozone (reaction is instantaneous with fatty acid methyl esters), purge the solution with nitrogen, evaporate the solution of ozonide on Lindlar hydrogenation catalyst, and then introduce the catalyst plus ozonide into the injection port (225°C) of the gas chromatograph. (Argon, not hydrogen, is used as the carrier gas.) The products are aldehydes:

$$\underset{\text{RCH}\underline{\quad\quad}\text{CHR}'}{\overset{\text{O}\underline{\quad}\text{O}\underline{\quad}\text{O}}{\mid\qquad\mid}} \longrightarrow \text{RCHO} + \text{R}'\text{CHO}$$

Another procedure advanced by Beroza and Bierl (60) was used to simplify and extend GC microozonolysis to a wide variety of compounds. The compound is dissolved in carbon disulfide (CS_2) or pentyl acetate and ozone is passed into the solution at −70°C. Triphenyl phosphine is then added to reduce the ozonide to aldehydes (61), and the solution is injected directly into the gas chromatograph. The triphenyl phosphine does not interfere in the analysis of the fatty acid methyl esters.

The setup used is shown in Figure 4-16. The simple ozonizer (62) is a three-way stopcock having its single leg C covered with aluminum foil D (outer electrode, encased with rubber tube E for insulation and grounded via F and G) and a 5-in. length of 21-gauge needle stock tubing A (inner electrode) held within the same leg by an injection-port septum B (W-10 septum, Applied Science Labs.). When oxygen (10 ml/min) passes through the leg and a vacuum tester H (Tesla coil type) is applied to the needle stock tubing, sufficient ozone (0.01 meq of ozone per minute) is generated for microanalyses. The ozone passes through the needle stock tubing into 0.1 ml of the sample K (as little as 0.005 ml has been used for small samples) in solution contained in a reaction tube J made from a medicine dropper by sealing off its small end. Teflon tubing L ($\frac{1}{16}$-in. o.d.) conducts gases emerging from the reaction tube into an indicator solution N (5% KI in 5% H_2SO_4 plus starch) which turns blue when ozone enters.

In an analysis, the sample (usually 25 μg) in 0.1 ml of solvent is held at −70°C (xylene–Dry Ice bath), and oxygen is passed into the solution. The vacuum tester is applied to the inner electrode until the blue color indicating excess ozone appears (15 sec). The three-way stopcock is reversed to replace the oxygen with nitrogen. The cold bath is removed and the reaction tube is slipped off the rubber stopper I; about 0.5 mg of powdered

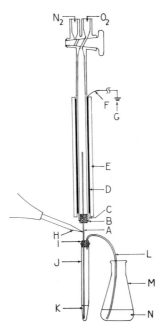

Fig. 4-16. Setup for ozonization of compounds. *A*, Needle stock tubing; *B*, rubber stopper (injection port septum); *C*, single leg of three-way stopcock; *D*, aluminum foil; *E*, rubber tubing (insulation); *F*, copper wire connected with *D*; *G*, ground; *H*, high-voltage source (vacuum tester); *I*, rubber stopper (injection port septum); *J*, reaction tube; *K*, solution of compound; *L*, Teflon tubing; *M*, 10-ml erlenmeyer flask; *N*, indicating solution (4 ml). Cold bath ($-70°C$) is not shown (60).

triphenyl phosphine is dropped into the tube which is then stoppered and swirled to dissolve the powder. About 20 μl ($\frac{1}{5}$) of the resulting solution is injected into the gas chromatograph.

With the flame ionization gas chromatograph used in our work, CS_2 was found to be the best of the solvents tested because it obscured a minimum of the chromatogram and produced a minimum of interference. Peaks of straight-chain aldehydes and ketones above C_4 were visible. For aldehydes and ketones between C_2 and C_5, pentyl acetate was employed.

Some typical chromatograms are shown in Figure 4-17, and the results of ozonolysis of different types of compounds (e.g., alcohols, ethers, esters, halides, aromatics, heterocyclics) are shown in Table 4-2. Retention time comparisons with available known compounds showed that the expected products were obtained in each case. Yields of products were between 70 and 95%. Even though the ozonolysis is not stoichiometric,

TABLE 4-2

Gas Chromatographic Analysis of Ozonolysis Products

Compound	Products identified[a]
$C_6H_5CH{=}CHP(O)(OCH_2CH_3)_2$	C_6H_5CHO, +
$C_6H_5CH{=}CHC(O)NHCH_2CH_2CH_3$	C_6H_5CHO
$(CH_3CH_2CH_2)_2C{=}CHCH_2Br$	$(CH_3CH_2CH_2)_2C{=}O$
$CH_3(CH_2)_4CH{=}CHC(O){-}N(\text{morpholine})$	$CH_3(CH_2)_4CHO$
$C_6H_5C(CH_3){=}CH_2$	$C_6H_5C(O)CH_3$
$CH_3CH{=}CHC(O)N(C_2H_5)_2$	CH_3CHO[b]
$(CH_3)_2C{=}CHC(O)N(CH_2CH_3)_2$	$CH_3C(O)CH_3$, +
$CH_3O{-}C_6H_4{-}CH{=}CHCH_3$	CH_3CHO, $CH_3O{-}C_6H_4{-}CHO$

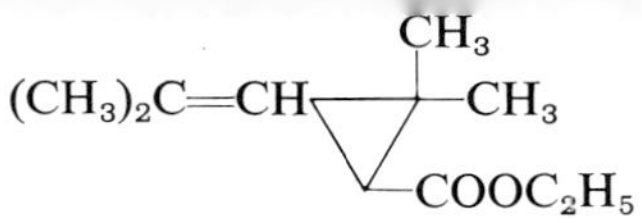

$(CH_3)_2C=CH$— [cyclopropane ring with CH_3 and $COOC_2H_5$] $\quad CH_3CCH_3$ (=O), +

$CH_3CH=CHCOCH_2CH_2Cl$ $\quad CH_3CHO^b$

$CH_3CH=CHCH=CHCOOC_4H_9$ $\quad CH_3CHO$, +

$CH_3CH_2CH=CH(CH_2)_2OH$ $\quad CH_3CH_2CHO$, +[c]

$CH_2=CH(CH_2)_2OH$ $\quad +^c$

$(CH_3)_2C=CHCH_3$ with OH $\quad CH_3CCH_3$ (=O), +

$CH_3(CH_2)_7CH=CH(CH_2)_7COOCH_3$ $\quad CH_3(CH_2)_7CHO$, +[d]

$CH_3(CH_2)_4CH=CHCH_2CH=CH(CH_2)_7COOCH_3$ $\quad CH_3(CH_2)_4CHO$, +[d]

$CH_3CH_2CH=CHCH_2CH=CHCH_2CH=CH(CH_2)_7COOCH_3$ $\quad CH_3CH_2CHO$, +[d]

$CH_3CH_2(CH=CHCH_2)_6CH_2COOCH_3$ $\quad CH_3CH_2CHO$, +

[cyclohexene] $\quad +$

$CH_3(CH_2)_7CH=CH(CH_2)_7CH_2NH_2$ $\quad CH_3(CH_2)_7CHO$, +

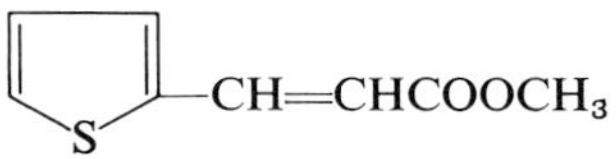

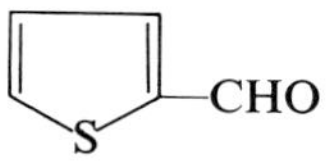

[thiophene]—$CH=CHCOOCH_3$ $\quad$ [thiophene]—CHO

---

[a] + signifies that another major product was detected but that a known compound was not available to check its identity.
[b] Analyzed only in pentyl acetate to determine small fragments.
[c] These two compounds give product with same retention time. The product expected from both is $HO(CH_2)_2CHO$.
[d] Also gave $OCH(CH_2)_7COOCH_3$.

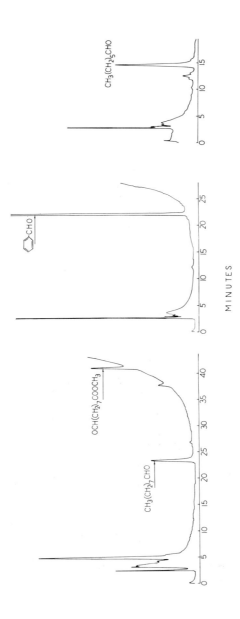

Fig. 4-17. Chromatograms of ozonolysis products of 1 μg of methyl oleate (left), 5 μg of methyl cinnamate (center); and 5 μg of 1-octene (right) (CS$_2$ solvent) (60). Analytical column was 3.6 m, 6-mm o.d. copper column containing 5% Carbowax 20M on 60/80 mesh GasChrom P; it was held at 50°C for 6 min and then programmed at 6.4°C/min to 200°C.

results are sufficiently reproducible to make it likely that the procedure can be used for quantitative analysis.

The direct ozonolysis of linseed and tung oils (glyceryl esters) is illustrated in Figure 4-18. The peaks obtained show the alkylidene groups present, i.e., groups from the hydrocarbon end of the fatty acid to the first double bond:

$$CH_2OOC(CH_2)_7CH{=}CH(CH_2)_7CH_3 \qquad\qquad \longrightarrow C_9 \text{ aldehyde}$$
$$|$$
$$CHOOC(CH_2)_7CH{=}CHCH_2CH{=}CH(CH_2)_4CH_3 \qquad \longrightarrow C_6 \text{ aldehyde}$$
$$|$$
$$CH_2OOC(CH_2)_7CH{=}CHCH_2CH{=}CHCH_2CH{=}CHCH_2CH_3 \longrightarrow C_3 \text{ aldehyde}$$

The glyceryl residue is too large to traverse the GC column and does not appear on the chromatogram. The ozonolysis of isomerized linseed oil produces an interesting series of alkylidene group products. Inasmuch as only straight-chain acids are present, the ozonolysis of isomerized linseed oil provides a convenient series of retention time standards for the straight-chain aldehydes. The major fatty acid in tung oil, eleostearic acid, has three conjugated double bonds with a C_5 alkylidene group. Tung oil also contains oleic and linoleic acids. Since the ozonolysis products of this oil are consistent with the presence of these acids, it appears that the alkylidene analysis of conjugated structures causes no problems. [Conjugated double bonds are ozonized with greater difficulty than isolated double bonds (63).]

Though double bonds were readily cleaved, certain structures were found to resist ozonolysis under the conditions described. These were triple-bond compounds and α,β-unsaturated nitriles. Undoubtedly other structures resistant to the conditions described will also be found.

Ketones rather than aldehydes are obtained when alkyl substituents are on double-bonded carbon atoms. Reaction 11 is typical.

$$\begin{array}{ccc} CH_3 & & CH_3 \\ | & & | \\ RC{=}CHR' & \longrightarrow & RC{=}O + R'CHO \end{array} \qquad (11)$$

Subtractive loops that remove aldehydes and not ketones (see Section V-D) may be interposed in the systems to distinguish between aldehydes and ketones and thus determine whether or not such an alkyl substituent is present. By analyzing the aldehyde fragments obtained from unsaturated esters on two columns of different polarity and by interrupting the ozonolysis of polyunsaturated esters so that unsaturated as well as saturated fragments were produced, Kleiman, Spencer, and Earle (64) were able to locate unsaturation in esters with multiple double bonds more precisely.

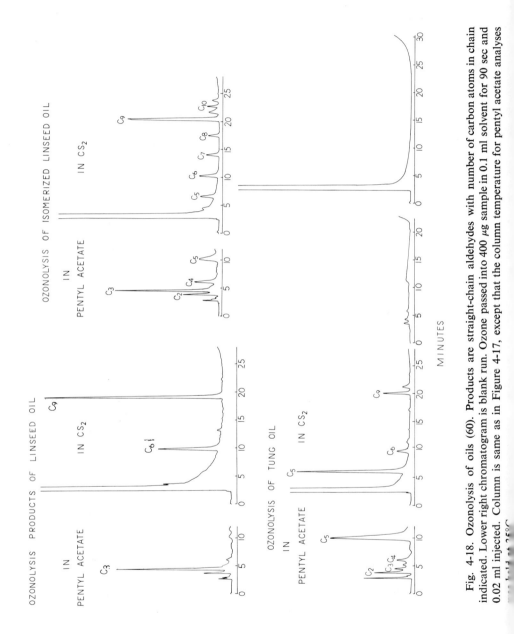

Fig. 4-18. Ozonolysis of oils (60). Products are straight-chain aldehydes with number of carbon atoms in chain indicated. Lower right chromatogram is blank run. Ozone passed into 400 μg sample in 0.1 ml solvent for 90 sec and 0.02 ml injected. Column is same as in Figure 4-17, except that the column temperature for pentyl acetate analyses

124

The methods of Davison and Dutton and Nickell and Privett have the advantage of requiring only one chromatogram to observe all products. However, the Beroza and Bierl procedure requires smaller amounts (as little as 1 μg) and may also be used to analyze oils and highly volatile compounds not suitable for analysis by the other procedures.

After the positions of double bonds have been located, it may be necessary to determine the configuration (*cis* or *trans*) of these bonds. A method to accomplish this for polyunsaturated fatty acid esters has been advanced by Privett and Nickell (65); although it utilizes thin-layer chromatography with selective argentation rather than gas chromatography, it is worth noting. Similarly, Scholfield, Davison, and Dutton (66) separated the geometric isomers of unsaturated fatty acid esters by liquid chromatography on a silver-saturated cation-exchange resin column; the positions of the double bonds were then located by the ozonolysis–pyrolysis technique (56).

Ozonolysis preceding gas chromatography has also been used in the identification of carbonyl compounds (67–70); oxygen containing about 2% ozone is passed through solutions of the 2,4-dinitrophenylhydrazones of aldehydes and keto acid methyl esters; the resulting carboxylic acids and keto acid esters are analyzed by gas chromatography.

V. Subtractive Processes

The removal of a compound or compounds before or during gas chromatography (often from a mixture) may be considered a subtractive process. Subtractions are generally recognized by comparing chromatograms of the sample made before and after the application of a given treatment or exposure to a chemical and noting the peaks that disappear completely or partially. Such a subtractive process is a most useful adjunct in making group separations, in determining the presence of functional groups, and in verifying the identity of a compound.

Many chemical and physical reactions can serve as subtractive processes; a number will be obvious to the analyst from his knowledge of qualitative organic analysis. For example, chromatograms of a compound or mixture of compounds in an organic solvent made before and after extraction with aqueous acid, base, or buffer can indicate whether compounds are acidic, basic, neutral, or amphoteric; an ion-exchange resin may also be used for this purpose. Esters may be saponified, derivatives or complexes formed, or compounds degraded, and the shifting or disappearance of the peaks noted on chromatography after treatment. In one study (71) carbonyl compounds were absorbed by a solution of hydroxylamine hydrochloride, and potassium permanganate was used to remove unsaturated compounds and aldehydes from test solutions before gas chromatography. Useful though such reactions may be, only those techniques designed to be used in close association with or as part of the GC process can be considered within the scope of this chapter.

The inclusion in the GC pathway of a chemical that reacts with certain types of compounds is a simple subtractive technique. A reaction loop is packed with a chemical, frequently on an inert GC support, and inserted in the system either before or after the GC column. The loop shown in Figure 4-19 is convenient to use (22); it is made from a 6-in. length of $\frac{1}{4}$-in. o.d. stainless steel tubing coiled once, and its ends are fitted with a 400-C-316 Swagelok fitting as shown; the nut is attached to one end in the usual way and the male plug, after drilling through the plug part of the fitting, is silver soldered to the other end. The loop can usually be attached with little distortion of the GC column. It is apparent that the length may be varied by using a straight tube, more coils, or different-size coils in accordance with requirements. The loop may be maintained at a temperature higher than the GC column (even if within the oven) by wrapping it with heating tape and then with insulation, or by some similar heating arrangement.

Compounds may be retained chemically or physically and may often be released subsequently by appropriate manipulation of temperature. The sample may be passed sequentially through a column (to separate components of a mixture) and through a reaction zone. One or more peaks may then be diverted into cold traps; by returning the trapped products to the original column, chemical changes that took place may be determined. Operations of this kind have been carried out rapidly and conveniently in a closed system with four-way valves (17). Compounds are thus removed or are changed to others with different retention times

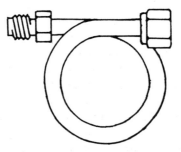

Fig. 4-19. Subtraction loop (22).

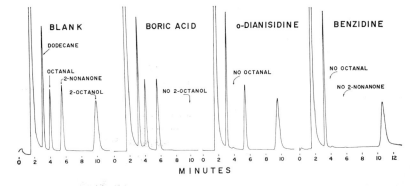

Fig. 4-20. Chromatograms obtained with subtraction loops (22).

either for identification purposes or to eliminate their interference in the analysis of other compounds. Structural information may also be deduced from the fact that a compound is partially subtracted (lesser yield) or retarded (increased retention time).

Some workers have acquired data rapidly with a dual-column, dual-recording instrument having a subtraction agent on one column but not on the other. A valve that can operate without leakage for long periods would make it possible to use a single column and divert a sample through either the subtraction loop or a blank loop at will.

One difficulty when using subtractive agents is that only limited information is available on their action with compounds other than the ones they are intended to subtract (e.g., those with a given functional group) and on the effect of neighboring groups in the molecule. Accordingly, the subtractive technique must be regarded as one of limited specificity and reliability at this time.

Table 4-3 gives a summary of subtractive processes and typical references. Chromatograms illustrating the use of subtractive loops are shown in Figure 4-20.

A. Hydrocarbons

Molecular sieves, especially Molecular Sieve 5A (Linde Co.), have been widely used in the petroleum field to separate normal alkanes and olefins from mixtures of hydrocarbons containing branched, cyclic, and aromatic structures (73,89–94). Retention of molecules has been attributed to the physical adsorption of molecules of small diameter within the uniform-

TABLE 4-3

Summary of Subtractive Reactions

Subtracting agent	Functional group	Compound or type	Effect	Typical refs.
Molecular sieves	Straight-chain	n-Alkanes, olefins[a]	Subtracted	72–76
Ag, Cu, Hg salts, H_2SO_4, Br_2	C=C	Olefins[a]	Subtracted	75, 77–80
4% Ag_2SO_4 in 95% H_2SO_4, $PdSO_4$ in H_2SO_4, $Hg(ClO_4)_2$, $AgNO_3$ on alumina	C=C	Olefins + aromatics[a]	Subtracted	78, 79, 81–83
Maleic anhydride	C=C—C=C	Conjugated dienes	Subtracted	84, 85
Ethylene glycol maleate polyester	C=C—C=C	Anthracene	Subtracted	86
Chloromaleic anhydride	C=C—C=C	Conjugated dienes	trans preferentially subtracted	87
Versamide 900	R—X	Labile halogen compounds	Subtracted	88
Boric acid	—OH	Satd. prim. and sec. alcohols	Subtracted	22, 50
		Tert. alcohols, unsatd. alcohols	Not subtracted[b]	22
		Allylic (except with terminal double bond)	Not subtracted[b]	22
		Others	Subtracted	22
		Phenols	Detained, peak spread	22
		Salicylaldehyde	Subtracted	22
	—CH——CH—\\/O	Epoxide	Not subtracted	50
Zinc oxide[c]	—COOH	Most carboxylic acids	Subtracted	22, 56
		Acids, alpha-substituted	Delay, broad tailing peaks	22

(continued)

TABLE 4-3 (*continued*)

Subtracting agent	Functional group	Compound or type	Effect	Typical refs.
Zinc oxide[c]	—OH	Alcohols and phenols	20–50% subtracted but retention time and peak shape are good	22
o-Dianisidine[c]	—CHO,—C(=O)—	Aldehydes	Subtracted	22
		Ketones except cyclohexanone	Not subtracted	22
		Cyclohexanone	20–50% subtracted	22
	—CH—CH— with O bridge	Epoxides	Some subtracted	22
Benzidine[c]	—CHO,—C(=O)—	Aldehydes	Subtracted	22
		Most ketones	Subtracted	22
Hydroxylamine	—CHO,—C(=O)—	Aldehydes, ketones	Subtracted	50
Phosphoric acid	—C(H)—C(H)— with O bridge	Epoxides	Subtracted	22, 50
		Tert. alcohols	Not subtracted	22, 50

[a] Used only in hydrocarbon analysis.
[b] See text.
[c] Types of compounds that are unaffected are not listed.

size pores of the molecular sieve rather than to chemisorption because adsorption does not occur above the critical temperature (90). The accuracy of a normal hydrocarbon determination in the C_7–C_{20} range using this subtraction was 2–3% (74). The *n*-paraffins sorbed on a molecular sieve at 350°C may be eluted by heating the sieve to 550°C and then determining their amounts by gas chromatography with temperature programming (76).

Hersh (90) reported adsorption characteristics of a variety of compounds, including some other than hydrocarbons, on Molecular Sieves

4A and 5A. Lower normal alcohols, aldehydes, and acids were subtracted with Molecular Sieves 5A and 13X at 100°C (95); however, compounds with a carbonyl or hydroxyl group adsorbed on the sieve may delay passage of other such compounds (96). Other molecules of small diameter that are subtracted are H_2O, CO_2, CO, H_2S, SO_2, NH_3, ethylene oxide (90), HCN, $(CN)_2$, and CO_2 (97), NO (98), N_2 (99), and O_2 (100). Because molecular sieves remove so many types of compounds, they have been used to purify GC gases; however, this lack of specificity has limited their usefulness as subtraction agents in structure determinations to the removal of normal alkanes and olefins from hydrocarbon mixtures.

Other means have been used to subtract olefins and aromatic hydrocarbons from hydrocarbon mixtures. Packings containing sulfuric acid, silver, copper, and mercury salts retain olefins; some packings additionally retain aromatics (78). The addition of bromine to volatile alkenes by passing them over a charcoal–bromine packing caused them to become nonvolatile in a GC system (80); ethylene and acetylene could be selectively subtracted by varying the temperature of the packing (101). Butadiene was subtracted with a maleic anhydride column at 100–140°C (84,85). A polyester of maleic anhydride and ethylene glycol was effective in subtracting anthracene from a mixture of anthracene, phenanthrene, and carbazole (86). *Cis–trans* conjugated dienes were distinguished by passing them through a chloromaleic anhydride column; the *trans* isomer reacts more readily and more of it is removed (87,102).

B. Alcohols

It has been reported that primary and secondary alcohols are subtracted by boric acid on packing materials, while tertiary alcohols usually are dehydrated and pass the packing, presumably as olefins (50,103,104). To evaluate the potential utility of this reaction for structure analysis, a variety of alcohols, phenols, thiols, and compounds of other types were passed through a 6-in. boric acid loop [filled with a mixture of 1 part (w/w) powdered boric acid and 20 parts of 5% Carbowax 20M on 70/80 Anakrom ABS (Analabs, Hamden, Connecticut)] to determine which compounds were subtracted or otherwise affected (22). Comparisons were made with a check loop containing the Carbowax 20M packing but no boric acid. In general it was considered best to test the compounds with the loop following the analytical column to avoid bleeding of the subtraction agent or reaction products into the column; however, a compound that passed a loop was also checked with the loop placed ahead of the column. A difference in the retention time of the compound with the loop before and after the column immediately indicated that the compound was altered in passage through the loop. For example, with the loop ahead of the analytical column, a tertiary alcohol was dehydrated to the olefin, which traversed the column; the retention time observed was that of the olefin. With the loop following the column, the tertiary alcohol traversed the column as such, and the olefin was formed just ahead of the detector; the observed retention time was therefore that of the alcohol.

Our studies (22) verified the results of previous workers and revealed some additional facts. Primary and secondary alcohols were held up in the loop, probably as borate esters, but eventually they did bleed off as extremely broad peaks. Compounds that were subtracted partially (less than 50%) or not at all exhibited little, if any, peak broadening or delay in retention time. Tertiary alcohols were not subtracted, nor were alcohols having an internal double bond alpha to a secondary hydroxyl group; some with primary hydroxyl groups (allylic OH) were dehydrated. The peaks of phenols were broadened, and their retention times were increased but less than twofold. Salicylaldehyde was an exception; it was subtracted. Unhindered carboxylic acids (no alpha substitution) were partially subtracted and their peaks were broadened but not delayed. Efficiency of subtraction and extent of dehydration were not affected by temperature (75–200°C) or sample size (0.6–300 μg).

Using a precolumn microreactor containing 15% boric acid on diatomaceous brick, Osokin, Feldblum, and Kryukov (50) noted that some epoxides were isomerized to aldehydes.

C. Acids

Zinc oxide (ZnO) has been used to remove straight-chain carboxylic acids in the analysis of ozonolysis products (56). We found ZnO powder physically mixed in a ratio of 1:10 by weight with LAC-2R-446–phosphoric acid packing (used in the chromatography of the free acids (105)) gave good results in the 6-in. loop (22). Carboxylic acids were subtracted completely except for those considered hindered (alpha-substituted). Hindered acids tailed badly, their retention times were increased as much as 50%, and their peak heights were reduced at least threefold, more often tenfold. In any event, all carboxylic acids were recognized, either by their complete removal or by the drastic change in their peak shape. The only other compounds affected were alcohols and phenols. As much as 50% of the primary alcohols were subtracted, while less than 20% of the phenols and secondary and tertiary alcohols were removed; surviving peaks were not altered in shape or retention time. The ZnO loop has a total capacity of 1.5–2 mg of acid and its condition should be checked periodically with an acid that is known to be subtracted.

A base-treated precolumn [20% Apiezon M on 80/100 mesh Celite 545 which had been impregnated with 5% NaOH (w/w)] has been used to subtract fatty acids from flavor extracts of sterilized concentrated milk (106).

D. Aldehydes, Ketones, and Epoxides

Allen (107) has reported that FFAP, a commercially available stationary phase,* subtracts aldehydes; its mechanism of action is unknown. Bierl, Beroza, and Ashton (22) investigated this finding further. They found that a 6-in. loop containing 20% FFAP on 60/80 mesh acid-washed Chromosorb P subtracts most aldehydes from complex mixtures in the temperature range 75–200°C; aldehyde samples ranging from 0.6 to 60 μg were efficiently removed. For aldehydes with branching at the alpha carbon the loop is less efficient, and as much as 20% passes unchanged. FFAP also subtracts epoxides (22).

Osokin, Feldblum, and Kryukov (50) have used diatomaceous brick with polyethylene glycol 1000 impregnated with hydroxylamine hydrochloride (20%) and aqueous alcoholic sodium hydroxide (6%) to subtract aldehydes and ketones by formation of oximes. Alcohols and an epoxide were not affected.

The well-known reaction of aldehydes with amines to form Schiff bases led Bierl, Beroza, and Ashton (22) to try various amines in an attempt to find general and effective subtraction agents for aldehydes. Of the compounds tested, o-dianisidine proved best. A wide variety of aldehydes, including alpha-substituted ones, were quantitatively removed by a 6-in. loop containing 5% of the amine on Anakrom ABS with the last half inch being uncoated support to prevent bleeding of the o-dianisidine. With the exception of cyclohexanone (20–50% retained), ketones were not subtracted. Epoxides, C_{12} and higher, were partially or completely subtracted; epoxides of lower molecular weight passed the loop unchanged. Ethers, esters, alcohols, phenols, olefins, and hydrocarbons were unaffected by the loops. Bleeding of the o-dianisidine became excessive above 175°C.

A 6-in. loop of 20% benzidine on Chromosorb P efficiently subtracts aldehydes, most ketones, and epoxides at 100–175°C. Alpha-substituted ketones are partially subtracted. Removal of esters, ethers, and alcohols is slight. Because benzidine causes some delay, spreading, and retention of

---

* Varian Aerograph, Varian Associates, Walnut Creek, California.

unreactive compounds, subtractions of less than 40% with this loop cannot be considered reliable evidence for ketones, aldehydes, or epoxides in an unknown system.

Although the o-dianisidine and benzidine loops may differentiate an aldehyde from a ketone, subtraction of an unknown compound by either loop does not rule out the possibility that the compound might be an epoxide. To remove the uncertainty, a short post-column plug (100–200 mg) of 5% phosphoric acid (85%) on Chromosorb W was found (22) to be an effective subtraction agent for most epoxides (>80%; usually over 95% subtracted). In tests of other compounds that might also be subtracted by the phosphoric acid loop (other than basic compounds), it was found (22) that reactive ethers (benzyl ethers), tetrahydrofuran, and crotonaldehyde were reduced in peak height and suffered delays in retention time up to 30%; however, the subtraction of these compounds seldom exceeded 80%. Inasmuch as phosphoric acid is not a mild reagent, subtraction with it does not necessarily infer that an epoxide is present; rather it is useful as a confirmatory test for an epoxide. For example, some tertiary alcohols were dehydrated by the phosphoric acid loop (22,50). As a precolumn, the loop removed the alcohol peak from its original position because the alcohol was changed to an olefin; as a post-column the loop had no apparent effect on the alcohol peak.

A few epoxy compounds were not subtracted by this phosphoric acid loop; only 50% subtraction was observed with 2,3-(epoxyethyl)benzene. Although 2,3-epoxy-2-ethyl-1-hexanol was not subtracted by a post-column loop, the large decrease in the retention time with a precolumn loop indicated the compound had undergone some reaction (22). Osokin, Feldblum, and Kryukov (50), using a precolumn microreactor with 15% phosphoric acid on diatomaceous brick at 100°C, reported that 1,2-epoxy-2-methylpentane was isomerized to 2-methylpentanal, with a resultant increase in the retention time.

VI. Functional Group Analyses and Class Reactions

Many reactions discussed thus far could be considered functional group analyses. Subtractive reactions and hydrogenations, for example, fit within this category. The title of this section is therefore arbitrary and is actually a catch-all to embrace many of the remaining reactions.

In early GC work with polar compounds, emphasis was given to derivative formation. Compounds such as alcohols, phenols, and amines were often retained by the GC supports and quantification of small amounts tended to be unreliable. [Such adsorption or "selective retardation" of polar compounds on an active solid support has been proposed as a subtractive technique for the trace analysis of compounds such as esters, ketones, or alcohols (108).] More recently, the need for derivatizing polar compounds has been lessened through the availability of high-quality supports with adsorptive properties reduced by thorough silanization and the introduction of newly devised porous polymer bead packing materials (the various Porapak types). If adsorption problems are encountered, they are frequently minor; several injections of the material being analyzed (presumably to cover adsorptive sites) often overcome these difficulties. Accordingly, many polar compounds are now chromatographed directly, without derivatization. However, derivatization can often be used to indicate the presence of certain functional groups in a molecule or to improve the sensitivity of analysis by using one of the highly sensitive detectors to analyze derivatives containing responsive groups. The conversion of sterols to halogenated acetates, which can be observed with the electron capture detector, has been used to demonstrate the presence of hydroxyl groups (109). Similarly, Larkham and Pagington (110), working with complex mixtures such as tobacco smoke condensate, showed that reaction of the mixtures with chloroacetyl chloride followed by gas chromatography with a halogen-sensitive detector could be used to identify the alcohols present; Karmen (111) indicated that a detector specific to halogens could be helpful in the qualitative identification of acids and aldehydes as their 2-chloroethyl derivatives. Inasmuch as the subject of derivatization has recently been reviewed by Cavagnol and Betker (112), it will not be discussed here.

A. Vessel Reactions

Reactions that are too slow to go to completion within a loop or flow-through tubular reactor may be carried out in a small closed vessel located just ahead of the injection port. When the reaction is complete, the carrier gas is diverted through the vessel, and the volatile reaction product is measured by gas chromatography. This procedure has been useful mainly for quantitative measurement of functional groups. Thus, active hydrogen content has been determined by reacting lithium aluminum hydride with a compound and sweeping the evolved hydrogen into a gas chromatograph having a nitrogen carrier gas and a thermal conductivity detector (113, 113a). Active hydrogen was determined similarly by Lysyj and Greenough

(114), who also determined the hydrogen liberated from boron (hydridic hydrogen) following acid treatment. In another vessel reaction, Hoffmann and Lysyj (115) determined the amount of primary amino group; they added the compound and sodium nitrite solution to the vessel, purged it with the helium carrier gas, added acetic acid to generate nitrous acid, and after 1 min swept the evolved nitrogen into a gas chromatograph equipped with a thermal conductivity detector. The amount of nitrogen generated is a measure of primary amino group present. Franc and co-workers (116,117) used this reaction in combination with several other cleavage reactions for the systematic analysis of the nitrogen bonds in organic compounds.

Some reactions require as much as 15–60 min to go to completion. These include the Zeisel method for determining alkoxy groups (118–121), aromatic carboxyl group determination by decarboxylation with quinoline and cupric carbonate and measurement of the carbon dioxide generated (122), cleavage of alkyl or aryl sulfides by Raney nickel to form hydrocarbons (123), and determination of alkimino groups (121,124). In each instance, the final measurement was made by GC analysis.

Lower aldehydes and ketones are conveniently removed from natural products as their 2,4-dinitrophenylhydrazones. The carbonyl compounds are freed for GC analysis by heating the derivatives for 10 sec at 250°C with α-ketoglutaric acid in a capillary tube having its one open end exiting into the injection port of the gas chromatograph (125–129). The reaction is reportedly not satisfactory for furfuraldehyde or aldehydes larger than hexaldehyde (130). Other means of regenerating carbonyl compounds from their 2,4-dinitrophenylhydrazones have also been used. Jones and Monroe (131) obtained good results with a mixture of oxalic acid and p-dimethylaminobenzaldehyde in place of α-ketoglutaric acid. Ozonolysis of 2,4-dinitrophenylhydrazones has already been mentioned (Section IV).

B. Syringe Reactions

Hoff and Feit (132,133) chemically classified compounds by syringe reactions. Vapors of an organic compound were allowed to react with a chemical classification reagent in a hypodermic syringe, and the products were injected directly into a gas chromatograph with the syringe. They detected carbonyl compounds, differentiated between aldehydes and ketones, converted alcohols to acetates or nitrites, detected unsaturation, hydrogenated unsaturates, and distinguished ethers, olefins, aromatic hydrocarbons, and paraffins. Their technique has recently been extended to include relatively nonvolatile compounds (134).

C. Other Class Reactions

A number of class reactions of value for compound identification are given in Table 4-4.

TABLE 4-4

Precolumn Class Reactions

Type of compound	Reaction used	Typical refs.
Alcohols	Conversion to nitrites	135, 136
	Dehydration to olefins	136, 137
	Reduction to hydrocarbons	136
Acids	Esterification by flash-exchange with potassium ethyl sulfate	125, 138
	Esterification by decomposition of tetramethylammonium salts in injection port	139–144
Malonic acids	Decarboxylation to monocarboxylic acid in injection port	145
α-Hydroxy carboxylic acids	Periodic acid oxidation to carbonyl compounds in injection port	146, 147
Amino acids	Oxidation to aldehyde by ninhydrin	148
Phenols	Formation of methyl ethers by decomposition of tetramethylammonium phenolates in injection port	149
Purines, pyrimidines, barbiturates	Methylation in injection port by decomposition of tetramethylammonium salt	150, 151
Amides	Conversion to nitriles in phosphoric acid column	152
Esters	Saponification in a wet KOH column for analysis of the free alcohols	153
Salts of acids, mercaptans, and amines	Conversion to the free acid, mercaptan, or amine in an acidic or basic precolumn	154, 155
Tertiary amines	Conversion to olefins by Hofmann exhaustive methylation	156
Carbamates	Hydrolysis to phenols on short phosphoric acid precolumn	156a

VII. Fragmentation Reactions

Decomposition of compounds before gas chromatography to obtain characteristic fragments which produce a distinctive "peak pattern" or "fingerprint" chromatogram is another source of structural information. The most widely used application of this principle is in pyrolysis gas chromatography, discussed in Chapter 3. Other means of fragmentation have also been used.

Juvet and co-workers (157,158) have proposed the use of photolytic degradation as a structure-determining tool. From 1 to 3 mg of sample saturated with mercury is forced into the bottom of a quartz capillary tube which is then immersed in liquid nitrogen, and the tube is degassed by connecting it to a vacuum pump. After several freeze–thaw cycles, which include the admission of helium between cycles, the tube is sealed and irradiated at 25°C for 0.9 sec to 30 min. The irradiated sample in the tube is introduced into the gas chromatograph with a solid sampler injector. Although many products are obtained, the authors state that the functional groups present are deduced from "homologous" and common peaks present and that results are reproducible provided no more than 1–2% of the sample is decomposed. Data are given on saturated, mono-functional alcohols, aldehydes, ketones, ethers, and esters.

Fragmentation by electrical discharge (159–161), neutron activation (162), and "electron pyrolysis", with tritium in tritium water as the electron source (163) are other methods reported to give characteristic decomposition patterns.

Catalytic dehydrogenation of monoterpenes (19,20) and alkaloids (164) before GC gives distinctive dehydrogenation "spectra" as a result of the isomerization and decomposition that takes place; correlation of the peaks of the products with parent structures can be helpful for identification and structure elucidation.

Scholz, Bednarczyk, and Yamauchi (165) characterized polymers by oxidative degradation in a short precolumn followed by gas chromatography of the fragments. They suggested that ozonization, hydrolysis, or photolysis of polymers could be similarly employed. Minyard and Jackson (166) showed that the action of reagents such as Na_2CO_3, CuO, $CdCl_2$, $AlCl_3$, or $K_2Cr_2O_7$ in a precolumn reactor at 240°C with a number of common chlorinated pesticides gave "fingerprint" chromatograms which could be used to obtain more positive identification of the pesticides in residue extracts. It has been suggested (167) that the characteristic olefin peak patterns obtained in the gas chromatography of sterol methane-sulfonates can be used in identification of sterol systems.

VIII. Confirmation of Identity

When a compound has been tentatively identified, confirmation of its structure is usually necessary. Combinations of the precolumn reactions discussed in this chapter can often be used for this purpose. Spectrometric methods, such as are discussed in Chapters 5 and 6, are of great value. [The device described in Section I-C is also very useful for trapping material coming off the GC column and transferring it to infrared cells or to a mass spectrometer (7).]

The technique of peak shifting by on-column derivative formation can also be used for confirming identity; injection of the material is followed immediately by a second injection of a reagent which will form a volatile derivative. In this way alcohols have been converted to trimethylsilyl ethers (168) or acetates (169), alkaloids and steroids to acetates and propionates (170), and naphthylamines to acetamides (171).

To verify the identity of a compound, its distribution in a two-phase system (with solvents equilibrated before use) may often be checked against that of the authentic compound by determining the p value (172–175). The compound in 5 ml of nonpolar phase (usually the upper one) is analyzed quantitatively. The solution is then extracted by shaking for 1 min with 5 ml of the polar phase, and the nonpolar phase is again analyzed. The ratio of the second analysis to the first is called the p value [amount of compound in the nonpolar phase (second analysis) divided by the total amount in the system (first analysis)]. Each compound has a characteristic p value with a given solvent pair, and determinations with several solvent pairs may be used to further verify the identity. Results within 0.02 have been obtained because the same GC system, the same syringe, the same mode of injection, and the same solvents in the same amounts are used. This check for identity is mentioned because it may be used with the same GC system being employed, and there is no need to separate the compound from a mixture if its peak separates well from the other constituents. Thus, if 10 compounds on a chromatogram are identified (presumably by retention time), the p values for all 10 can be determined by a second analysis of the mixture after extraction with an equal volume of an immiscible phase. A check for quantitative linearity of the system may be made quickly by diluting the solution of the compound with an equal volume of solvent and analyzing; response (peak area or height) should be one-half that of the undiluted solution. If it is not, a calibration curve must be constructed. The p-value technique is based on a constant partition ratio regardless of concentration. It is reliable for dilute solutions of compounds that do not have a strong tendency to associate. It tends to work well with compounds that can be analyzed well in gas chromatography.

IX. Summary

Chemical reactions may advantageously be combined with gas chromatography to help determine the chemical structure of compounds, especially those available only in limited amounts. Several procedures and apparatus are described in detail, and others are simply cited to signal their availability. Reactions may be carried out prior to or during gas chromatography. Analyses applicable at the microgram level are stressed.

Carbon-skeleton chromatography may be used to strip functional groups from a molecule and thus provide information on the carbon skeleton. Principles, apparatus (including one that fits in the injection port), notes on the proper procedure for different compounds, and applications of the technique are described. Its use with the retention index or a similar system offers a good starting point for structure determination.

Hydrogenation is another technique useful for determining chemical structure; with the appropriate setup, it is easily carried out within the gas chromatographic pathway. Hydrogenation is applicable to a wide variety of compounds, and in many instances, it is quantitative.

Location of double bonds in many types of compounds may be carried out quickly and easily by subjecting them to ozonolysis. The resulting aldehyde or ketone fragments are determined by gas chromatography.

Subtractive reactions may be used to remove or distinguish between compounds having different functional groups, such as alcohols, acids, aldehydes, ketones, and epoxides. A convenient means of performing such reactions is to include in the gas chromatographic system a small length of packing which reacts with the desired functional group of a compound and prevents passage of that compound.

Analyses for specific functional groups may be carried out in a vessel located ahead of the injection port or in a syringe and the products led into a gas chromatograph for qualitative or quantitative determination.

The structures of compounds tentatively identified should be verified by spectrometry, derivative formation, reactions discussed here, or by other means. The many precolumn reactions reported attest to the usefulness and versatility of this relatively new technique.

References

1. M. Beroza and R. A. Coad, *J. Gas Chromatog.*, **4**, 199 (1966).
2. M. Beroza and R. A. Coad, in *The Practice of Gas Chromatography*, L. S. Ettre and A. Zlatkis, Eds., Interscience, New York, 1967, pp. 461–510.
3. V. G. Berezkin, *Analytical Reaction Gas Chromatography*, translated from Russian, Plenum Press, New York, 1968.
4. A. Wehrli and E. Kováts, *Helv. Chim. Acta*, **42**, 2709 (1959).
5. R. Teranishi, R. E. Lundin, W. H. McFadden, and J. R. Scherer, in *The Practice of Gas Chromatography*, L. S. Ettre and A. Zlatkis, Eds., Interscience, New York, 1967, pp. 407–459.
6. S. Dal Nogare and R. S. Juvet, Jr., *Gas-Liquid Chromatography*, Interscience, New York, 1962.
7. B. A. Bierl, M. Beroza, and J. M. Ruth, *J. Gas Chromatog.*, **6**, 286 (1968).
8. M. Beroza, *J. Gas Chromatog.*, **2**, 330 (1964).
9. M. Beroza, *Nature*, **196**, 768 (1962).
10. M. Beroza, *Anal. Chem.*, **34**, 1801 (1962).
11. M. Beroza and R. Sarmiento, *Anal. Chem.*, **35**, 1353 (1963).
12. M. Beroza and F. Acree, Jr., *J. Assoc. Offic. Agr. Chemists*, **47**, 1 (1964).
13. M. Beroza and R. Sarmiento, *Anal. Chem.*, **36**, 1744 (1964).
14. M. Beroza and R. Sarmiento, *Anal. Chem.*, **37**, 1040 (1965).
15. A. I. M. Keulemans and H. H. Voge, *J. Phys. Chem.*, **63**, 476 (1959).
16. R. J. Kokes, H. Tobin, Jr., and P. H. Emmett, *J. Am. Chem. Soc.*, **77**, 5860 (1955).
17. R. Rowan, Jr., *Anal. Chem.*, **33**, 658 (1961).
18. I. C. Nigam, *J. Chromatog.*, **24**, 188 (1966).
19. I. Mizrahi and I. C. Nigam, *J. Chromatog.*, **25**, 230 (1966).
20. T. Okamoto and T. Onaka, *Chem. Pharm. Bull. (Tokyo)*, **11**, 1086 (1963).
21. R. G. Brownlee and R. M. Silverstein, *Anal. Chem.*, **40**, 2077 (1968).
22. B. A. Bierl, M. Beroza, and W. T. Ashton, *Mikrochim. Acta*, in press.
23. M. Beroza, *J. Org. Chem.*, **28**, 3562 (1963).
24. R. M. Silverstein, R. G. Brownlee, T. E. Bellas, D. L. Wood, and L. E. Browne, *Science*, **159**, 889 (1968).
25. E. G. Perkins, *J. Am. Oil Chemists Soc.*, **44**, 197 (1967).
26. M. Beroza, B. A. Bierl, and J. M. Ruth, *J. Am. Oil Chemists Soc.*, **45**, 441 (1968).
27. M. Beroza, F. Acree, Jr., R. B. Turner, and B. H. Braun, *J. Econ. Entomol.*, **59**, 376 (1966).
28. T. M. Valega and M. Beroza, *J. Econ. Entomol.*, **60**, 341 (1967).
29. R. I. Asai, F. A. Gunther, and W. E. Westlake, *Residue Rev.*, **19**, 57 (1967).
30. J. J. Broderick, *Am. Perfumer Cosmet.*, **80**, 39 (1965).
31. G. Schomburg, *Z. Anal. Chem.*, **200**, 360 (1964).
32. R. C. Crippen and C. E. Smith, *J. Gas Chromatog.*, **3**, 37 (1965).
33. P. M. Adhikary and R. A. Harkness, *Biochem. J.*, **105** (3), 40P (1967).
34. J. A. Wenninger, R. L. Yates, and M. Dolinsky, *J. Assoc. Offic. Anal. Chemists*, **50**, 1304 (1967).
35. J. P. Minyard, J. H. Tumlinson, A. C. Thompson, and P. A. Hedin, *J. Agr. Food Chem.*, **14**, 332 (1966).

36. J. P. Minyard, J. H. Tumlinson, A. C. Thompson, and P. A. Hedin, *J. Agr. Food Chem.*, **15**, 517 (1967).
37. P. A. T. Swoboda and C. H. Lea, *J. Sci. Food Agr.*, **16**, 680 (1965).
38. C. J. Thompson, H. J. Coleman, R. L. Hopkins, and H. T. Rall, *J. Gas Chromatog.*, **5**, 1 (1967).
39. I. Klesment, *J. Chromatog.*, **31**, 28 (1967).
40. E. Kováts, in *Advances in Chromatography*, Vol. 1, J. C. Giddings and R. A. Keller, Eds., Dekker, New York, 1966, pp. 229–247.
41. L. S. Ettre, *Anal. Chem.*, **36**(8), 31A (1964).
42. L. S. Ettre, in *The Practice of Gas Chromatography*, L. S. Ettre and A. Zlatkis, Eds., Interscience, New York, 1967, pp. 381–387.
43. H. J. Dutton and T. L. Mounts, *J. Catalysis*, **3**, 363 (1964).
44. T. L. Mounts and H. J. Dutton, *Anal. Chem.*, **37**, 641 (1965).
45. H. J. Dutton, in *Advances in Tracer Methodology*, Vol. 2, S. Rothchild, Ed., Plenum Press, New York, 1965, pp. 123–134.
46. M. Beroza and R. Sarmiento, *Anal. Chem.*, **38**, 1042 (1966).
47. R. G. Ackman, *J. Am. Oil Chemists Soc.*, **40**, 564 (1963).
48. A. B. Littlewood and W. A. Wiseman, *J. Gas Chromatog.*, **5**, 334 (1967).
49. P. L. Gordon, *J. Inst. Petrol.*, **51**, 398 (1965).
50. Y. G. Osokin, V. S. Feldblum, and S. I. Kryukov, *Neftekhimiya*, **6**, 333 (1966).
51. I. Hornstein, P. F. Crowe, and R. Hiner, *J. Food Sci.*, **32**, 650 (1967).
52. I. Hornstein, P. F. Crowe, and J. B. Ruck, *J. Gas Chromatog.*, **5**, 319 (1967).
53. J. Lecerf and J. Bezard, *Rev. Franc. Corps Gras*, **13**, 455 (1966).
54. O. S. Privett, in *Progress in Chemistry of Fats and Other Lipids*, Vol. IX, R. T. Holman, Ed., Pergamon Press, London, 1966, pp. 91–117.
55. O. S. Privett and E. C. Nickell, *J. Am. Oil Chemists Soc.*, **43**, 393 (1966).
56. V. L. Davison and H. J. Dutton, *Anal. Chem.*, **38**, 1302 (1966).
57. H. J. Dutton, V. L. Davison, and E. D. Bitner, *J. Am. Oil Chemists Soc.*, **44**, 133A (1967).
58. E. D. Bitner, V. L. Davison, and H. J. Dutton, *J. Am. Oil Chemists Soc.*, **44**, 360A (1967).
59. E. C. Nickell and O. S. Privett, *Lipids*, **1**, 166 (1966).
60. M. Beroza and B. A. Bierl, *Anal. Chem.*, **39**, 1131 (1967).
61. R. A. Stein and N. Nicolaides, *J. Lipid Res.*, **3**, 476 (1962).
62. M. Beroza and B. A. Bierl, *Anal. Chem.*, **38**, 1976 (1966).
63. A. Greiner, *J. Prakt. Chem.*, **13**, 157 (1961).
64. R. Kleiman, G. F. Spencer, and F. R. Earle, *J. Am. Oil Chemists Soc.*, **44**, 362A (1967).
65. O. S. Privett and E. C. Nickell, *Lipids*, **1**, 98 (1966).
66. C. R. Scholfield, V. L. Davison, and H. J. Dutton, *J. Am. Oil Chemists Soc.*, **44**, 648 (1967).
67. P. Ronkainen and S. Brummer, *J. Chromatog.*, **28**, 253 (1967).
68. P. Ronkainen and S. Brummer, *J. Chromatog.*, **28**, 259 (1967).
69. P. Ronkainen, S. Brummer, and H. Suomalainen, *J. Chromatog.*, **28**, 443 (1967).
70. P. Ronkainen, S. Brummer, and H. Suomalainen, *J. Chromatog.*, **28**, 270, (1967).
71. G. E. Howard and A. Hoffman, *J. Sci. Food Agr.*, **18**, 106 (1967).
72. N. Brenner and V. J. Coates, *Nature*, **181**, 1401 (1958).
73. E. R. Adlard and B. T. Whitham, *Nature*, **192**, 966 (1961).

74. E. M. Barrall II, and F. Baumann, *J. Gas Chromatog.*, **2**, 256 (1964).
75. I. H. Williams, *Anal. Chem.*, **37**, 1723 (1965).
76. G. C. Blytas and D. L. Peterson, *Anal. Chem.*, **39**, 1434 (1967).
77. W. B. Innes and W. E. Bambrick, *J. Gas Chromatog.*, **2**, 309 (1964).
78. W. B. Innes, W. E. Bambrick, and A. J. Andreatch, *Anal. Chem.*, **35**, 1198 (1963).
79. D. L. Klosterman and J. E. Sigsby, Jr., *Environ. Sci. Technol.*, **1**, 309 (1967).
80. N. H. Ray, *Analyst*, **80**, 853 (1955).
81. D. J. McEwen, *Anal. Chem.*, **38**, 1047 (1966).
82. D. E. Seizinger, *Instr. News*, **18** (1), 11 (1967).
83. R. L. Hoffmann, G. R. List, and C. D. Evans, *J. Am. Oil Chemists Soc.*, **43**, 675 (1966).
84. J. Janák and J. Novák, *Chem. Listy*, **51**, 1832 (1957).
85. J. Janák and J. Novák, *Collection Czech. Chem. Commun.*, **24**, 384 (1959).
86. Y. Suzuki, D. Ishii, and T. Takeuchi, *Kogyo Kagaku Zasshi*, **69**, 1916 (1966); *Chem. Abstr.*, **66**, 85632t (1967).
87. E. Gil-Av and Y. Herzberg-Minzly, *J. Chromatog.*, **13**, 1 (1964).
88. M. Rogozinski, *J. Gas Chromatog.*, **2**, 163 (1964).
89. P. W. Sherwood, *Brennstoff-Chem.*, **40**, 354 (1959).
90. C. K. Hersh, *Molecular Sieves*, Reinhold, New York, 1961.
91. E. R. Adlard and B. T. Whitham, in *Gas Chromatography*, N. Brenner, J. E. Callen, and M. D. Weiss, Eds., Academic Press, New York, 1962, pp. 371–390.
92. P. A. Schenck and E. Eisma, *Nature*, **199**, 170 (1963).
93. R. Stock and C. B. F. Rice, *Chromatographic Methods*, Reinhold, New York, 1963, pp. 114–143.
94. T. P. Maher, *J. Gas Chromatog.*, **4**, 355 (1966).
95. N. Brenner, E. Cieplinski, L. S. Ettre, and V. J. Coates, *J. Chromatog.*, **3**, 230 (1960).
96. L. S. Ettre and N. Brenner, *J. Chromatog.*, **3**, 235 (1960).
97. R. E. Isbell, *Anal. Chem.*, **35**, 255 (1963).
98. M. Lefort and X. Tarrago, *J. Chromatog.*, **2**, 218 (1959).
99. I. A. Murdoch, *Analyst*, **86**, 856 (1961).
100. G. W. Heylmun, *J. Gas Chromatog.*, **3**, 82 (1965).
101. H. Wirth, *Monatsh. Chem.*, **84**, 751 (1953); *Chem. Abstr.*, **48**, 2522b (1954).
102. E. Gil-Av and Y. Herzberg-Minzly, *Proc. Chem. Soc.*, **1961**, 316.
103. F. W. Hefendehl, *Naturwiss.*, **51**, 138 (1964).
104. R. M. Ikeda, D. E. Simmons, and J. D. Grossman, *Anal. Chem.*, **36**, 2188 (1964).
105. L. D. Metcalfe, *Nature*, **188**, 142 (1960).
106. R. G. Arnold, L. M. Libbey, and E. A. Day, *J. Food Sci.*, **31**, 466 (1966).
107. R. R. Allen, *Anal. Chem.*, **38**, 1287 (1966).
108. R. Jeltes and R. Veldink, *J. Chromatog.*, **32**, 413 (1968).
109. R. A. Landowne and S. R. Lipsky, *Anal. Chem.*, **35**, 532 (1963).
110. T. W. Larkham and J. S. Pagington, *J. Chromatog.*, **28**, 422 (1967).
111. A. Karmen, *J. Lipid Res.*, **8**, 234 (1967).
112. J. C. Cavagnol and W. R. Betker, in *The Practice of Gas Chromatography*, L. S. Ettre and A. Zlatkis, Eds. Interscience, New York, 1967, pp. 71–127.
113. M. N. Chumachenko and L. B. Tverdyukova, *Proc. Acad. Sci. USSR, Chem. Sect. (English Transl.)*, **142**, 77 (1962); *Dokl. Akad. Nauk SSSR*, **142**, 612 (1962).
113a. M. N. Chumachenko, L. B. Tverdyukova, and F. G. Leenson, *Zh. Anal. Khim.*, **21**, 617 (1966).

114. I. Lysyj and R. C. Greenough, *Anal. Chem.*, **35**, 1657 (1963).
115. E. R. Hoffmann and I. Lysyj, *Microchem. J.*, **6**, 45 (1962).
116. J. Franc, V. Kovář, and F. Mikeš, *Mikrochim. Acta*, **1966**, 133.
117. J. Franc and F. Mikeš, *J. Chromatog.*, **26**, 378 (1967).
118. D. L. Miller, E. P. Samsel, and J. G. Cobler, *Anal. Chem.*, **33**, 677 (1961).
119. T. Mitsui and Y. Kitamura, *Microchem. J.*, **7**, 141 (1963).
120. F. Ehrenberger, *Z. Anal. Chem.*, **210**, 424 (1965).
121. M. M. Schachter and T. S. Ma, *Mikrochim. Acta*, **1966**, 55.
122. T. S. Ma, C. T. Shang, and E. Manche, *Mikrochim. Acta*, **1964**, 571.
123. J. Franc, J. Dvořáček and V. Koloušková, *Mikrochim. Acta*, **1965**, 4.
124. N. D. Cheronis and T. S. Ma, *Organic Functional Group Analysis by Micro and Semimicro Methods*, Interscience, New York, 1964, pp. 622–626.
125. J. W. Ralls, *Anal. Chem.*, **32**, 332 (1960).
126. J. W. Ralls, *Anal. Chem.*, **36**, 946 (1964).
127. M. E. Mason, B. Johnson, and M. C. Hamming, *Anal. Chem.*, **37**, 760 (1965).
128. M. E. Mason, B. Johnson, and M. C. Hamming, *J. Agr. Food Chem.*, **15**, 66 (1967).
129. I. G. Mokhnachev, S. V. Kamenshchikova, and V. S. Kovtun, *Isv. Vysshikh. Uchebn. Zavedenii, Pischevaya Tekhnol.*, **1966**, No. 5, 58.
130. R. P. Collins and K. Kalnins, *Lloydia*, **28**, 48 (1965).
131. L. A. Jones and R. J. Monroe, *Anal. Chem.*, **37**, 935 (1965).
132. J. E. Hoff and E. D. Feit, *Anal. Chem.*, **35**, 1298 (1963).
133. J. E. Hoff and E. D. Feit, *Anal. Chem.*, **36**, 1002 (1964).
134. K. M. Fredericks and R. Taylor, *Anal. Chem.*, **38**, 1961 (1966).
135. F. Drawert and G. Kupfer, *Angew. Chem.*, **72**, 33 (1960).
136. F. Drawert, R. Felgenhauer, and G. Kupfer, *Angew. Chem.*, **72**, 555 (1960).
137. F. Drawert and K. H. Reuther, *Chem. Ber.*, **93**, 3066 (1960).
138. I. R. Hunter, *J. Chromatog.*, **7**, 288 (1962).
139. E. W. Robb and J. J. Westbrook III, *Anal. Chem.*, **35**, 1644 (1963).
140. N. E. Hetman, H. G. Arlt, Jr., R. Paylor, and R. Feinland, *J. Am. Oil Chemists Soc.*, **42**, 255 (1965).
141. D. T. Downing, *Anal. Chem.*, **39**, 218 (1967).
142. D. T. Downing and R. S. Greene, *Anal. Chem.*, **40**, 827 (1968).
143. R. II. Leonard, *J. Gas Chromatog.*, **5**, 323 (1967).
144. J. J. Bailey, *Anal. Chem.*, **39**, 1485 (1967).
145. N. E. Hoffman and I. R. White, *Anal. Chem.*, **37**, 1541 (1965).
146. N. E. Hoffman, J. J. Barboriak, and H. F. Hardman, *Anal. Biochem.*, **9**, 175 (1964).
147. N. E. Hoffman and P. J. Conigliaro, *Develop. Appl. Spectry.*, **4**, 299 (1965).
148. A. Zlatkis, J. F. Oró, and A. P. Kimball, *Anal. Chem.*, **32**, 162 (1960).
149. H. G. Henkel, *J. Chromatog.*, **20**, 596 (1965).
150. J. MacGee, *Anal. Biochem.*, **14**, 305 (1966).
151. G. W. Stevenson, *Anal. Chem.*, **38**, 1948 (1966).
152. D. J. Gaede, *Dissertation Abstr.*, **27**(8), 2612B (1967).
153. J. Janák, J. Novák, and J. Sulovsky, *Collection Czech. Chem. Commun.*, **27**, 2541 (1962).
154. G. F. Thompson and K. Smith, *Anal. Chem.*, **37**, 1591 (1965).
155. H. D. LeRosen, *Abstr., 148th Natl. Mtg. Am. Chem. Soc., Chicago, Ill., Aug.– Sept. 1964*, p. 26B.

156. H. B. Hucker and J. K. Miller, *J. Chromatog.*, **32**, 408 (1968).
156a. M. C. Bowman and M. Beroza, *J. Agr. Food Chem.*, **15**, 894 (1967).
157. R. S. Juvet, Jr. and L. P. Turner, *Anal. Chem.*, **37**, 1464 (1965).
158. R. S. Juvet, Jr., R. L. Tanner, and J. C. Y. Tsao, *J. Gas Chromatog.*, **5**, 15 (1967).
159. R. C. Cavenah and T. Johns, *Analyzer*, **7**(3), 2 (1966). *Chem. Abstr.*, **67**, 60707d (1967).
160. J. C. Sternberg and R. L. Litle, *Anal. Chem.*, **38**, 321 (1966).
161. J. C. Sternberg, I. H. Krull, and G. D. Friedel, *Anal. Chem.*, **38**, 1639 (1966).
162. S. P. Cram and J. L. Brownlee, Jr., *J. Gas Chromatog.*, **5**, 353 (1967).
163. H. Schildknecht and O. Volkert, *Z. Anal. Chem.*, **216**, 97 (1966).
164. C. Radecka and I. C. Nigam, *J. Pharm. Sci.*, **56**, 1608 (1967).
165. R. G. Scholz, J. Bednarczyk, and T. Yamauchi, *Anal. Chem.*, **38**, 331 (1966).
166. J. P. Minyard and E. R. Jackson, *J. Agr. Food Chem.*, **13**, 50 (1965).
167. W. J. A. VandenHeuvel, *J. Chromatog.*, **26**, 396 (1967).
168. S. H. Langer and P. Pantages, *Nature*, **191**, 141 (1961).
169. J. K. Haken, *J. Gas Chromatog.*, **1**(10), 30 (1963).
170. M. W. Anders and G. J. Mannering, *Anal. Chem.*, **34**, 730 (1962).
171. D. M. Marmion, R. G. White, L. H. Bille, and K. H. Ferber, *J. Gas Chromatog.*, **4**, 190 (1966).
172. M. Beroza and M. C. Bowman, *J. Ass. Offic. Agr. Chemists*, **48**, 358 (1965).
173. M. C. Bowman and M. Beroza, *J. Ass. Offic. Agr. Chemists*, **48**, 943 (1965).
174. M. C. Bowman and M. Beroza, *Anal. Chem.*, **38**, 1427 (1966).
175. M. C. Bowman and M. Beroza, *Anal. Chem.*, **38**, 1544 (1966).

CHAPTER 5

Gas Chromatography and Mass Spectroscopy

J. Throck Watson,* *USAF School of Aerospace Medicine, Aerospace Medical Division (AFSC), Brooks Air Force Base, San Antonio, Texas*

*Present address: Faculté des Sciences, Université de Strasbourg, Strasbourg, France.

I. Introduction

When gas chromatographic data seem dubious, the investigator is forced to seek an additional analytical technique which will more completely characterize the substance and finally result in its identification. Ideally, such an augmentative technique would be operationally or directly combined with the gas chromatograph.

The vaporous nature of gas chromatographic effluents renders them especially amenable to further analysis by mass spectroscopy. In addition, both gas chromatography and mass spectroscopy have modest sample-size requirement (typically in the microgram range). Once the problems of interfacing are resolved, the combined gas chromatograph–mass spectrometer (GC–MS) provides a facile, sensitive, and convenient system for the separation and identification of complex mixtures.

II. Introduction to Mass Spectroscopy

A. Principles of Operation

Characteristic mass spectral information is obtained from a compound by ionization and fragmentation of the sample vapor followed by analysis of the resulting ions according to mass. The schematic (Fig. 5-1) shows a mass spectrometer as a series of five regions which sequentially operate on the molecules and ions of the compound under investigation. A portion of the sample vapor in the reservoir is allowed to flow into the ionizing region

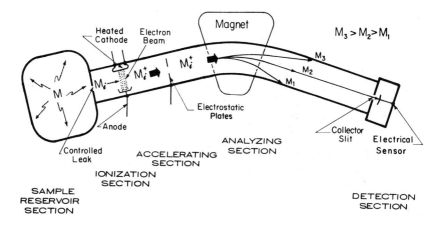

Fig. 5-1. Schematic of magnetic single-focusing mass spectrometer.

through a controlled leak which maintains the ion source pressure between 10^{-7} and 10^{-5} torr (mm Hg). This portion may be as low as 0.01% for a standard inlet to as high as 90% for some GC–separator–MS systems. An electron beam provides the energy for ionization and fragmentation; the electrons emanate from a heated cathode and are attracted (with an energy of 20–70 eV) through a small well-defined volume to an anode. The sample molecules (M) diffuse randomly throughout the ion source where some reach the electron beam and are ionized. The positively charged ions are extracted from the ion source and can now be controlled and focused by electrostatic plates between the ionizing region and the ion accelerating region. The ions are accelerated toward and through an electrostatic plate maintained at a large negative potential (e.g., -4000 V) relative to the ionizing region. The ions thus acquire 4000 eV of kinetic energy, and hence a characteristic velocity (KE $= \frac{1}{2}$ mv^2), before they enter the magnetic field for separation or mass analysis.

Charged particles moving through a magnetic field at right angles to the lines of magnetic flux are accelerated in a direction orthogonal to the initial velocity of the ions and to the lines of magnetic flux, resulting in a curvature of trajectory. This accelerating force is dependent on the velocity, and therefore the radius of curvature depends on the mass of an ion. The electrostatic and magnetic fields can be adjusted to focus the ion current of given mass-to-charge ratio (m/e) onto the slits of an electrical sensor (see Fig. 5-1), while the other ions are curved too much or too little and collide with the walls of the analyzer housing. The ions M_1 and M_3 in Figure 5-1 can be focused to the collector by readjusting either the magnetic field or the accelerating potential.

The relationship between the charged species and the electric and magnetic fields is given by

$$m/e = H^2R^2/2V \qquad (5\text{-}1)$$

where H is the magnetic field strength, V is the accelerating potential, and R is the radius of curvature. The magnetic mass spectrometer is used here merely for simple illustration; a complete discussion on the operating principles of these and other types of mass spectrometers (e.g., time of flight, radiofrequency, etc.) can be found elsewhere (1–3).

If a mass spectrometer produced only molecular ions in the ion chamber, it would be an excellent instrument for determinations of molecular weight, but would not give information regarding structure. Actually, the electron beam has sufficient energy (typically 70 eV) to cause both ionization of the molecule and fragmentation of chemical bonds. In general, fragmentation of the molecular ion (M^+) yields ions whose mass gives characteristic information about the expelled atom or group of atoms. The abundance of the various ions is governed by the structure and stability (potential energy) of the original and final charged and uncharged species. Thus, the mass spectrum, indicating the relative intensities and m/e values for the molecular and fragment ions, yields considerable characteristic information about the various portions of the molecule. These bits of information may be pieced together by deductive reasoning to "reconstruct" the structure of the original molecule.

The MS data output generally consists of a tracing of ion intensity vs. time. Time, of course, is related to the scan rate and can be converted to m/e. Figure 5-2a is an actual data trace of n-butanol recorded from a magnetic focusing mass spectrometer. The mass scale was ascertained empirically by recognizing the mass peaks of H_2O, N_2, and O_2 (m/e 18, 28, and 32, respectively) on a background scan (not illustrated). The broad, poorly shaped peaks midway through the mass spectrum are the so-called "metastable peaks" produced by ions which decompose in transit from the accelerator section to the analyzer (magnetic) section. The recorded m/e value for the diffuse "metastable peak" is related to the mass of the

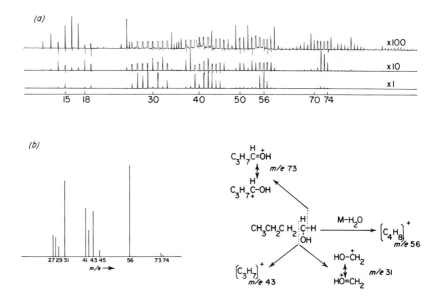

Fig. 5-2. (a) Tracing of actual recorded mass spectrum of n-butanol (courtesy of J. A. McCloskey, Baylor University, Houston, Texas). (b) Abbreviated plot of mass spectral data for n-butanol and suggested origin of some fragment ions (4). GC–MS inlet, 50°C; ion source, 250°C; 70 eV.

parent and daughter ions, and thus these "metastable peaks" are helpful in elucidating single-step modes of decomposition (1).

An abbreviated plot of the mass spectral data for *n*-butanol is shown in Figure 5-2*b*. The accompanying chemical formulas indicate possible modes of fragmentation which produce ions consistent with the mass spectrum. Ionization of the intact molecule gives an ion of mass 74, while fragmentation of this molecular ion involving loss of either a hydrogen or a propyl group results in resonance-stabilized ions of mass 73 and 31, respectively. Additionally, the molecular ion loses H_2O through a rearrangement resulting in a predominant ion of mass 56. The spectrum of butanol can be interpreted further (4), but this brief introduction indicates that a careful examination of both the molecular and fragment ions in a mass spectrum advances valuable clues in the investigation of molecular structure.

There are several reference books which discuss specific modes of fragmentation as well as the general approach to spectra interpretation for various compounds (1,5–11). The early catalogs of mass spectral data have been summarized recently by McFadden (12) and are now published as a single volume (13) which lists the compounds by molecular weight and by each of the four most abundant ions. A more recent catalog (14a) and its supplement (14b) also list the mass spectral data by order of the most intense peaks. Mass spectral data for approximately 6000 compounds listed according to molecular weight are now available in a set of three volumes (15). Furthermore, a new journal (16) will be devoted to the continuing distribution of mass spectra in a manner comparable to the early system of "uncertified mass spectral data."

B. Conventional Mass Spectroscopy

Resolving power is a parameter of extreme importance since it directly affects the quality of the mass spectrum. Adjacent peaks should be discernible, and, of course, greater resolving power allows easier and more reliable designations of mass number for the various ions. Medium-resolution (generally accepted as 500–2500) instruments produce better quality conventional mass spectra than the low-resolution (150–500) spectrometers. The spectrum in Figure 5-2*a* was obtained on a single-focusing magnetic mass spectrometer (LKB) with a resolving power of ~1000.

While a mass marker (accessory) conveniently indicates the *m/e* values on the recorded spectrum, the investigator may prefer to ascertain the mass scale experimentally. In the past, three methods have been used to determine the mass scale in conventional mass spectroscopy: (*a*) The masses are "counted" from an unambiguously known peak (e.g., m/e 28 from air) up to the molecular ion (1). (*b*) A second parameter such as time is recorded simultaneously [commonly utilized in time-of-flight mass spectrometry (18)]; unknown masses are then calculated mathematically by comparison with a set of known peaks. (*c*) A so-called "overlay" or "reproducible" mass scale can be used for reference.

When rapid scanning is necessary, as in most GC–MS instruments, all three methods, particularly (*b*) and (*c*), are not very reliable above masses 200–300. "Counting" (*a*) depends on the quality of the spectrum as well as the type of compound, and a well-resolved peak is desired at every few mass units. The most reliable technique of establishing a mass scale—the addition of a calibration compound whose mass spectrum has charac- teristic peaks well distributed over the entire mass range—requires scan- ning the mass spectrum twice, once with only the unknown compound and again while also bleeding the calibration compound vapors into the ion source. Alternatively, the calibration spectrum may be subtracted from the total, but this is time consuming and occasionally ambiguous.

C. High-Resolution Mass Spectroscopy

Modern high-resolution mass spectrometers (19) generally achieve resol- ution greater than 1 part in 10,000. The effectiveness of this resolving power is illustrated in Figure 5-3, which is a densitometer tracing of a

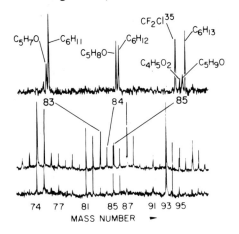

Fig. 5-3. Densitometer recordings of portions of mass spectra recorded on photo- graphic plate (32). Bottom: Calibration compound only (perfluorodimethylhexane). Middle: Superimposed spectra of methyl stearate and calibration compound. Top: Part of middle spectrum at slower densitometer speed.

small portion of a photographic plate (Mattauch-Herzog type of mass spectrograph) containing the mass spectra of a mixture of methyl stearate and perfluorodimethylhexane (20,32). The top trace is a much slower scan of the peaks at the three m/e values: 83, 84, 85. Although the densitometer was not designed for such fine lines, the slower scan indicates that at nominal m/e 85 there are four contributing species of different elemental composition. Note that the ion containing fluorine and chlorine lies at the low mass edge of the quadruplet, while the other species lie at progressively higher m/e as their hydrogen content increases. This "splitting" of the charged species at any given nominal mass is due to the characteristic mass defect observed for various elements, so that ^{19}F (18.9984) and ^{16}O (15.9949), for example, have an exact atomic mass slightly under the nominal integral mass, while ^{1}H (1.0078) is slightly over integral. (By convention, $^{12}C = 12.0000$.)

In high-resolution mass spectroscopy the exact mass scale can be determined conveniently and reliably because at any nominal mass, the ions of a suitably chosen calibration compound will be resolved from the sample ions. The calibration compound is carefully selected to have an elemental composition different from that of the compounds under investigation, e.g., a fluorocarbon is chosen if fluorine-free organic compounds are to be studied. Another convenience which results from this feature of high-resolution mass spectroscopy is that in the GC–MS combination interference from bleeding packed columns is avoided because the interfering ions are easily separated (as discussed in Section VIII-A).

The most important capability of a high-resolution instrument is to permit measurement of the mass of a charged particle with a high degree of accuracy. When the mass of an ion can be measured with high confidence to within a millimass unit (mmu), the number of elemental compositions which must be considered is substantially reduced from the very large number of compositions that fall within one nominal mass unit. This unique information, not easily obtained with low-resolution instruments, becomes particularly important in studies involving complex molecules of unknown structure (21). Consider for example, a portion of the conventional mass spectrum of a derivative of eburnamenine (20) shown in Figure 5-4. If only the conventional mass spectral data were known and no other descriptive data were available, over 200 different empirical formulas (2) would have to be considered at the outset for a molecular ion of nominal mass 308. However, since measurement of the molecular ion mass gives the value 308.188 $\pm$ 0.001, $C_{20}H_{24}N_2O$ is suggested as the only plausible empirical formula (20). The high-resolution mass spectral data for selected ions are also tabulated in Figure 5-4. The

High-resolution mass spectral data for selected ions:

Experimentally found	Calculated	Elemental composition
308.1878	308.1889	$C_{20}H_{24}N_2O$
None	238.1359	$C_{17}H_{18}O$
238.1249	238.1232	$C_{16}H_{16}NO$
None	238.1106	$C_{15}H_{14}N_2O$

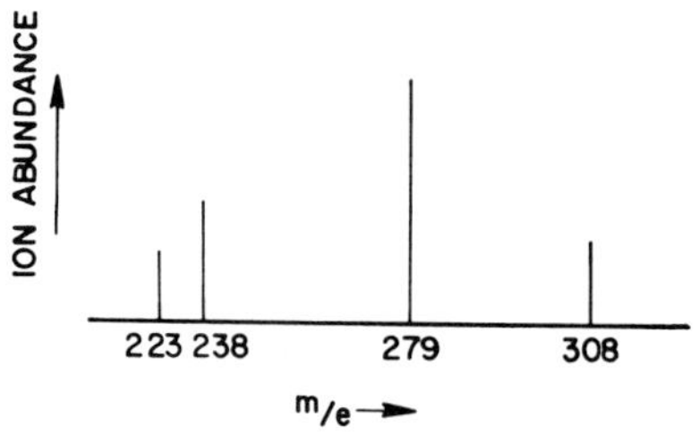

Fig. 5-4. Comparison of some of conventional and high-resolution mass spectral data of an alkaloid derivative of eburnamenine (see text) (20).

molecular ion (M$^+$) is experimentally "found" at m/e 308.1878 which corresponds to the elemental composition $C_{20}H_{24}N_2O$ (calcd.: 308.1889). [The term "found" denotes the numerical value of the mass-to-charge ratio computed from the position of the line on the photographic plate (22) of a high-resolution mass spectrograph (CEC-21-110). The term "calcd." refers to a summation of the atomic weights ($^{12}C = 12.0000$) of a set of elements. This summation should fall within a specified range of the mass "found," in this case ± 3 mmu.] The mass of the fragment M-70 is "found" to be 238.1249 and corresponds to $C_{16}H_{16}NO$ (calcd.: 238.1232). This information indicates that the M-70 fragment originates by loss of C_4H_8N from the molecular ion rather than, for example, C_5H_{10} or $C_3H_6N_2$.

As has been demonstrated, high-resolution mass spectral data can be both very useful and very awkward. While an accurate mass measurement can confirm the elemental composition of an ion, a long list of seven-digit mass numbers or even elemental compositions soon becomes relatively unintelligible. For this reason, Biemann, Bommer, and Desiderio (22) suggested the "elemental map" as a systematic format for presentation of high-resolution mass spectra.

Nominal mass number	CH	CHO	CHN	CHNO	CHN$_2$	CHN$_2$O
110			7/12(0)+			
115	9/7(1)+					
124			8/14(1)+++++			
130			9/8(1)+			
145			10/11(−2)+			
167			12/9(0)+			
180			13/10(−1)+			
223				15/13(0)+		
238				16/16(1)+++++++		
254		19/10(2)+				
279						18/19(1)++++++
308						20/24(0)+++++

Fig. 5-5. Element map format of high-resolution mass spectral data of an alkaloid discussed in text (20).

Briefly, the element map arranges the elemental compositions of the various ions vertically in order of increasing nominal mass and horizontally according to the heteroatom content. As shown in Figure 5-5, the first column contains the hydrocarbon species; the other columns list species which contain hydrogen and carbon but which also contain additional elements as indicated by the headings, e.g., CHO, CHN, CHNO. The number of carbons for each entry is indicated by the number preceding the solidus and the hydrogen content is given by the number following. The error of the mass measurement in millimass units is shown in parentheses, and the relative abundance of the ion is indicated on an exponential scale by the number of pluses following the parentheses. [This relationship is usually conveniently indicated by the number of asterisks printed on the computer output (22).]

III. Interfacing Considerations

In combining a gas chromatographic unit with a mass spectrometer, one often encounters unfortunate incompatible features. The desirable maximum operating pressure in the ion source of most mass spectrometers is typically in the region of 10^{-6} to 10^{-5} torr. On the other hand, most conventional gas chromatographs are operated conveniently with the column exit at ambient pressure, and, of course, the sample is necessarily diluted with the carrier gas in the process of elution. In the mass spectrometer, high pressure leads to excessive ion–molecule collisions and a deterioration of resolution, and therefore cannot be tolerated. In some cases (23,24), elevated pressures (approaching 10^{-5} torr of helium, uncorrected*) have significantly altered the characteristic fragmentation pattern of compounds. Such spectra indicate considerable bias against ions of higher mass. For example, n-octacosane was introduced through a conventional inlet into the ion source (Atlas CH4) in which the pressure was varied by helium flow through the GC inlet. As the total pressure approached 10^{-5} torr, uncorrected (23), the molecular ion virtually disappeared relative to the ions at lower mass. Helium ions are apparently not involved since similar results are obtained at both 20 and 70 eV. (The ionization potential of helium is 24.4 V.) On the other hand, this effect is not so significant in a differentially pumped instrument in which the ion source and the mass analyzer are evacuated by separate pumping systems and are connected only by the very small slit through which the ion beam

---

* Most ion gauges are calibrated for air. The reading should be multiplied by 8 to get an approximate corrected pressure.

is transmitted. At a helium pressure of about 2×10^{-4} torr in the ion source and 2×10^{-6} torr in the analyzer, the molecular ion of pristane (mass 268) is decreased by only 20% relative to the major fragment at mass 183 (25). Furthermore, the resolution under these conditions was degraded to only 20,000 from 30,000 obtained under normal low-pressure conditions (no helium). Of course, the purpose of differential pumping is to preserve resolution at high ion source pressures. The nature of intensity distortion at elevated helium pressure has not been fully explored; further investigation is needed to explain the mechanism of this phenomenon.

The ion source pressure in any mass spectrometer is established primarily by the volume flow rate entering the spectrometer in relation to the pumping rate. Most mass spectrometers can operate effectively while accepting volume flow rates up to about 0.2 ml/min. Thus in GC–MS combinations employing packed columns, only about 1% of effluent can be channeled into the mass spectrometer, while in some cases nearly all the effluent from an open tubular column may be introduced into the spectrometer without inviting deleterious pressure effects.

In order to retain the characteristic distribution of relative ion abundances, the sample pressure in the ion source should be constant while scanning the mass spectrum. The very nature of the GC output, namely a changing sample concentration per unit time, violates this classical mass spectroscopic requirement. The relative peak heights in a mass spectrum will be significantly distorted if the required scanning time is comparable to the duration of the emerging GC fraction. An extreme example of this intensity distortion is illustrated in Figure 5-6, which shows two mass spectra taken at different scanning rates from a 66-sec GC peak. The upper plot represents a spectrum scanned in 1.7 sec at the apex of the GC peak. In this case, the characteristic relative intensities of the ions are retained in the recording, as verified by comparison with a reference spectrum of ditrimethylsilylsebacate (23). However, in a second GC–MS analysis, the mass spectrum was scanned in 44 sec with obvious distortion of ion abundances in the resulting record (lower plot, Fig. 5-6). The M-15 fragment ion at m/e 331 is characteristically the most abundant, but in this recording its intensity is greatly diminished, as are the intensities of the ions at lower mass; the ion intensities in the middle portion of the spectrum are enhanced, of course, because they were recorded when the sample concentration was at the maximum.

Attempts have been made to correct slowly recorded (and therefore distorted) mass spectra from a GC–MS system. By simultaneously monitoring the total ion current (which is dependent on sample concentration in the ion source), the recorded ion intensities can be normalized or corrected.

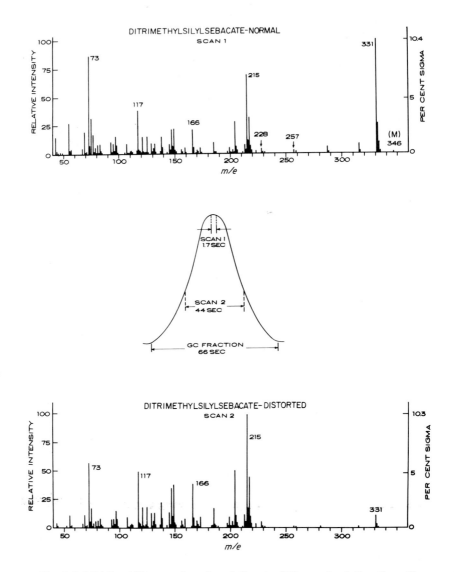

Fig. 5-6. Middle: MS scan times in relation to GC sample elution time. Top: Mass spectrum as recorded with essentially constant sample pressure. Bottom: Spectrum of same compound distorted by changing sample pressure during scanning (23).

For example, this can be accomplished on the Bendix Time-of-Flight mass spectrometer by shunting the total ion current to a second gate (or output) on alternate scan cycles (26). Changes in total ion current are then used to correct the intensities recorded during the scan. Similar corrections have been accomplished graphically from GC peak shape and known delay times and scan rates (27). However, the most practical correction technique yet offered produces spectra which are normalized to an artificially constant total ion current, regardless of changes in sample concentration during the scan. Kennet (28) designed a device for recording ratios (peak intensity vs. total ion intensity) by attenuating the photomultiplier output with a potentiometer mechanically coupled to the recorder monitoring the total ion current (derived from the double ion source of an Atlas CH4 mass spectrometer). In this arrangement (28) the potentiometer is geared to the main slide wire shaft of the total ion current recorder so that the zero-resistance position of the potentiometer corresponds to a total ion current level of 10% full scale. Thus, the recorded spectra will be independent of changes in sample concentration as long as the total ion current recorder indicates between 10% and 100% of full scale. Furthermore, the signal-to-noise ratio was improved by an order of magnitude when ratio recording at a rate of 6 sec per octave after incorporating a noise filter (frequency attenuator) (28).

Rapid scanning devices have been used in mass spectrometry which allow scanning of the spectrum (masses 50–500 at a resolution of at least 500 with 10% valley) in 1–3 sec (23,29,30). If one selects GC conditions so that the duration of the average GC peak is approximately 5–10 times the required scanning time, all but the very sharp peaks should yield usable mass spectra. If the spectra are used to distinguish isomers (e.g., *cis* and *trans*), no intensity distortion can be tolerated. On the other hand, slight distortion should have little effect on most routine interpretations.

Fast-scanning mass spectrometers require fast-response amplifying and recording systems. The electron multiplier is the most acceptable detector–amplifier since it provides wide-band amplification with a low noise level. Furthermore, multiplier–electrometer combinations with frequency response of 10^3–10^4 Hz are particularly advantageous in GC–MS work and in high-resolution mass spectrometry as well. Recorders should have a frequency response of approximately 1000 Hz; oscillographic recorders with four to six galvanometers in parallel (at different attenuations) are most useful. Recently, magnetic tape recorders have been used successfully in the rapid acquisition of data, as will be discussed later. McFadden (12,31) has reviewed the characteristics of various detectors, amplifiers, and recorders.

While the rapid scanning device was a necessary innovation for the mass spectrometer in GC work, the subsequent introduction of the mass spectrograph to GC–MS demonstrated the advantage of its photographic recording system which integrates the variation in intensity (due to changing sample concentration) throughout the exposure time (32). Oscilloscopic photography (33) utilizing a time-of-flight instrument has also been used to avoid distortion in otherwise slowly recorded mass spectra. Unfortunately, this technique results in data records (3×4 in. photos) of limited mass range that are difficult to use because of the large relative error in measuring the various masses and intensities. For these reasons, very few workers prefer to use the system of oscilloscope photography.

Fortunately, there are compatible features of gas chromatography and mass spectroscopy. If a sample is sufficiently volatile to be separated by gas chromatography, it will be sufficiently volatile for mass spectroscopy. The sensitivities of the two instruments are approximately comparable, although the fact that in gas chromatography it is necessary only to detect the fractions (i.e., indicate the appearance or presence of a compound) gives it greater sensitivity (on a numerical basis) compared to mass spectroscopy (34).

In mass spectroscopy, on the other hand, the sensitivity value will vary depending on the amount of information required from the mass spectrum. For example, 10–100 times as much sample is required for identification by mass spectroscopy as opposed to mere detection of the most intense ion. Nevertheless, the two techniques are quite compatible in their ability to glean considerable information from less than microgram quantities of sample. Most important of all, direct introduction of a GC fraction is probably the most convenient method of fully realizing the available MS sensitivity (34,35).

IV. The Use of "Separators"

A. Operating Principles

The early GC–MS combinations were inspired by workers who were faced with copious samples of mixtures of fairly simple hydrocarbons, fixed gases, or other volatile compounds (for reviews of early work see references 17, 20, 23, 32, and 34). As the potential of the GC–MS combination was realized by those working with complex, high-boiling, natural products, it was obvious that good quality mass spectra would be desired to afford identification of these compounds. Since good quality data were usually produced under optimum pressure conditions, it was necessary to devise a system which would utilize a large portion of the GC effluent and still maintain acceptable operating conditions in the mass spectrometer. This requirement led to the development of devices which operated on the GC effluent by preferentially removing the carrier gas molecules (32,36–39) or the sample molecules (40). The improved GC–MS connections were designed to channel a gas stream enriched in sample concentration to the mass spectrometer and still maintain the integrity of the GC output, i.e., no mixing, absorption, condensation, decomposition, etc.

These sample-enrichment devices have acquired the generic name "separators" probably as a result of the GC–MS technique proposed by Ryhage (36) which incorporated a "separation jet" designed earlier by Becker (41) for separation of gaseous mixtures of istopes.

The discussion of operating principles of the various separators will be more meaningful if the necessary terminology is reviewed and summarized. Flow conditions through orifices and channels are already recognized in mass spectroscopy (and other vacuum technology) as important considerations. Constrictions or restrictions in the form of orifices or capillaries are used to effect the necessary pressure adjustments in the gas flow through the GC–separator–MS system. In a manner analogous to electrical circuitry, the gas flow suffers a pressure drop in overcoming the resistance (or impedance) offered by the dimensions of the vacuum conduit (42). The resistance (or impedance) to flow is given by

$$R = kl/d^4 \qquad\qquad (5\text{-}2)$$

where l is the length, d the diameter, and k the characteristic proportionality constant for the constriction or restriction. It is preferable to construct flow restrictions from measured lengths of capillary of known diameter; this procedure allows "fine tuning" of the pressure drop because of the first-power dependence of resistance on length rather than the fourth-power (inverse) dependence on the diameter of an orifice.

Viscous leaks are often used to restrict the rate (or reduce the pressure) without affecting the composition (on a mass basis) of the flowing gas. The primary requirement for viscous flow is that the mean free path, L, of the molecules be small compared to the diameter, d, of the orifice. When the condition $L < d$ is met, viscous flow exists and the volume flow rate is proportional to the *square* of the *total* pressure differential of the gas mixture (43). No separation occurs under these conditions.

Molecular flow (44) describes the process of effusion or movement of gases through an orifice at pressures for which L is approximately an order of magnitude greater than d. Under these conditions, other parameters being constant, the flow rate of the gas is *directly* proportional to the pressure drop through the orifice and inversely proportional to the square root of the molecular weight. Since some of the mathematical expressions for effusion are similar to those that describe Graham's law, the two processes are occasionally confused (45); Graham's law applies to the diffusion of gases.

The enrichment factor is a measure of the improvement in sample concentration effected by the separator. This factor is generally defined as the ratio of sample concentration entering the mass spectrometer to that leaving the gas chromatograph. An experimental enrichment factor can be measured by comparing the ratio of sample to carrier (determined mass spectrometrically) entering the spectrometer through the separator with the ratio of sample to carrier entering through a conventional inlet (20,32). In such a determination it is important to maintain viscous flow through the conventional inlet so that there is no discrimination against the true sample composition.

The term *efficiency* is used to designate that fraction of the injected GC sample which finally reaches the MS ion source after the enrichment process. The partial loss of sample is unavoidable if the GC sample effluent is to be enriched by preferential removal of the lighter carrier gas. If it is disturbing that some sample is lost, it should be recognized that if no separator were used with packed columns (flow rates of about 20 ml/min) a maximum of 0.2 ml/min or about 1% of the sample would be admitted to the mass spectrometer. Experimentally, the efficiency can be determined by monitoring the total ion current produced when equal quantities of the sample are alternately introduced into the ion source by the GC–separator inlet and a direct heated inlet. The integrated value of total ion current from the direct sample introduction is taken as 100% (37) in the comparison with the value from the GC–MS operation. This method is valid only if both techniques of sample introduction are equally effective in directing the sample vapor into the ionization chamber. The method suggested by Krueger and McCloskey (46) probably gives the most realistic appraisal of efficiency because the mode of sample introduction with and without the separator is virtually the same as will be described in Section IV-B-2-b.

B. Types of Separators

1. Diffusion: The Jet Orifice

The separation jet (41) provides the principal diffusive means of preferentially removing helium from GC effluents (36). The principle of the separation jet method is illustrated in Figure 5-7a. The gaseous mixture of

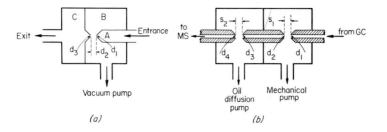

Fig. 5-7. (*a*) Schematic of separation jet designed by Becker (41). (*b*) Schematic of operational two-stage jet orifice separator used in GC–MS (47). Typical dimensions (in mm): $d_1 = 0.1$, $s_1 = 0.15$, $d_2 = 0.3$, $d_3 = 0.24$, $s_2 = 0.5$, $d_4 = 0.3$.

organic compound and carrier gas (helium) enters through the convergent nozzle, A, with a small orifice, d_1, from which the gas emerges in the form of an expanding jet. The orifice of the nozzle is aligned at a short distance, d_2, from a small orifice, d_3, in the wall separating chambers B and C. Optimal adjustments (41,47) can be made on the dimensions, d_1, d_2, and d_3, which are generally the order of 0.1–0.3 mm. Because of lower forward momentum and greater diffusivity, the lighter component of the gas is removed (at right angles to the direction of the jet stream) into the peripheral volume B, which is evacuated by a vacuum pump. The gas stream entering chamber C is thus enriched in the heavier component. Chamber C in Figure 5-7a represents the channel that carries the enriched sample gas to the mass spectrometer if the separator functions as the connection between the GC and MS units. Thus the enrichment index of the heavy component in the core gas may be measured mass spectrometrically in evaluating the separator system.

The separator for GC–MS work introduced by Ryhage (36) utilizes two separation jet assemblies in series (Fig. 5-7b) to obtain an enrichment factor of approximately 100 and an efficiency of up to 60% (47). The first chamber in the separator is evacuated by a forepump (pumping speed of 2.6 liters/sec) which maintains a pressure of approximately 0.1 torr for a helium flow of 30 ml/min. The second chamber is connected to an oil diffusion pump (pumping speed of 150 liters/sec at 10^{-6} torr). To achieve optimum performance of the mass spectrometer, the pressure in the second separator chamber should be less than 10^{-3} torr. The dimension of the first nozzle, d_1, ensures the correct pressure drop across the first separation jet, but it also affects the operation of the GC column and the forepumping system. For optimum operation of the separator, a small ratio (< 1) of mean free path to orifice dimension is necessary. Viscous flow is more difficult to attain in the second separation unit where the

pressure is quite low due to at least 90% removal of the carrier gas in the
first unit where the mean free path is 0.001–0.01 mm. The dimensions of
the second separation jet must therefore be larger than those of the first
(47).

The use of only one separation jet in combination with the high flow
rates of packed columns results in low efficiency if the throttling is sufficient to reduce the pressure to the MS range. If the pressure drop from
chromatograph to spectrometer is accomplished in two stages, the overall
efficiency is better. When two separation jets are operated in series on a
helium flow rate of approximately 30 ml/min, there is about 16% sample
loss in the first unit and 40% loss (of the original sample) in the second
(47). On the other hand, a single separation jet is all that is needed to
efficiently enrich the effluent from an open tubular column which may have
a helium flow rate in the order of 0.5–2 ml/min (48). If such a low GC flow
rate were sent through two separation jets in series, there would be too
much sample loss at no appreciable gain in vacuum.

The separator system (containing either one or two separation jets), is
generally made of stainless steel, and can be heated sufficiently to prevent
any condensation or absorption of the effluent from a GC column up to
300°C. In general, it is desirable to maintain the separator and lines leading
to the ion chamber at a temperature comparable to that of the GC column.

Ryhage has made use of a "separation factor," S, for his molecule
separator (47) where

$$S = \frac{n_e(1 - n_0)}{n_0(1 - n_e)} \tag{5-3}$$

n_0 and n_e are the mole fractions of the heavier component before and
after the gas mixture has passed the separator. For small amounts of
sample (0.1 μg of methyl stearate) the concentration of sample in the
carrier gas is approximately $1:10^4$ at a helium flow rate of 30 ml/min and if
the separator improves the sample concentration by a factor of 100, the
sample concentration is still very low ($1:10^2$). Thus the mole fraction of the
lighter component in both cases is still essentially unity and the expression
(Eq. 5-3) for the "separation factor" can be reduced to

$$S = n_e/n_0 \tag{5-4}$$

It was found that at a flow rate of 30 ml/min, 65% of the GC sample
reached the ion source with an analyzer pressure of 1.2×10^{-6} torr (uncorrected), corresponding to a helium volume flow rate of 0.25 ml/min
into the mass spectrometer. These data (47) can be used to calculate an
approximation of the "separation factor" ($65:0.25/100:30 = 78$). This

value characterizes the separator jet only for the conditions stated. The enrichment capability and the efficiency vary as a function of helium flow rate, helium pressure in the ion source, molecular weight, and retention time on the GC column (47). The shorter the retention time, the higher the ratio of sample to carrier gas at the outset; this results in a larger portion of the fraction finally reaching the spectrometer. The efficiency or sensitivity of the separator system employing two separation jets in series increases with helium flow rate to a maximum at 30 ml/min, but decreases at helium flow rates above 30 ml/min. The efficiency of the single jet separator increases with helium flow rate, but at 45 ml/min the pressure in the analyzer is approximately 2×10^{-6} torr (uncorrected). Relatively high pressure in a mass spectrometer which is not differentially pumped results in a deterioration of resolution. In the GC–MS system described by Ryhage (47), the resolution decreases from about 900 to 600 for an increase in helium pressure from 5×10^{-7} to 7×10^{-6} torr.

The separation jet molecule separator has been used successfully with both packed (36) and open tubular columns (48) in the analyses of several groups of natural compounds (some of which will be discussed in Section VIII). However, in considering use of the separator, the compounds should not be sensitive to hot metal surfaces (catalytic degradation). The very small orifices are prone to clogging, although care can be taken to avoid introduction of particulate material. The commercially available separation jet assemblies (under patent for a specific mass spectrometer) are machined so that the critical alignment and spacing of the orifices are mechanically established. Of all the separators reported, the jet orifice type is probably the least feasible to fabricate in the laboratory because of the necessary mechanical precision. The separation jet is very reliable in affording good sample enrichment and efficiency once the system is optimized.

2. Effusion: The Sinter

The process of effusive transport of gases through an orifice occurs only under the condition of molecular flow which requires the diameter, d, of the orifice be no greater than 1/10 the mean free path, L, of the gas at the higher pressure. An increase in pressure violates the condition $10d \leq L$ and the rate gradually shifts over to viscous flow. The factor of 10 is an empirical one that seems to offer an adequate margin of safety (45) and at $L = 2d$ the error is generally only a few percent.

In the case of a mixture of two gases under conditions of molecular flow, each gas will effuse independently of the other because of the absence of collisions near the orifice. The flow rate of each gas is determined by the difference between its *partial pressure* across the orifice through which effusion occurs (49). Thus,

$$F_1 = k_a \bar{v}(P_{1f} - P_{1b}) \qquad (5\text{-}5)$$

where P_{1f} and P_{1b} are the partial pressures (in front and back), $\bar{v}$ is the mean molecular velocity, and F_1 is the flow rate of the lighter gas through orifices characterized by k_a. The ratio of the flow rates is

$$\frac{F_1}{F_2} = \frac{v_1}{v_2} \cdot \frac{P_{1f} - P_{1b}}{P_{2f} - P_{2b}} = \left(\frac{M_2}{M_1}\right)^{1/2} \frac{P_{1f} - P_{1b}}{P_{2f} - P_{2b}} \qquad (5\text{-}6)$$

since the mean velocities are inversely proportional to the square of the molecular weights M_1 and M_2.

For the special case in the separator where the exit volume from the orifice will be maintained at vacuum, $P_b \ll P_f$, and the ratio of the flow rates reduces further to

$$\frac{F_1}{F_2} = \left(\frac{M_2}{M_1}\right)^{1/2} \frac{P_{1f}}{P_{2f}} \qquad (5\text{-}7)$$

Equation 5-7 also shows that the flow rate varies directly with the partial pressure of the gas and indicates preferential rate of removal for the lighter gas. The concentration of microgram quantities of compounds in the carrier gas of a ⅛-in. o.d. packed column will be in the range of $1:10^5$ to $1:10^4$ as they emerge from the gas chromatograph as a peak approximately 1 min wide. The rate of removal of the carrier gas (helium) will be approximately 10^5–10^6 times greater than that of a sample of molecular weight 300, and the ratio of remaining sample to remaining carrier gas will rise rapidly at a modest partial expenditure of sample.

a. The Fritted Glass Separator.   A fritted glass separator utilizing the effusion principle was introduced by Watson and Biemann (32). The main idea of the design was to provide a short, relatively inert path from the gas chromatograph to the mass spectrometer which would reduce the pressure in the spectrometer by removing most of the excessive carrier gas. As shown in the schematic (Fig. 5-8) of a later version of the separator (50), the apparatus consists essentially of a fritted (porous) glass tube which is housed in an evacuation chamber. The fritted glass tube serves as the valve or throttle ordinarily used to reduce the amount of gas entering the MS ion source. As shown in Figure 5-8, the GC effluent enters through a constriction which reduces the pressure so that molecular flow will be established through the wall of the fritted glass tube. The carrier gas (helium) is preferentially removed through the many small (diameter

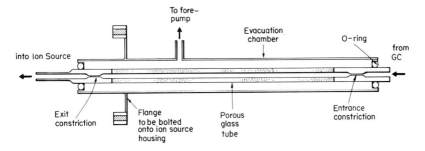

Fig. 5-8. Schematic of the fritted glass separator (50).

~ 1 μm) pores in the fritted glass, and the residual gas, enriched with the (heavier) sample molecules, continues through the exit constriction into the MS ionization chamber. The entrance constriction primarily controls the flow rate from the GC column and the mean pressure inside the fritted glass tube. The exit constriction controls the sensitivity (efficiency) of the apparatus which is limited by the maximum permissible operating pressure of the mass spectrometer.

In a typical Watson-Biemann separator, the porous section is 20 cm long, 7 mm o.d., 4 mm i.d., and has an average pore diameter of about 1 μm ("ultrafine" porosity). At this writing, the fritted glass separator is commercially available (Special Glass Apparatus Dept., Corning Glassware, Corning, N. Y.) with a glass evacuation chamber similar to the original design (32). The solid-walled capillaries fused to the ends of the fritted tube in the commercial apparatus almost always require modification to meet the vacuum requirements of a specific mass spectrometer. This may be accomplished with a constriction in both capillaries formed by heating the glass until it collapses to the desired orifice (an empirically established dimension which leads to acceptable enrichment and efficiency) (32,50). An alternative (and preferable) method of effecting the necessary pressure drop involves fusing measured lengths of smaller capillary (e.g., 1 cm and 4.6 cm of 0.375 mm capillary for the entrance and exit restrictions, respectively) to the ends of the standard glass chamber (51). In either case, conditions of viscous flow should be maintained through both restrictions.

An estimate of the actual efficiency of the separator requires a knowledge of the flow rate into the separator, the enrichment factor, and finally the flow rate of the gas stream entering the spectrometer. For example, the efficiency of a Watson-Biemann (prototype) separator (20) was estimated from the following data. The helium flow rate through the column and into the separator was 20 ml/min. The enrichment factor was experimentally determined to be 50. The flow rate into the spectrometer was estimated to be 0.2 ml/min. Since the sample-to-helium ratio in a GC peak is generally less than 1 part in 100, the total gas flow is essentially all helium. Thus, $0.2/20 \times 100 = 1\%$ of the helium and 50% of the sample reach the ion source. If a better evaluation of separator efficiency is desired, the total ion current technique described in Sections IV-A or IV-B-2-b may be used.

The efficiency of the glass separator is affected by the residence time of the sample in the separator and decreases as a greater amount of gas passes through the wall of the fritted section. Obviously, if the exit constriction were virtually closed, almost all the helium and most of the sample would be drawn through the wall of fritted glass. In such a case, the enrichment factor would be enormous, but the system would have very little utility. Thus, it behooves the investigator to establish a viscous flow rate through the exit constriction which is limited only by the MS operational requirements. For example, the efficiency of one fritted glass separator (23) was found to be 20% when the attending pressure in the spectrometer was 2×10^{-6} torr as measured with a Penning gauge calibrated for air; the efficiency was approximately doubled as the pressure was allowed to reach 1×10^{-5} torr. Most mass spectrometers suffer a significant loss in resolution when operated at such high pressures, but good mass spectra can be obtained at pressures of approximately 10^{-5} torr (20,50) if the instrument employs differential pumping.

A recent quantitative study of the fritted glass separator (52) reemphasizes the importance of maintaining viscous flow through those stages of the GC–separator–MS system which serve only to transfer the gas mixture, and molecular flow in those sections which effect enrichment (i.e., the fritted surface and the ion source itself). Each section of the system which imposes molecular flow on the gas mixture also effects an enrichment of the heavy component to an extent given by the enrichment factor

$$E = (M_s/M_c)^{1/2} \qquad (5\text{-}8)$$

where M_s is the molecular weight of the sample (heavier component) and M_c is the molecular weight of the carrier gas (52). The total enrichment effected by the system then is the product of the individual enrichment factors describing each encounter of molecular flow. Thus, the overall enrichment of the gas through the conventional single-stage fritted glass separator and into the ion source is given as

$$E' = M_s/M_c \qquad (5\text{-}9)$$

if the flow into the mass spectrometer is much less than that from the gas chromatograph and if the sample is very dilute as it is in GC effluents.

Furthermore, ten Noever de Brauw and Brunnée (52) derive a maximum enrichment factor by assuming that no sample is lost through the separator:

$$E_{\max} = (Q_{GC}/Q_{MS})(M_s/M_c)^{1/2} \qquad (5\text{-}10)$$

where the new terms are Q_{GC}, gas flow (throughput) from the gas chromatograph, and Q_{MS}, gas flow into the mass spectrometer. Thus, this $E_{\max}$ is a function of the adjustable ratio of throughputs, Q_{GC}/Q_{MS}, but in practice this limit is never reached, of course, because sample loss is unavoidable. However, $E_{\max}$ serves as a useful guide or approximation. Experimentally, the conventional single-stage glass frit was found to achieve about 72% of the enrichment anticipated from equation 5-9 (52).

A two-stage design in which two sections of fritted glass tubing were connected via a viscous leak (52) increased the effective enrichment by a factor of $(M_s/M_c)^{1/2}$ to give

$$E'' = (M_s/M_c)^{3/2} \qquad (5\text{-}11)$$

Experimental values of the enrichment factor reached 85% of those expected from equation 5-11 (52). More specifically, a compound of molecular weight 250 would enjoy an enrichment factor of about 400 at an efficiency of 25–50% when the ratio of throughputs was $Q_{GC}/Q_{MS} = 100/1$ through the two-stage design (52).

The two-stage model of the fritted glass separator was designed to cope with GC flow rates of 20–60 ml/min, but it will efficiently handle flow rates down to 10 ml/min. Both the enrichment and efficiency of the two-stage separator deteriorate because the gas kinetic conditions for molecular or viscous flow in the specified stages are not maintained (52). The single-stage separator model is preferable for flow rates between 1 and 10 ml/min because optimum flow conditions are easy to establish; also the low volume gives a lower time constant for sweepout and thus better response to sharp peaks from open tubular columns, etc.

The fritted glass separator has been used in the analysis of many different types of complex molecules (20,50) with no chemical alteration in the GC effluent due to interaction with the porous glass surface. However, severe sample adsorption has been noted in studies (52,53) which require the separation and identification of trace quantities of heavy alcohols and other polar compounds.

In one case (53), attempts to remedy the sorption problem included the following silanization procedure. The fritted glass separator (obtained from Perkin-Elmer) was washed successively with chloroform, methanol, distilled water, concentrated hydrochloric acid, aqua regia, distilled water, methanol, and chloroform, and dried overnight at 180°C. The separator was cooled in a desiccator and treated for 20 min with a freshly prepared 5% (v/v) solution of dimethyldichlorosilane (DMCS) in toluene. It was then rinsed with toluene, washed with anhydrous methanol to react with residual Si—Cl groups, and dried overnight at 180°C. However, this careful DMCS silanization treatment did not eliminate the sorption of small quantities of sample (53).

Another silanization procedure, treatment with bistrimethylsilyacetamide (BSA), was more successful. This method involves treating the separator with the silanizing agent under operating conditions of temperature and pressure (54). The exit capillary of the separator and the evacuation port of the chamber are connected to a mechanical pump and the separator heated to 150°C. A 0.1-ml aliquot of BSA is slowly injected into a 5 ft × 0.1 in. stainless steel transfer line and the fritted glass separator through a temporary injection port. The excess reagent and by-product (acetamide) are removed through both the pores of fritted glass and the exit capillary. This technique results in a silanized sinter which permits GC–MS sampling of most alcohols, aldehydes, ketones, and esters down to 10^{-9} g without serious absorption effects.

This silanization procedure has also been successfully carried out on a fritted glass separator which was mounted in the ion source of a mass spectrometer. All electronics, i.e., filament, heaters, ion guages, etc., were turned off, but the vacuum pumps were left on. The excess reagent and by-product were readily pumped off by baking the source for a few hours. No complications resulted from this *in situ* silanization treatment (54); however, many MS specialists would prefer the use of a separate vacuum system.

The operating temperature of the fritted glass separator module shown in Figure 5-8 is limited to approximately 250–275°C in the vicinity of the O-ring vacuum seal. At these temperatures, Viton O-rings become brittle and silicone rubber O-rings become soft and malleable; both conditions promote vacuum leaks. Fortunately, a maximum temperature of 250°C is not a severe limitation and is compatible with GC column temperatures up to 300°C. A more vacuum-tight system may be effected by incorporating Kovar seals with Swagelok fittings for connections at either end of the separator.

b. The Porous Stainless Steel Separator. Another separator analogous in design to the fritted glass separator utilizes a porous stainless steel tube for the preferential removal of the carrier gas (38,46). The principal component of this separator (Fig. 5-9) is the 4-in. length of porous stainless

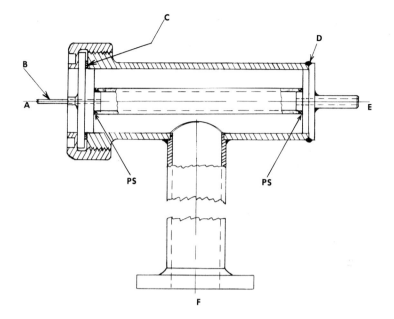

Fig. 5-9. Schematic of porous stainless steel separator (46). *A*, From gas chromatograph; *B*, $\frac{1}{16}$ in. standard stainless steel tubing, silver soldered; *C*, 0.02 in. flat silver gasket; *D*, heli-arc weld; *E*, to mass spectrometer; *F*, to vacuum; *PS*, pressure seal.

steel tube (available from the Pall Trinity Micro Corp., Cortland, N. Y. or Mott Metallurgical Corp, 272 Huyshope Ave., Hartford, Conn.) which is housed in a stainless steel evacuation chamber. Both ends of the porous tube are pressure sealed against a lip in the chamber housing as indicated (*PS*) in Figure 5-9. The GC effluent enters the separator through an appropriate length (approximately 50 cm of 0.25 mm i.d.) of stainless steel capillary which effects the initial pressure drop. The final pressure adjustment is accomplished with a micrometer valve or another suitable length of stainless steel capillary. The apparatus is attached to the MS ion source housing by means of a modified probe assembly.

The porous stainless steel separator operates on the effusion principle discussed in the section describing the fritted glass separator. The discriminating interface, the porous stainless steel wall, is available in various grades which have an average pore size as small as 0.1 μm (46). These smaller dimensions should allow molecular flow at the relatively high pressure resulting from helium flow rates of approximately 30 ml/min into the separator. The porous stainless steel separator has achieved an enrichment factor of 100 and an efficiency of 40% while accepting a GC flow rate of 30 ml/min (46). The efficiency was determined experimentally by introducing identical samples through the same inlet system except that in one case the separator was turned off. Such a procedure is possible in this design (46) since the separator is built into the lower end of a probe (see Fig. 5-18) which fits the standard vacuum lock assembly on the mass spectrometer (Section V-D). Under normal operating conditions the GC effluent flows into the separator in the probe and the enriched effluent enters the ionization chamber. In the efficiency determination, one standard aliquot of a given sample is introduced under these conditions. In order to subject the second standard aliquot (which will be taken as the base or 100% value in the calculation) to the same introduction pathway, the probe is withdrawn into the vacuum lock and isolated from the spectrometer. The electrical heater to the probe is turned off. The separator is isolated from its evacuation system, but it is maintained at its normal operating temperature. Since the separator is now ineffective, all the GC effluent passes through the exit capillary into the vacuum lock where it is evacuated by the auxiliary pumping system. After injection into the GC, the second standard passes the hot separator but condenses on the cool exit capillary which runs concentrically from the separator through the length of the probe. The separator is then evacuated and the entire probe assembly is moved back into the ion source through the vacuum lock. The exit capillary is reheated to volatilize the sample and the ion current is monitored to establish a basis for 100% efficiency of sample introduction through the GC–separator system. This technique is quite realistic since it allows for sample adsorption in the GC column, separator, and capillaries. Furthermore, this technique establishes identical geometries of sample introduction in both cases.

The separator is constructed of rugged materials, requires no critical alignment or adjustments, and can be easily dismantled. It is heatable and bakeable to high temperatures (300°C), has excellent heat-transfer properties, will not plug with small particles, and is easily cleaned by pouring solvents through the tube. The porous stainless steel separator does not interact with large polar molecules (38) if it is silanized with dimethyldichlorosilane by a procedure similar to that described for the fritted glass separator (Section IV-B-2-b). Except for cytidine monophosphate, the trimethyl silyl ether derivatives of all major nucleotides and nucleosides traverse the silanized stainless steel with no observable interaction (46).

c. The Teflon Separator.   A unique effusion separator introduced by Lipsky, Horvath, and McMurray employs a thin Teflon membrane as the discriminating interface (37,55). The basic component (Fig. 5-10) consists

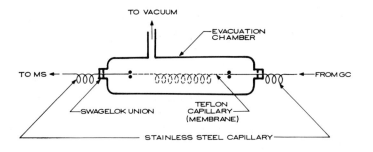

Fig. 5-10. Schematic of the Teflon membrane separator (37).

of a 7-ft length of thin-walled (0.005 in.) Teflon capillary (0.020 in. o.d.) which is held in an evacuation chamber by means of Swagelok fittings. A 4-ft coil of stainless steel capillary (0.010 in. i.d.) is connected to each end of the Teflon capillary to effect the necessary pressure reductions. The separator unit is evacuated by a two-stage mechanical pump.

Temperature control of this system is critical. At 250°C the Teflon capillary is essentially a direct connection between the gas chromatograph and the mass spectrometer, allowing no loss of either helium or sample (56). The optimal operating temperature is approximately 280–290°C, depending on the volume flow rate of carrier gas. However, at temperatures near 350°C, "pinholes" may form in the Teflon and allow complete loss of sample and helium (56). At flow rates up to 20–30 ml/min, efficiencies of 40–70% are reported at an estimated enrichment factor of 200 (55) which resulted in an ion source pressure of about 10^{-4} torr (corrected for helium). The mass spectrometer used in this case was differentially pumped; an instrument with an open ion source might not operate effectively at such high pressures.

The exponential temperature dependence of helium removal through the separator (37,55) suggests a permeation process (57) through the Teflon membrane. However, such a mechanism would require that Teflon exhibit the unusual behavior of preferentially sorbing an inert gas in lieu of organic vapors. The importance of the similarity between the nature of the membrane barrier and the permeant has been reviewed by Lebovits (58). However, permanent gases sometimes exhibit permeabilities which vary directly with the pressure differential across the membrane and inversely with molecular weight without dissolving (in the conventional sense) in the membrane material. In some high polymers, diffusion and viscous flow are still related processes, and the heat of activation in both processes, due to the formation of holes, is approximately the same (59).

It is possible that the energy requirement for a phase transition resulting in enlarged pores in the Teflon membrane also follows an exponential dependence on temperature. If this were the case, the actual mechanism of gas removal could be effusive in nature, which would also explain the preferential removal of helium under the optimum flow conditions (55) and the further removal of both sample and helium at higher temperatures as the pores expanded even more. However, the formation of pores is unlikely, since the crystalline fusion point of Teflon FEP is 261°C (60) and thus the membrane material is essentially a superviscous liquid at 300°C. The exact mechanism of carrier gas removal by the Teflon separator is thus not clear at this time, but it probably results from a combination of the above effects.

A delay time of 20–30 sec between the emergence of the GC fraction and its appearance in the mass spectrometer has been observed (56). Furthermore, the efficiency of the separator depends on the nature of the compounds as well as the molecular weight (55). Both of these phenomena probably result from sample interaction with the Teflon surface (Section V-B) which in some cases may alter the overall GC efficiency (61). Such a loss in GC efficiency or an increase in peak tailing would be prohibitive in the analysis of complex mixtures.

The temperature dependence of helium permeation gives some flexibility in coping with various flow rates, but on the other hand, the absolute magnitude of the temperature may limit its use for some compound types. The components are reasonably rugged and conveniently assembled; there is no required delicate alignment or critical adjustment of pressure within the separator. The apparatus has good efficiency, but it should be used with instruments which can tolerate reasonably high ion source pressures.

The Teflon separator has been used successfully in the analysis of fatty acid esters, amino acid derivatives, terpenes, steroids, and alcohols (37,55). No memory of outgassing effects have been noted after many analyses. In general, the Teflon is chemically inert, but it should be remembered that Teflon may interact with certain compounds. For example, halide exchange reactions (62) may occur if Teflon is used at elevated temperatures in the analysis of reactive trimethylsilyl compounds such as trimethylsilyl esters, amides, amines, and halides.

d. The Porous Silver Separator.   The porous silver separator, which also operates primarily on an effusive principle, utilizes a porous silver membrane as the discriminating interface (39). The GC effluent is introduced into a disk-like chamber formed by a "sandwich" of circular sheets of a porous silver membrane which fit on top and bottom of the annular brass ring as shown in Figure 5-11. The "sandwich" is sealed between two stainless steel flanges (Fig. 5-11) which form the evacuation chamber when connected to a mechanical pump. The pores or channels in the silver membrane have an average diameter of 0.2 μm so that an operating pressure of about 1 torr in the evacuation chamber will establish the conditions of molecular flow. Thus, helium is preferentially removed perpendicular to the circular sheets of silver membrane and the remaining enriched sample is drawn into the mass spectrometer via the exit capillary. In this particular design, the entrance and exit capillaries are positioned by line-of-sight within a few millimeters of one another. Slight enrichment has been observed even when the silver membranes are removed (63), which suggested some similarity to the jet orifice device (47). However, the dominating mechanism involves preferential removal of helium by effusion through the porous silver.

The silver separator is a new development and, at this writing, has been used only in analysis of reasonably simple hydrocarbons and oxygenated

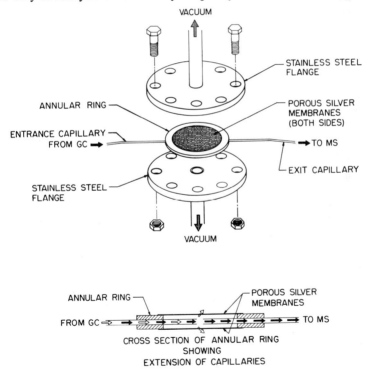

Fig. 5-11. Schematic of the porous silver separator (39).

compounds having up to seven carbons. The effect of the silver separator on the GC peak shape has not been documented, but simple solvents, i.e., acetone and benzene, when eluted within 10 sec of one another, do not produce significant "cross-talk" in their recorded spectra (39). An enrichment factor of 100 or more is reported (39) for compounds having a molecular weight of at least 200. This enrichment factor was determined (at an undesignated efficiency) as the ratio of MS sensitivity *with the functioning separator* versus the sensitivity for the same sample entering the MS *through the dismantled separator*. The ion source pressure was the same in both cases.

The porous silver membrane has several attractive features which qualify it for use as a discriminating interface in the GC–MS system. The average pore size of 0.2 μm ensures molecular flow conditions at conveniently established separator pressures and the high porosity (80% dead volume) accommodates reasonably high column flow rates (40 ml/min). The silver surface should be inert, but this has not yet been empirically substantiated by a survey of compound types. Possibly a tubular form of the silver membrane would be desirable, since the sample molecules might have a greater probability of reaching the exit capillary at the end of a tubular chamber rather than an exit in the center (high pressure point) of a disk-like chamber. If so, the tubular design would enhance the efficiency but not the enrichment. The present system is simple in design, conveniently dismantled for modification or replacement of parts, and bakeable to 450°C.

Blumer (63a) has recently described a separator which utilizes a small disk of silver membrane (6 mm o.d., 2 mil thick; pore diameter: max 3 μm) as an interface for effusion. The very thin membrane permits much higher flow rates per unit cross-sectional area and, thus, the separator unit can be miniaturized to minimize surface interaction and dead volume. Blumer suggests a design which may be constructed from readily available materials and which lends itself to facile modification in the laboratory. This separator design should be useful for small bore (e.g. 2.2 mm i.d.) packed columns as well as for open tubular columns.

3. Adsorption and Diffusion: The Permeable Barrier

The permeation or true diffusion of gases in solids is usually a highly specific process, depending both on the solubility and on the mobility of the gas molecules or atoms in the solid (59). For example, the solubility of hydrogen in palladium (in the form of atomic hydrogen) is very high at elevated temperatures and the rate of diffusion is high also, while there is hardly a perceptible permeation of other gases through palladium. True diffusion generally exhibits a large positive temperature coefficient, while gas flow through capillaries shows comparatively small negative temperature coefficient.

Another approach to the GC–MS connection has been demonstrated in the separator introduced by Llewellyn and Littlejohn (40). In this case, the mechanism of GC sample enrichment involves preferential selection of the organic components. The organic compounds are adsorbed or dissolved in a thin elastomer membrane which is the actual interface between the GC and MS units. The total amount of gas passing through the membrane is given by the permeability, which is a function of the solubility and the diffusion rate of the substance in the elastomer material. More specifically, the relationship is:

$$P = SDA[(P_1 - P_2)/d]t \qquad (5\text{-}12)$$

where S is the solubility, D the diffusion rate, A the area, $(P_1 - P_2)$ the pressure differential across the membrane, d the thickness of the membrane, and t the time (40). The permeation is measured in torr-liter/sec which are the units of throughput (42). Conductance values (40) of various fixed gases (Table 5-1) for a silicone rubber membrane ($A = 5$ cm^2, $d = 2.5 \times 10^{-3}$ cm) at 30°C are almost insignificant compared with most organic compounds (e.g., n-octane). Helium has a high diffusion rate in the elastomer membrane material, but its very low solubility precludes any significant permeability. Thus, an elastomer membrane as the GC–MS interface provides a barrier for the carrier gas.

The two-stage separator of Llewellyn and Littlejohn (Fig. 5-12) is designed to receive the GC effluent at ambient pressure in the small volume above the first elastomer membrane ($A = 5$ cm^2). At carrier gas flow rates of 30–50 ml/min, the first membrane adsorbs approximately 90% of the organic component. The dissolved organic material diffuses

TABLE 5-1

Conductance of Silicone Rubber Membrane for Various Gases at 30°C (40)

Nitrogen	0.12 ml/min
Helium	0.18
Hydrogen	0.60
Argon	0.60
Octane	180

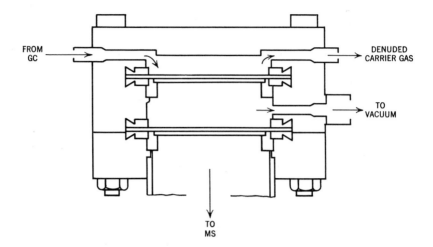

FROM GC

DENUDED CARRIER GAS

TO VACUUM

TO MS

Fig. 5-12. Schematic of the permeable barrier separator (40).

through the elastomer to the second-stage chamber which is evacuated via a conductance adjusted to about 10 ml/sec. The small amount of helium that manages to permeate the first membrane is removed from the intermediate volume, along with some of the organic component. The remaining sample (typically 50% of that injected into the chromatograph) permeates the second elastomer membrane and enters the spectrometer. A helium partial pressure as low as approximately 10^{-7} torr can be maintained in the ion source as established by measurement on a residual gas analyzer (40).

There is a time lag, τ, associated with mass transfer through the elastomer membrane. When the material comes into contact with the membrane, an adsorption process takes place followed by a movement into the bulk of the material. The rate of movement depends on the size of the molecule and the temperature. Thus, the molecule requires some average time to traverse the thickness of the membrane (40). This lag time is related to the diffusion coefficient and membrane thickness as follows:

$$\tau = d^2/6D \tag{5-13}$$

It is possible to design a membrane which will have a short lag time. For example, a substance having $D = 10^{-6}$ cm^2/sec in a silicone rubber membrane of thickness $d = 2.5 \times 10^{-3}$ cm would have a τ of 1 sec. Such fast response should prevent any distortion of the GC peaks. Since D is also a function of temperature, the choice of thermal environment and dimensions of the membrane are important considerations in the design of this separator. Conditions which promote short response times (high diffusion rates) are also favorable for high percentage transmission (permeability).

Over a fairly wide range, the operating temperature of the membrane is not critical. To obtain up to 70% transmission, the membrane should be kept at, or below, the boiling point of the material being chromatographed (40). If the temperature is too low, adsorption and "tailing" (longer τ) will occur. At very high temperatures (250–300°C), a small amount of organic material is released from the membrane, but this is usually less than column bleed. Naturally, the permeation of carrier gas increases slightly with temperature (a ΔT of 50°C will double the gas load), so that in general the membrane should be maintained close to the GC column temperature. It should also be pointed out that very high temperatures decrease the solubility of organic substances, with a resulting decrease in permeability.

The elastomer membranes are firmly fixed to rigid supports and sealed into the separator assembly by vacuum flanges and gaskets. Silicone rubber was chosen as the elastomer membrane (rather than Mylar, nylon, etc.) because of suitable permeability values and reasonable temperature stability. It is possible to saturate the interface membrane at an organic flow rate of ~ 100 μg/sec. This represents a high GC concentration and in practice, it is generally possible to keep the mass flow rate well below that value.

Llewellyn and Littlejohn (40) report enrichment factors of approximately 10^5 and an efficiency of approximately 50% for the two-stage permeable barrier separator. During operation, the pressure in the volume between the two membranes is such that neither viscous or molecular flow conditions are fully attained. The efficiency of this separator could be further enhanced if it were modified slightly so that the intermediate volume could be evacuated under conditions of molecular flow. For example, evacuation of the intermediate volume through a section of porous stainless steel (pore diameter: 1–5 μm) would increase the probability of the heavier organic component reaching the second permeable membrane since the lighter carrier gas (especially helium) would be preferentially removed through the porous exit.

The simplicity of this separator design and operation offers several advantages. The components may be easily dismantled and replaced. The GC connection is conveniently maintained at ambient pressure and the separator is not particularly sensitive to average GC column flow rates. A significant advantage of the permeable barrier separator is its applicability to several different carrier gases. Also, the general restricted transmission of fixed gases through the membrane barrier uniquely qualifies it for use in specialized cases. For example, this separator could be very helpful for monitoring trace contaminants in a sealed physiological environment also containing relatively large amounts of carbon dioxide and water vapor (64).

C. Electrical Analog of Separator

The operation of the various separators is often readily understood in terms of the general principles of physics and the kinetic behavior of gases. However, some investigators have had difficulty in fabricating new models of the separators reported in the literature. A better understanding of the practical and technical details may be facilitated if the operating principles and requirements of separators are explained or represented on a familiar basis, such as that of electrical circuitry.

An elementary analysis of an electrical circuit requires a determination of its characteristic resistances (connected in series or parallel) by measuring the electron flow (current) through the various elements at the applied voltage (driving force). The kinetic behavior of gases is such that, following proper definitions, the flow characteristics of fluids in a continuous system of conduit can be represented by an electrical circuit (42). The flow of gas in a conduit is analogous to the flow of electrons in a conductor. Gas flow is impeded by the physical dimensions (e.g., the diameter of an orifice or constriction) of a conduit just as electron flow is impeded by electrical resistance. Gas flow is proportional to a pressure differential (ΔP) for a driving force just as current or amperage carried by an electrical circuit is proportional to the applied voltage (ΔV). In vacuum terminology (42), however, the flow resistance is customarily expressed as the conductance (or the reciprocal of the resistance), which is denoted by the symbol G and defined as

$$G = Q/(P_2 - P_1) \tag{5-14}$$

where ($P_2 - P_1$) is the pressure drop and Q is the gas volume flow. The gas flow or throughput, Q, has the units torr-liter/sec, and hence the units of G are liter/sec. In summary, the analogous descriptive terms of the electrical and vacuum systems are:

Electrical terms	Gas flow terms
$I = \Delta V/R$	$Q = \Delta P \cdot G$
ΔV = voltage drop, V	ΔP = pressure drop, torr
R = resistance, ohms	G = conductance, liter/sec
I = current, C/sec	Q = throughput, torr-liter/sec

Under conditions of molecular flow, G is independent of the viscosity and inversely proportional to the molecular weight of the flowing gas. Viscous flow conductance varies directly with $\bar{P}$, the average pressure in the element under consideration, and is inversely proportional to the viscosity of the gas (42). It is important to realize that for either viscous or molecular flow, G is independent of ΔP (65).

Hayes (65) has suggested an electrical analog of the separator system which provides a convenient and useful means of studying and analyzing possible modifications of an existing system. An electrical analog for the complete GC–MS system incorporating a fritted glass separator is shown in Figure 5-13. Carrier gas and sample enter the system at the column inlet and flow together through the column and the separator entrance orifice. The dotted line signifies a length of fritted glass tube with pore dimensions small enough that molecular flow is established. The helium carrier gas is preferentially removed through the many small pores (collectively represented as having the conductance G_{frit}), and the remaining sample and helium flow through the exit orifice into the mass spectrometer.

An analysis of the separator characteristics or proposed modifications can be accomplished "on paper" if all the conductances noted in Figure 5-13 are known. The conductances can be evaluated by actual pressure and throughput measurements. For example, column conductance, G_{col}, is simply determined by dividing Q_1 by the pressure drop through the column ($P_{\text{inlet}} - P_1$). Reference to Figure 5-13 indicates that the next conductance, G_1, can be calculated using measurements of P_1, P_x, and Q_1. Similarly, G_{frit} and G_2 can be calculated from measurements of P_y, P_{frit}, and Q_2. Q_{frit} is equal to $Q_1 - Q_2$. Pumping speed is treated as a conductance.

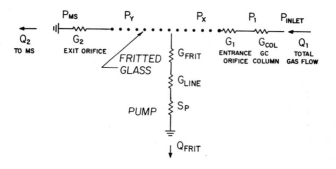

Fig. 5-13. Electrical analog of fritted glass separator representing the various throughputs, Q, and conductances, G, at the pressures, P (65).

The operating pressures can be measured on a modification of the fritted glass evacuation chamber in which the entrance and exit orifices are replaced by a suitable arrangement of valves and McLeod gauges (65). No significant pressure drop is found along the inside of the fritted glass tube (P_y is approximately 10% lower than P_x) (65).

The calculations which may be made with the measured conductances allow a better understanding of the operating conditions and/or modifications of the separator. Consider the question that often arises concerning the use of a larger mechanical pump on the separator system. What would be the benefit, for example, if a pump having a resultant pumping speed, S_p, of 1.25 liters/sec at 0.5 torr were connected to the fritted glass separation unit instead of the usual 0.3 liter/sec pump? The applicable conductances (65) are arranged in the formula (42) for the net conductance of elements connected in series:

$$\frac{1}{G_{net}} = \frac{1}{G_{col}} + \frac{1}{G_1} + \frac{1}{G_{frit}} + \frac{1}{G_{line}} + \frac{1}{S_p} \qquad (5\text{-}15)$$

Inserting typical values:

$$\frac{1}{G_{net}} = \frac{1}{3 \times 10^{-4}} + \frac{1}{10^{-2}} + \frac{1}{2 \times 10^{-2}} + \frac{1}{G_{line}} + \frac{1}{S_p} \qquad (5\text{-}16)$$

Normally, the dimensions of the pumping line are such that $G_{line} = 0.8$ liter/sec and the typical rotary pump has $S_p = 0.3$ liter/sec. If changes are made such that $G_{line} = 10$ liters/sec and $S_p = 1.2$ liters/sec, it is clear from equation 5-16 that the net conductance will not be affected (Q_1 and P_1 will be unchanged). Consider further the conditions just inside the separator entrance orifice (the last three terms of equations 5-15 and 5-16); the net conductance through the evacuation chamber is normally 0.0183 liter/sec and with the proposed changes it should reach 0.0196 liter/sec. The 7% increase in net separator pumping path conductance would decrease the chamber operating pressure slightly, but since the separator is already operating in the molecular flow range, a further pressure reduction would not promote better enrichment. If any change were noted, it would probably be a slight loss in efficiency and a slightly improved source pressure. If the source pressure is too high, the only effective means of lowering it is by reducing the conductance of the separator exit orifice, G_2. Similarly, it is seen that the use of a diffusion pump on the separator would have little effect.

The analog representation is useful in investigating the efficacy of other modifications in the separator, such as varying the length of the fritted glass section. The electrical analog also provides a basis for mathematical calculations of efficiency as a function of sample molecular weight and conductance values (65).

This discussion of the use of an electrical analog system has, for explanatory purposes, dealt with the fritted glass separator (65). It should be emphasized that a similar representation can be devised for any of the other separator systems. Calculations based on the analog of the fritted glass separator confirm the empirically derived conclusions of Watson's dissertation (20). This and preliminary results from prototype modifications (65) demonstrate the validity of the analog method of analysis of a separator system. Once the characteristic conductances of other separator systems are measured, the corresponding analog representations will be useful in the optimization and evaluation of the GC–MS operating conditions.

V. General Considerations of the GC–Separator–MS System

A. The Gas Chromatograph

The peculiarities of the GC–MS operating conditions require occasional deviations from normal GC column operation. For example, in cases where the GC column is connected directly (without a split) to the MS, the exit of the column is at subambient pressure. That GC columns operated in this manner suffer no serious loss in efficiency was indicated in a study of the low-pressure behavior of open tubular columns (66).

There has been little emphasis on modifying GC columns or making them more compatible for combination with a mass spectrometer. A preliminary study of multichannel open tubular columns (67) suggests some attractive column characteristics. These columns are designed to provide a relatively large surface area of a thin film of liquid phase, require a low volume of carrier gas under certain conditions, and effect minimal pressure drop. The columns were constructed by longitudinally packing $\frac{1}{8}$ in. o.d. stainless steel tubing with 16 strands of 0.01 in. twisted tin-coated copper wire. The columns (25 ft lengths) were coated with stationary phase in the manner generally employed with open tubular columns. These columns exhibited very low pressure drop at normal flow rates. Preliminary results of this study (using simple solvents, e.g., n-propyl, benzene, hexylacetate, toluene) indicate advantages of low flow rates (0.5-2 ml/min) and low column temperatures which (if also applicable in analysis of more polar compounds) will diminish interference from column bleeding and high gas flow into the separator-MS system.

TABLE 5-2

Comparison of GC Columns for GC–MS Operations (68)

Column	Wall-coated open-tubular (Capillary)			Porous-wall (support-coated) open-tubular	Packed	
Outside diameter, in.	—	—	—	—	$\frac{1}{8}$	$\frac{1}{4}$
Inside diameter, in.	0.010	0.020	0.030	0.020	0.065	0.155
Length (commonly used), ft.	300	200	500	100	40	20
Total number of theoretical plates $\times 10^3$	75	35	30	35	20	6
Plates per foot	250	175	60	350	500	300
Sample size, largest component, μg	5	20	100	100	1000	5000
Helium flow, ml/min	1.5	6	10	3	15	60

The operating capabilities and parameters of some GC columns are summarized (68) in Table 5-2. While Table 5-2 is not meant to be exhaustively comprehensive, it does indicate the column types which are suitable for GC–MS work. In general, one is interested in fairly *high sample loads* at relatively *low flow rates* for GC–MS operation. The 0.010 in. i.d. open tubular columns give the highest efficiencies for separation in the smallest oven space, but their low sample capacity limits their use in GC–MS analysis. The porous-wall (support-coated) open tubular columns have more attractive features except that they exhibit adsorptive tailing of polar compounds on nonpolar phases. The 0.030 in. i.d. open tubular columns are not commonly used, but they have a reasonably high sample load to helium flow ratio and still retain the separations possible with open tubular columns of smaller diameter but at the expense of greatly increased length. Packed columns are preferred for work with higher boiling compounds such as steroids, or when analysis of traces at the ppm level or lower is desired.

In order to obtain quantitative information in GC–MS analyses, a stream splitter must be installed upstream from the separator. It is necessary to split the flow to a parallel conventional GC detector because the total MS ion current depends on the ionization efficiency of each compound and, furthermore, the sample transmission of the separator varies inversely with molecular weight. If one plans to temperature program the gas chromatograph, however, it is desirable that a flow regulator be installed on the GC inlet to compensate for the increased flow resistance in the column at higher temperatures (68). The flow rate into the separator is constant, so that if the GC were temperature programmed at a constant inlet pressure it is possible that the flow to the conventional GC detector would terminate or even reverse as the flow through the column decreased throughout the program.

In the analysis of complex mixtures, it is quite helpful to know the distribution of components in the GC peak so that the MS scan may be initiated at the proper moment. Normally, a preliminary chromatogram is obtained on an auxiliary chromatograph under similar conditions except that there is no connection to the mass spectrometer. This preliminary chromatogram is then used to plan the time and frequency of the MS scans. A much better method involves recording a "preview" chromatogram (23) which indicates the GC sample concentration about to enter the ion source. The preview chromatogram results from sampling the GC vapors before they reach the column exit. This can be accomplished by inserting a small capillary or needle into the column packing material from the exit end of the column; the further the needle is inserted into the column, the more advance notice one has of the ensuing GC peaks. This small fraction of GC vapor is sent to a flame ionization detector which gives a preview trace virtually identical to the one which results from the total ion current monitor when the GC effluent enters the spectrometer. The preview trace gives the operator a selected anticipation time (typically 5–30 sec) which is extremely helpful in selecting points on the actual chromatogram for activating the MS scan. This technique is especially valuable in cases of unresolved shoulders and complex mixtures which have a high frequency of peaks.

Alternatively, selection of MS scans (recordings) can be made while monitoring the spectra of GC fractions as they enter the mass spectrometer. Use of this technique will be described in Section VII-A.

It is sometimes advisable to prevent the GC solvent peak from entering the MS unit since the accompanying sudden increase in ionization may upset the space charge in the ion source and defocus the instrument. In cases where the column exit is maintained at ambient pressure, the GC effluent can merely be vented to the atmosphere for the duration of the solvent peak (40). However, when the column has no split and is connected directly to the separator, it may be necessary to install a low-volume valve which will allow the effluent to be shunted to the auxiliary pumping system for the duration of the solvent emergence (23,69).

B. The Separator

The primary purpose of the GC–MS combination is to provide a facile means of obtaining good quality mass spectra which are usable or suitable for identification of GC fractions. It is necessary that sufficient sample reach the MS ion source to result in many discernible peaks of reasonable intensity in the recorded spectrum. However, this must be accomplished without increasing the operating pressure to levels which promote apparent mass discrimination or peak broadening (deterioration of resolution in the MS) as discussed in Section III. The separator should thus be used *as needed* to maintain acceptable conditions in the mass spectrometer in spite of the high flow rates from the GC column.

There is often a tendency to think that a separator system of high efficiency and enrichment is an absolute requirement in a GC–MS combination. This may not be the case. The basic components of the MS system should be examined before considering a separator. The magnitude of GC column flow rates will also influence the necessity of a separator. A differentially pumped spectrometer can operate with a much higher ion source pressure than one which employs a single pump to evacuate both the ion source and the analyzer section. The size and/or the arrangement of the main pump is another consideration; in some cases the main pump can keep up with the full flow from an open tubular column (about 0.5 ml/min) or a portion of the flow from a packed column without inviting deleterious pressure effects in the analyzer. The complexity of the molecular structure is important also. High molecular weight steroids require mass spectra of much higher quality for unambiguous identification than, for instance, low molecular weight hydrocarbons. In the latter case, one can possibly afford to abuse the optimum operating conditions in the mass spectrometer and still accomplish his analyses.

It should be remembered that sputter-ion pumps fail to maintain good pumping speeds in the presence of inert gases. These pumps cannot accommodate the usual flow rates from either packed or open tubular columns. Certainly, in these instances a separator of high enrichment is necessary to reduce the volume flow rate of helium into the MS unit. To circumvent this problem in one case (70), the carrier gas for open tubular column was changed to hydrogen which was readily removed by titanium sublimation pumping without a separator. It might also be mentioned that if a separator is necessary, and if hydrogen is acceptable as the carrier gas, the enrichment process will be enhanced due to the lower molecular weight of hydrogen (71).

If the GC–MS system is to be used in the analysis of large quantities of readily accessible materials, then a 1% split of the effluent to the spectrometer (no separator) provides adequate sample partial pressure in the ion source. For example, 1% of the effluent from a $\frac{1}{8}$-in. packed column with a 1000-μg sample slug carries 1 μg of a 10% GC component to the spectrometer over a period of 30–60 sec. This is plenty of material for a mass spectrum. However, in the analysis of trace components (less than 0.5% injected total) or precious samples, it is preferable to use a separator so that 40–90% of the eluted component enters the ion source.

The choice of a separator should be based on convenience, simplicity, and compatibility of the separator material with the sample. The individual separators have already been described, but some general comments on fabrication materials are in order.

For some time, many investigators have tried to avoid exposure of GC samples to hot metal surfaces (32,50) which have been thought to promote degradation of sensitive compounds, e.g., steroids, alkaloids, sugars, etc. However, it has been demonstrated recently that silanized glass, stainless steel, and aluminum in the form of GC columns are practically identical in providing an inert surface, but copper has a very definite deleterious effect on sensitive samples (61). These materials might be even less harmful when used in the separator if lower temperatures are used, but there are still some instances where metal surfaces seem to promote thermal decomposition of samples, e.g., the degradation of trimethylsilyl ethers of 1,2-diglycerides (72). In general, stainless steel seems to be reasonably innocuous, especially when it is silanized (38). The fritted glass surface is also improved significantly (54) by the silanization procedure spelled out in Section IV-B-2. The general rule seems to be: "When in doubt, silanize!" and silanize all components that the sample will encounter en route from the GC to the MS.

Teflon is often used in GC construction because of its high thermal stability and unreactive nature. However, Teflon has seriously affected the GC peak shapes in some analyses of alkaloids and steroids (61). Short lengths (6 in.) of $\frac{1}{16}$-in. i.d. Teflon tubing caused considerable tailing on peaks arising from both polar and nonpolar compounds. Increasing the lengths of Teflon tubing to 20 in. resulted in a prohibitive loss of resolution. Furthermore, when the Teflon section was heated to 300°C, loss of the basic amine components occurred. This loss was presumably due to interaction with reactive groups on the Teflon surface, since subsequent injections afforded normal response even at the elevated temperature (61). It has also been reported that compounds which contain reactive silyl functional groups undergo an exchange reaction with perfluorocarbon components such as Teflon valves and seals (62). Metal separators offer the advantages of ruggedness and ease of mechanical adjustment. Glass components are inexpensive and easy to inspect for contamination, but they are fragile and require some "in-house" glass-blowing ability.

One particular separator should not be selected over another because the enrichment factor is 60 instead of 40 or because the efficiency can be increased by 10%. As a matter of fact, a factor of 2 is seldom a significant advantage in mass spectroscopy.

Finally, there are some operating parameters of the various separators which render them more suitable in certain types of MS. The fritted glass and porous stainless steel separators are best suited for instruments which are differentially pumped or at least have a well-pumped ion source. The two-stage jet orifice separator is suitable for open source instruments since it seems to result in a relatively greater pressure drop in the GC effluent. The permeable barrier separator results in a very good pressure drop which makes it compatible with any MS type (even one evacuated by a sputter-ion pump). The porous silver separator appears to reduce the pressure of the GC effluent sufficiently to be compatible with open source instruments.

C. The Mass Spectrometer

1. Conventional Mass Spectrometers

The Time-of-Flight (TOF) mass spectrometer was originally designed to study gas-phase reactions and therefore has the capability of an exceptionally fast scan, e.g., the entire spectrum can be scanned every 10^{-4} sec or faster. It should be emphasized, however, that this rate of scan has virtually no utility in GC–MS analyses and is generally applied only to special fast-reaction kinetic studies. For GC-MS studies, an analog amplifier is generally used to integrate the individual mass peaks and permit recorder scans at the practical rates of a few seconds per mass decade. The poor statistical representation found in each 10^{-4} sec scan is thus integrated out. Furthermore, if a simultaneous GC record is desired, a similar analog amplifier can be used to collect the total ion current on alternate cycles (26).

One important feature of the very fast scan rates (10^4/sec) is the facile presentation of a high-quality oscilloscope display. This also permits oscilloscopic photography (33) mentioned earlier, but the restricted dynamic and mass range of this technique have limited its popularity. On the other hand, the oscilloscope display is very useful in monitoring GC peaks, particularly fused or shoulder peaks whose maximum intensity may be slightly displaced from the actual point indicated by the conventional chromatogram.

Another common nonmagnetic mass spectrometer employs a quadrupole or monopole radiofrequency (rf) filter (3) for mass analysis of the ions. These instruments are attractive because they are relatively inexpensive, and are adequate for simple routine analyses. These rf instruments are capable of scanning the mass spectrum in a few milliseconds and thus also provide the convenience of visual "real time" observation of the mass spectrum on an oscilloscope. An inherent advantage of the quadrupole or monopole instruments is the linear mass scale in the output; i.e., on the final data trace, the distance between the peaks at m/e 350 and 351 is the same as that between m/e 40 and 41. Furthermore, this type of rf instrument is capable of maintaining quality output at higher pressures than other conventional mass spectrometers, a particular advantage in the GC–MS combination. This tolerance to higher pressures is due to the fact that the quadrupole mass spectrometer is not a beam instrument; that is to say, ions of tuned mass enjoy a stable existence in the quadrupole "well," and drift randomly, if necessary, down the inner region formed by the quadrupole rods to the detector. The only harmful effect of higher pressures ($P \sim 10^{-4}$ torr) is that once the mean free path of the gas is less than the tube length, molecular collisions effect a loss of ions to the poles, which results in a loss of sensitivity linearly proportional to the increase in pressure.

The problem of mass discrimination has plagued the rf mass spectrometer for some time. This so-called mass discrimination manifests itself in the diminished relative intensity of the high-mass ions from a given compound compared to the spectrum obtained on a magnetic instrument. There are indications from the mass spectrometer manufacturers that this problem is rapidly being corrected. Also, a significant improvement in the sensitivity of the quadrupole mass spectrometer may be realized soon without concurrent loss of resolution. Brubaker (73) has found that an arrangement of four auxiliary electrodes placed near the entrance to the quadrupole rods and maintained at a smaller ratio of dc to ac potential than that applied to the quadrupole effects a more efficient injection of ions into the analyzer volume. This effect has been simulated in a computer study and most of the effects have been verified by preliminary experimental results. It has been reported that at a resolution of 100, the sensitivity was increased 50-fold, while at a resolving power of 400, the sensitivity was 250 times greater than that possible in a quadrupole instrument without the auxiliary electrodes.

In considering any of these mass spectrometers which claim scan speeds of the complete mass spectrum in less than a few tenths of a second, one should remember the difficulty involved in establishing the mass scale (Section II-B). Generally, this difficulty increases with scan speed since the sampling of each group of ions becomes statistically unacceptable, leading to poorly represented peak shapes due to low signal-to-noise ratio.

The *magnetic single-focusing mass spectrometers* have a history of providing reliable, good quality low-resolution mass spectra which are suitable for structure elucidation. Older models of these magnetic instruments must be overhauled if they are to be used in the GC–MS combination. The requirements of fast amplifiers and oscillographic recorders have been reviewed elsewhere (12) recently. Since 1963, modern magnetic instruments have been designed for rapid scanning of the mass spectrum (m/e 50–500 in 1–3 sec), and have produced good quality data for manual reduction (29) or data processing (30).

The output from a magnetic instrument can also be displayed on an oscilloscope and a tracing of the total ion current (obtained with a partial ion collector) can be recorded to provide a permanent record of those portions of the gas chromatogram examined by the mass spectrometer. A complete discussion of the operations and utilization of a rapid, repetitive, magnetic scanning mass spectrometer is given in Section VII-A. Furthermore, the data discussed in Section VIII were obtained from magnetic deflection instruments.

2. High-Resolution Mass Spectrometers

The advantages of high-resolution mass spectral data and the resulting assignment of elemental composition was briefly discussed in Section II-C. Unfortunately, high-resolution mass spectroscopy is very expensive both in initial cost and in operation (12). An instrument itself may cost $150,000 while maintenance could run as high as $5,000 annually. The cost of data reduction and processing could approach $30,000 annually if the instrument is used continuously and efficiently. (This figure is estimated by assuming that 15 complete high-resolution mass spectra would be reduced to element map form per week; approximately 5 min of computer time is required per spectrum, computer time valued at $500 per hour.) If the mass spectra are recorded photographically, it is necessary to acquire a microdensitometer to measure the line positions on the photoplate with high accuracy. Because of this accuracy (0.5 μm) the accessory costs at least $12–15,000 and sophisticated systems run as high as $50,000.

Recently a manual system has been introduced commercially for about $3,000, but accuracy has been sacrificed for economy.

Regardless of how the data are recorded (see next paragraph), it is only feasible to reduce the data completely by computer to an intelligible format such as an element map. Furthermore, a high-resolution mass spectroscopy group might consist of four people: (*a*) a professional investigator who has had experience in mass spectroscopy, (*b*) a part-time electronics maintenance expert who has had experience in the peculiarities of mass spectroscopy, (*c*) a part-time programmer or computer consultant, and (*d*) a technician.

There are two basic designs of high-resolution instruments currently available. The design incorporating Mattauch-Herzog geometry (Fig. 5-14) separates and focuses all the ions on a focal plane. A photographic plate held at this focal plane records all the ions all the time, and thus avoids the restrictions imposed by scanning. This is a particular advantage in GC–MS work since the photographic emulsion is essentially an integrating device for the changing ion current during GC elution. An electron multiplier, positioned at one end of the photoplate assembly, is used for "tuning" the instrument, or for electrical recording if a scan of the mass spectrum is desired.

Another basic design is the Nier-Johnson which focuses the ions of one *m/e* value only on the slit of an electron multiplier (Fig. 5-15). Accurate

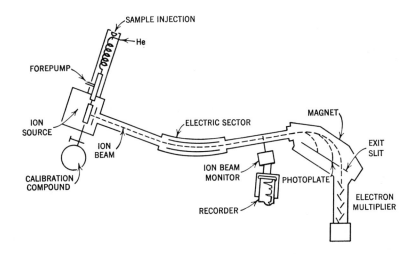

Fig. 5-14. Schematic of high-resolution mass spectrometer (Mattauch-Herzog geometry) with ion source modified for gas chromatography (50).

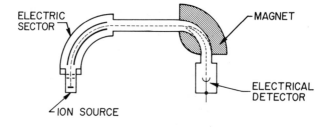

Fig. 5-15. Schematic of high-resolution mass spectrometer with Nier-Johnson geometry.

mass determinations can be made by a process of "peak matching" (74). This technique involves very accurate measurement of the ratio of the accelerating voltages necessary to focus first an ion of unknown m/e and then an ion of standard m/e on the slits of the electron multiplier. In general, it is not considered practical to attempt peak matching during a GC–MS run, and the spectra should be recorded. This is normally done on magnetic tape which facilitates further data processing. Unfortunately, the process of rapid scanning (m/e 60–600 in 10 sec) limits the effective resolving power of the instrument to approximately 10,000, and requires a sample of approximately 1 μg (75). This is so because of statistical limitations; at a resolution of 10,000 and a scan time of 10 sec, each 1/10,000 of the mass spectrum is examined by the electron multiplier for only 10^{-3} sec. If the increments of the spectrum are further reduced (resolution increased), the electron multiplier receives too few ions to establish a representative peak shape. The center of the peak cannot be accurately determined, and thus there is a deterioration in the accuracy of the mass measurement.

The Mattauch-Herzog design with photographic recording does not compromise accuracy in recording spectra of transient samples because it records all the ions simultaneously as a photograph. Data reduction from the photoplate can thus be performed under less taxing circumstances. Since the Mattauch-Herzog design also provides a facility for electrical recording, this instrument may utilize peak matching also. In summary, the Mattauch-Herzog double-focusing mass spectrometer can have the desired electrical features of the Nier-Johnson type and still allow the mass spectroscopist to enjoy the option of photographic recording when necessary. McFadden (12,31) has reviewed the comparisons and limitations of photographic and electrical detection systems.

High-resolution instruments are differentially pumped, that is, the pressures in the ion source and in the analyzer are essentially independent to one or two orders of magnitude. This design, which was included to minimize gas scattering of the ion beam over the relatively long distance to the detector, is especially advantageous in the GC–MS combination, since the ion source pressure can be increased by an order of magnitude or more without adversely affecting the pressure in the mass analyzer. Further advantages of the use of high-resolution instruments in GC–MS analysis, such as clearly resolving ions from column bleed and/or calibration compound will be demonstrated later in Section VIII-A.

D. The GC–MS Chassis

Widespread use and demand for GC–MS analyses have resulted in the commercial availability of several models of combined instruments. The combined instrument is generally a permanent fusion of the gas chromatograph and the mass spectrometer, neatly arranged on a single chassis as, for example, the model shown in Figure 5-16. Such an arrangement is

Fig. 5-16. Permanent arrangement of gas chromatograph with double-focusing mass spectrometer. *A*, injection port; *B*, GC oven; *C*, ion source; *D*, magnet; *E*, electrical detector of mass spectrometer.

excellent for a laboratory that has sufficient sample loads to justify the exclusive use of a mass spectrometer to monitor and examine GC effluents. The main advantage of a permanent arrangement is that both instruments are optimized for combined routine operation. That is, the MS vacuum system will be designed to cope with the anticipated higher gas flow into the ion source. The MS unit will also have fast amplifiers and high-frequency oscillographic recorders as standard equipment for scanning. The permanent GC–MS arrangement usually assures the incorporation of a good quality gas chromatograph with all of the latest refinements of good thermal environment and reproducible temperature programming.

Alternatively, there are often advantages of permanent incorporation of a short, unsophisticated GC column. Even though such an arrangement may be considered as merely a modification of the MS ion source or inlet system, it provides the great convenience of facile introduction of small samples and eliminates gross contamination of the principal compound. A compact version of a GC inlet which will not interfere with other routine uses, such as sample introduction through more conventional inlet systems, can be designed to mount permanently on the mass spectrometer (50).

If a laboratory has *several* good quality GC instruments and *one* mass spectrometer, it may be preferable to have an arrangement that will permit convenient, temporary attachment of a gas chromatograph. A detachable arrangement allows each instrument to be used independently under optimum conditions. Furthermore, the operator may choose the best GC system for any specific problem.

Many techniques of MS sample introduction have been reported in the past decade (34), and most modern mass spectrometers now have some facility for direct introduction of solid or high-boiling samples into the ion source via a vacuum lock assembly. Temporary attachment of the chromatograph to the spectrometer is sometimes accomplished through the standard vacuum lock hardware, as has been demonstrated with both the fritted glass separator (76) and the porous stainless steel separator (46).

The design of a "gas-flow" connection in the physical form of a direct inlet probe facilitates the temporary combination of the GC–separator–MS system. Use of the gas-flow probe attachment allows the actual attachment of the GC–separator components to be accomplished in only a few minutes since it is only necessary to evacuate the vacuum lock before the exit of the gas-flow probe is introduced directly into the ionization chamber in the center of the ion source (Fig. 5-18). The gas-flow probe transfers the enriched GC sample from a rollaway chassis (Fig. 5-17) which contains all the electrical equipment for a high performance gas

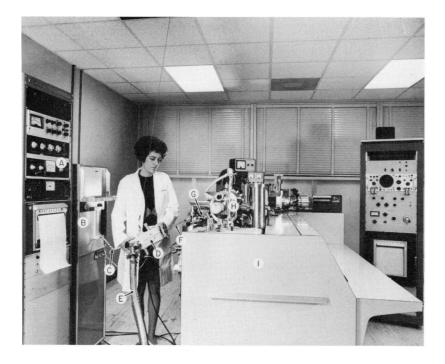

Fig. 5-17. Temporary arrangement of GC–MS. *A*, GC control unit; *B*, GC oven; *C*, capillary transfer line; *D*, separator enclosed in oven; *E*, separator vacuum line; *F*, gas-flow probe; *G*, vacuum lock; *H*, ion source; *I*, high resolution MS (17).

chromatograph, including a flame ionization detector. Several thermo-couples mounted throughout the heated components of this interface system can be monitored on meters in the rollaway chassis.

In one case (46) the gas-flow probe is fabricated from stainless steel of the dimensions required for the standard direct introduction accessory for, in this case, a CEC-21-110B high-resolution mass spectrometer. The separator and metering valve are built directly into the lower end of the probe as shown in the superficial drawing (upper left) in Figure 5-18. The upper right portion of Figure 5-18 shows a cross-sectional schematic of the probe passing through the vacuum lock into the ion source. A metal capillary which runs concentrically along the inside of the main body of the probe channels the enriched GC effluent into the ionization chamber.

The metal capillary and main probe body also function as part of the heating circuit (46). The inner capillary is pressure fitted to the main probe body near the exit. The heating circuit is completed appropriate through

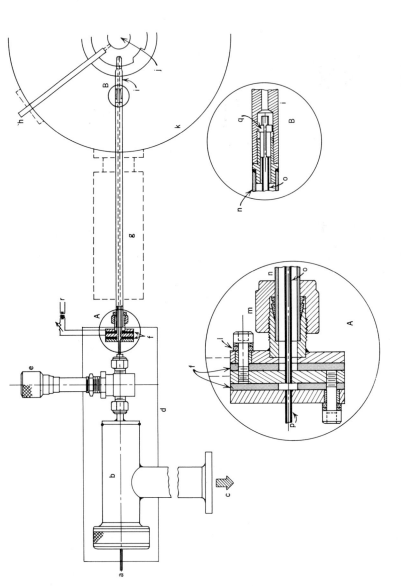

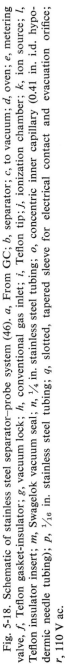

Fig. 5-18. Schematic of stainless steel separator–probe system (46). *a*, From GC; *b*, separator; *c*, to vacuum; *d*, oven; *e*, metering valve; *f*, Teflon gasket-insulator; *g*, vacuum lock; *h*, conventional gas inlet; *i*, Teflon tip; *j*, ionization chamber; *k*, ion source; *l*, Teflon insulator insert; *m*, Swagelok vacuum seal; *n*, $\frac{1}{4}$ in. stainless steel tubing; *o*, concentric inner capillary (0.41 in. i.d. hypodermic needle tubing); *p*, $\frac{1}{16}$ in. stainless steel tubing; *q*, slotted, tapered sleeve for electrical contact and evacuation orifice; *r*, 110 V ac.

194

electrical connections in the Teflon gasket insulator block (see enlargement at bottom of Figure 5-18). To insure uniform heating of the capillary surfaces it is necessary that the annular space between the capillary and heavier probe body be evacuated. This requirement demands that vacuum seals be established by a Swagelok connection with the main probe body and also by soldering the inner capillary to the appropriate metal disk in the Teflon gasket insulator block. Evacuation of the annular region is accomplished by the MS pumping system through a small orifice at the upper end of the probe. The other major components of the system have separate heating ovens; the longer length (2 ft of standard $\frac{1}{16}$ in.) of stainless steel tubing between the chromatograph and the separator is wound with heating tape.

The rollaway GC–separator system easily utilizes all the latest refinements of sophisticated gas chromatography and also permits the independent operation of the high-resolution mass spectrometer. When combined GC–MS operation is required, the GC–separator chassis can be rolled up and directly connected to the mass spectrometer ready for operation—in less than 5 min.

In all such GC–MS combinations the transfer path should be kept as short as possible. However, when temporary combinations are employed, it may be necessary to incorporate capillary connections 3–5 ft in length. It is important to channel the gas stream directly into the relatively small volume of the ionization chamber as shown in Figure 5-18. The sensitivity of the overall operation would be greatly diminished if the gas stream were only channeled to the inner periphery of the ion source, a relatively large volume, where a significant portion of the sample may be evacuated before it can diffuse to the ionization region.

From this discussion it is clear that a completely new, combined instrument is not required to furnish the most sophisticated combined techniques of separation and identification for occasional analyses. Modifications of GC and MS instruments for convenient combination are within the technical capabilities of most laboratories.

VI. Other Techniques for Examination of GC Fractions

The discussion of GC–MS combination has dealt primarily with a tandem, continuous system with some technique of recording the mass spectra during the emergence of the GC fractions. Such a dynamic system, while often the most convenient and flexible, is certainly not always required.

For several years, many investigators have practiced the simple technique of collecting specific fractions by condensation in a small capillary maintained at a lower temperature than the exit of the GC column. The GC fraction can then be removed from the capillary and introduced into the batch inlet of the mass spectrometer, or the short length of capillary can be inserted into the ion source of the spectrometer by means of the vacuum lock ordinarily used for solid samples (1,18,34,77). A short length of capillary containing GC column packing has also been suggested as an efficient means of fraction collection (78). This technique also involves introduction of the capillary containing the GC fraction and the column material into the spectrometer by the direct inlet system; the unit is carefully heated to remove the sample without causing "bleeding" from the liquid phase of the GC column material. However, a small capillary filled with activated coconut charcoal (70–100 mesh) is much more successful in collecting volatile compounds (79). Furthermore, except for adsorbed CO_2, the charcoal contributes no appreciable background spectrum.

An alternative to monitoring the continuous flow of GC effluents utilizes the technique of interrupted elution (80–83) of the GC column. This "stop–start" technique allows individual peaks to be trapped for analysis while the remaining GC system is successfully preserved under static conditions. One successful design of a "stopped-flow" GC–MS system (82) incorporates a thermal conductivity detector in series with the trapping cell so that all the GC fraction is available for analysis by the (time-of-flight) mass spectrometer. This arrangement avoids the necessity of a stream splitter, but requires that the detector be shut off when the carrier gas is stopped. After a GC fraction enters the trap through a manually operated manifold, the inlet pressure is relieved to minimize peak, creeping in the isolated column. The trap is evacuated to $\sim 10^{-4}$ torr and then removed from liquid nitrogen and heated to regenerate the sample as required for a usable MS signal. This system has been quite useful in the analysis of complex, high molecular weight mixtures (82).

Another version of a stop–start GC–MS system has been modified for unattended operation including analysis by both mass and infrared spectroscopy (83). An essential part of this unit is a 15 ft precolumn upstream from the injection port to prevent loss of components through backward movement of peaks during the stop–start procedure. During operation, individual GC fractions are channeled through an arrangement of solenoid-operated valves to a cold trap containing GC partitioning materials for trapping efficiency, while the carrier flow through the column is stopped. The trapped component is then regenerated by heating and transferred by backflushing with carrier gas to a 25-ml infrared cell which also has a leak to the mass spectrometer. The isolation and backflushing operation usually increases the concentration of the component by a factor as great as 50.

This technique has the advantage of permitting near-optimum recording of the mass spectrum. The sample pressure is nearly constant throughout the relatively long sampling period, and this facilitates operation of the spectrometer. The ion focusing and other parameters may be readjusted if necessary before the mass spectrum is scanned at a reasonably slow rate. Furthermore, substantial enrichment is accomplished by the trapping method. However, the system is not always practical for analysis of complex mixtures with many overlapping peaks and the operation is relatively slow.

VII. Data Reduction and Processing

A. Data Reduction

It was recognized some time ago that high-resolution mass spectral data required extensive data reduction for effective utilization in structure elucidation. Several approaches to the processing of high-resolution data are well established and will be mentioned briefly.

Desiderio and Biemann (22,84) initiated the use of computers in reduction of high-resolution mass spectra when they realized the importance of considering all the several hundred lines in a mass spectrum recorded on a photoplate. Venkataraghavan and McLafferty have extended the fundamentals of the two-dimensional element map to a topographical display which readily indicates the relative importance of the various ions (85). The topographic element map is automatically plotted in three dimensions by a computer.

Tunnicliff and Wadsworth (86) have developed a program for accurate mass determinations in which the relation between mass and line position on the photoplate is fitted to a power series. The positions of a set of standard masses, taken from an arbitrary zero, are raised to appropriate powers to determine a unique set of coefficients which characterize a given spectrum, and can be used to compute accurate mass values for the recorded unknown lines. This approach is fundamentally important in that all the calibration data are utilized *en masse* and thus a significant error in any single calibration line is effectively eliminated through a least squares fit of the data. Preliminary results of this program are encouraging; errors of accurate mass measurement are 1–2 ppm up to mass 500 (86).

The basic techniques of reduction of high-resolution mass spectral data have also been demonstrated for data recorded directly onto magnetic tape. Direct recording on magnetic tape offers the most practical method for considering all the high-resolution data from a Nier-Johnson instrument (75), especially from fast GC–MS scanning. Conventional slow scanning at fairly fast recorder charts produces an awkward record (87), and although inexpensive compared to a tape recorder computer, the records require considerable time to process.

The Mattauch-Herzog instruments record the complete high-resolution mass spectrum on a photographic plate under optimum conditions; the photoplate is then measured and the resulting data submitted for reduction. The extra steps of preparation and measurement of the photoplate have been cited as inconvenient and time-consuming, but the drawbacks to this plate-reading technique are slowly diminishing. Desiderio (88) recently described a technique of direct communication with the computer through an analog-to-digital converter which permits a complete scan of the photoplate in less than 3 min.

All the techniques of data acquisition and reduction of high-resolution mass spectra usually display the output in some form of "element map" with the option of a plotted conventional mass spectrum and tabulation of accurate masses and elemental compositions. Recently, systems of data acquisition and processing have also been demonstrated with an on-line computer located in the laboratory (89–91).

Most of the analyses by the GC–MS technique involve a medium-resolution mass spectrometer, partly because of the high cost of high-resolution instruments but also because most samples are amenable to conventional mass spectrometry. It is rather ironic that the GC–MS combination which provides such a potent analytical technique also produces an overwhelming number of data. If one analyzes a mixture with only 20 peaks on the gas chromatograph and scans each peak twice with the mass spectrometer to ensure each indicates a single component, then within an hour he produces at least 40 mass spectra. Even a cursory examination of the spectra requires another 2 hr. If the spectra are to be studied and compared, they must be mass marked (i.e., mass scale ascertained by actually counting, mass by mass, from an identifiable peak—such as 28 or 32 from air—at low mass up to the molecular ion). For many GC peaks, the spectra must be measured for relative abundance, and preferably replotted or tabulated. To manually reduce all these data to the often-preferred bar graph form could require as much as 25 man-hours of labor. As a result, the full advantage of the GC–MS system is not exploited because only a few of the spectra are chosen for reduction to a form suitable for careful comparison with library spectra. [It should be pointed out that most systematic interpretations of mass spectra (1) may be accomplished from the analog data record. Also, some approximate comparisons with tabulated spectra can be made from the raw data trace, but if (automatic) library retrieval techniques (92–94) are to be used, the unknown spectra will have to be carefully reduced and tabulated.] Furthermore, the inhibiting awareness of the burdening data reduction problem may prevent the operator from scanning the GC fractions with sufficient frequency to fully examine the composition of each peak. The full impact of the data problem is realized when several hundred mass spectra are generated by continuous daily operation of the GC–MS system.

The dynamic system of data acquisition and reduction demonstrated by Hites and Biemann (30) employs a single-focusing mass spectrometer with a continuous repetitive magnetic scan. A 3-sec scan of the spectrum (masses 20–500) is suitable for most transient GC fractions while a 1-sec interruption between scans is adequate for complete recovery of the magnetic field. The analog signal from the electron multiplier can be directed to any one or all of three monitoring and recording devices: a memory oscilloscope, an oscillographic recorder, and an analog-to-digital (A/D) converter for magnetic tape recording. The automatic feature is realized when the A/D converter receives and samples the electron multiplier output at the rate of 3000 data points per second. The recording system is synchronized with the MS scan cycle with sufficient precision to establish a mass scale on a time basis. After the mass spectrometer is tuned for optimum operation, the time axis is calibrated for nominal masses by introducing a compound of known spectrum. A more sophisticated version (92) employs on-line A/D conversion (IBM 1800) which permits continuous scanning of the mass spectrometer for the duration of the gas chromatogram. This technique relieves the operator of the responsibility of deciding when or whether to initiate an individual scan for a given GC fraction. A self-correcting program recalibrates or compensates for any slight drift of the time-calibrated mass scale from scan to scan; this calibration is much more stable when the computer controls the magnetic scan (95).

The status of the GC effluent entering the ion source can be conveniently displayed by monitoring a portion of the total ion current. By noting the position of the total ion current trace or the intensity of the mass spectrum on the oscilloscope, one can select the next complete MS scan for permanent recording. It is also possible, if desired, to record continuously any or all of the consecutive scans. The final processing (30) involves smoothing the raw digital data and then locating the time datum point corresponding to each peak center. The resulting time vs. intensity data are converted to mass vs. relative intensity (normalized to the most intense peak). These data are stored on magnetic tape in digital form and are then readily available for tabulation, plotting, or further processing.

This type of data processing is automatic only in that several steps of reduction, computation, and adjustment of the raw data are possible without human intervention. The investigator, freed from the routine of data handling, can now command the efficient production of the final representation of the data. It is a simple task for the computer to subtract two consecutive scans, normalize the net intensities, and plot the resulting mass spectrum which is then relatively free of the "cross-talk" from a poorly resolved GC peak or background. It is also a valuable asset to have several consecutive scans of a substance to check purity and/or uniformity of composition of the GC peak as it issues into the ion source.

Retaining all the mass spectral data from consecutive 4-sec scans no longer presents a data storage problem, since they are stored on magnetic tape or disks. These data can be quite useful in determining the general nature of the compound types represented by the chromatogram. The total ionization current of each scan plotted against spectrum (or scan) number resembles the conventional chromatogram. However, the un-normalized intensity of any selected m/e value may also be automatically plotted versus spectrum index (95). In this way, the contribution of selected compound types (e.g., methyl esters by plotting un-normalized m/e 74 intensities) can be displayed for comparison with the total ion plot, the "histogram" (95) or "monitogram" (76). This technique of plotting a specific ion abundance has greatly facilitated the discovery of minute components in sections of the histogram or monitogram which might ordinarily be dismissed as background (95). This technique is reminiscent of the early GC–MS work in which preselected m/e values were monitored in an effort to ascertain the composition of GC fractions. However, with the modern system the investigator has the overwhelming advantage of selecting which m/e values he wants to plot after completion of the GC–MS run, and more important, of course, is the fact that all the mass spectral data are immediately available for any given scan (95).

A reasonably simple and inexpensive data acquisition system has been developed for gas chromatography–mass spectrometry which records digitally on incremental magnetic tape (96). The system uses a mass marker (Hall element) to trigger the tape movement at increments which correspond to nominal m/e units. This approach offers the advantage that no on-line data reduction is necessary to determine the peak centers and intensities; there is only one intensity measurement per mass unit. At the

end of a scan the data are immediately available for background adjustment, tabulation, and/or plotting. Furthermore, since the marker determines the mass scale automatically, the need for a reference compound is
obviated. However, the entire operation is dependent on the reliability
and precision of the mass markers. In order to minimize errors in mass
assignment of sample ions, adjustments in the mass marker are necessary
depending on the elemental composition of the sample (compensation for
mass defect). The system typically scans m/e 10–800 in 3 sec with an error
of ± 0.3 mass units.

The output from a quadrupole mass spectrometer is conveniently
amenable to on-line data processing because the m/e values are a linear
function of scanning voltage. In principle, the determination of the mass
scale requires only the measurement of the voltage difference between any
two nominal m/e peaks because the separation of masses is constant
throughout the mass spectrum. Such an approach has been used successfully for the simplified operation of a computer-controlled quadrupole
mass spectrometer (97).

B. Computer-Aided Identification

The vast majority of GC–MS analyses, as well as straight MS analyses,
have a common goal—the rapid and unambiguous identification of the
compound(s). This is usually accomplished by "fingerprint matching" or
comparing the mass spectrum with available spectra of authentic compounds. The problem is in providing convenient access to a sufficiently
large library of mass spectra to achieve the "matching" or identification.

Attempts to establish accessible libraries of mass spectra have resulted
in data storage on magnetic tape (93). This has been accomplished by
manually punching into IBM cards the available mass spectral data from
a few thousand compounds, and transferring the data to magnetic tape
for storage and retrieval. Various routines or criteria have been suggested
(94) for selecting a member from the library spectra which matches the
unknown spectrum.

The method of searching for the six most intense peaks in the spectrum
has the shortcoming of emphasizing examination of the low-mass end of
the spectrum (94). Most of the unique fragments in a mass spectrum
appear at the high-mass end and are often not among the six most intense
ions in the spectrum. Examination of a given number of the most intense
peaks between certain mass numbers (94) was reasonably successful, but
it required human intervention to establish the m/e value of the molecular
ion. Pettersson and Ryhage (94) found that comparison of all peaks of

3% or greater relative intensity for agreement within $\pm 5\%$ was a reasonably successful method of searching the library for a spectrum that matched the unknown. However, this large number of comparisons taxed the capacity and speed of a small computer (IBM 401). It was then suggested that the two most intense peaks every 25 mass units be compared, and, if necessary, the intensity of other peaks could be examined for a final "match."

The approach to library searching designed by Hites and Biemann (92) seems to be the most comprehensive and successful. One of the first requirements involves reducing the data to "abbreviated" spectra by retaining only the two most intense peaks in every 14 mass units. In almost all cases, the abbreviated spectra still contain all the characteristic peaks which one selectively considers under usual circumstances. The standard procedure for abbreviation guarantees the inclusion of the molecular ion and the heaviest fragment ion (except that due to loss of hydrogen).

The computer program which compares the unknown spectrum with those available in the library is fairly complicated (92), but it establishes the criteria briefly outlined in the next few paragraphs.

The program makes a "rough cut" by noting the mass number of the most intense peak in the unknown spectrum and then ignoring all those spectra in the library which do not exhibit a peak at the given m/e of at least 25% of the intensity of this main peak in the unknown spectrum, and vice versa. It is expected that this conservative "rough cut" will not disqualify spectra which have suffered intensity distortion due to instrumental peculiarities or changes in sample concentration accompanying the emergence of a GC fraction. The program also avoids comparison of the ions in regions where one of the two spectra may not have been scanned.

The intensities of peaks at the same mass in the unknown and library spectra are examined and a ratio of the two intensities is computed. If a peak appears in one spectrum, but does not in the other, the ratio is taken as zero. The reciprocal is taken of all intensity ratios greater than unity so that in the final tabulation all the ratios will have values between zero and unity. Since intensity agreement, as well as disagreement, of the more abundant ions is more significant than that of minor ions, a weighting factor (1, 2, or 3) is applied to the intensity ratio of the ions depending on their level of relative contribution or abundance ($< 1\%$, $> 1\%$, or $> 10\%$, respectively). Finally, these weighted ratios are summed and averaged. This "average ratio" or "identity index" is probably the single most important parameter by which to gauge the degree of confidence in matching the unknown and library spectra; a perfect match would give an identity index of 1.00 (92,95).

Finally, the sum of the intensities of all peaks that are present in one spectrum but absent in the other is compiled. This number should obviously be small or minimal in the comparison of two spectra from the same compound and large when the spectra are from different compounds.

Upon completion of the gas chromatogram, the spectra are immediately available on digital tape for processing by the search program. The results include a plot of the unknown spectrum and the name of the compound whose mass spectrum most nearly matches that of the unknown compound (92). The results also list "No Finds" when all the intensity "average ratios" are below 0.200 and all the sums of missing intensities are above 2000, or it may list a questionable assignment if the best "average ratio" is less than 0.050 units higher than the next best "average ratio" of intensities. (It is possible to express the "missing intensities" as a percentage of the "matched intensities"; this may be more meaningful to some investigators.) In any event, the operator may examine the data that the computer used to select the final result. That is, the complete raw spectrum of any scan can be plotted along with the name, chi-square value (a statistical test of the data (92)), and sum of missing intensities for the three most likely candidates as selected by their intensity ratios. This allows the investigator to evaluate the identifications suggested by the computer and, if necessary, interpret the mass spectrum in the conventional manner since it may not be listed in the library.

In an actual investigation, a record in the form of a tracing of the total ion intensity of each spectrum vs. its index number (Fig. 5-19) is produced. This plot (histogram) resembles the gas chromatogram, and greatly facilitates the selection of the pertinent scans for plotting, comparative processing, etc. For example, after the GC–MS run represented by Figure 5–19, certain scans (indicated by the × in Fig. 5-19) were submitted by index number for processing (92). For the fourth, fifth, and sixth peaks in Figure 5-19, the results indicated 3-methylcyclohexanone for the consecutive scans 22, 23, and 24; "No Finds" for scan 34; ethyl caprylate for scans 35, 36, and 37. (The same result for two or three scans of a single GC fraction confirms the uniform identification.)

In contrast, scans 50 and 51 showed questionable assignments of methyl undecyl ketone and ethyl 2-methyl valerate. (It should be pointed out that matching the molecular ion of the unknown spectrum with the molecular weight of the suggested library compound is not a criterion of this program.) A "questionable" assignment suggests that the unknown spectrum and accompanying computations be carefully examined by the chemist. Examination of the detailed results for scans 50 and 51 (Fig. 5-20) indicates the reasons for the questionable assignments. The "average

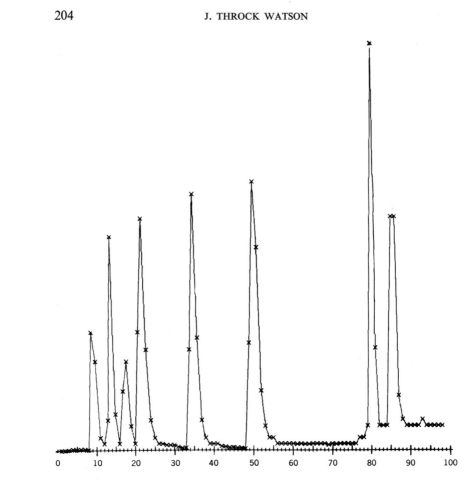

Fig. 5-19. Plot of total intensity (ordinate) of 100 consecutively scanned spectra plotted as indicated by index number (abscissa) (92).

ratios" are all low, the values of chi-square are high (especially in scan 51), and the missing peak intensities are also high. The fact that the results are quite different substantiates the suspicion that the sixth GC peak contains at least two components.

While the problem could have been simplified or solved by examining scans 49 and 52, further processing of scans 50 and 51 demonstrates the potential utility of the comparative treatment (by computer) of the data. Subtraction of scan 51 from 50 and 50 from 51 leads to the new spectra shown in Figure 5-21 (92). The spectrum compiled by subtraction of 51 from 50 (Fig. 5-21a) now more truly represents the component at the

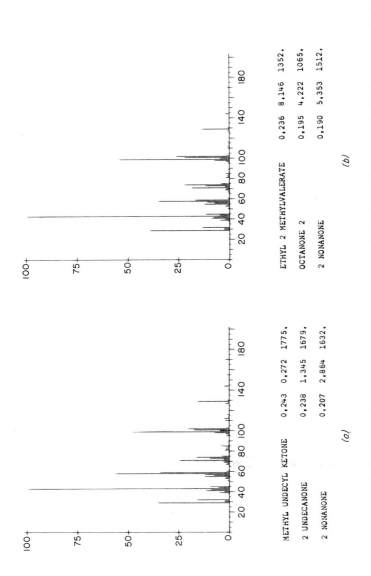

METHYL UNDECYL KETONE	0,243	0,272	1775,
2 UNDECANONE	0,238	1,345	1679,
2 NONANONE	0,207	2,864	1632,

(a)

ETHYL 2 METHYLVALERATE	0,236	8,146	1352,
OCTANONE 2	0,195	4,222	1065,
2 NONANONE	0,190	5,353	1512,

(b)

Fig. 5-20. Plotted spectra and search results (average ratio, chi square, missing intensities, respectively) from scans 50 (a) and 51 (b) (92).

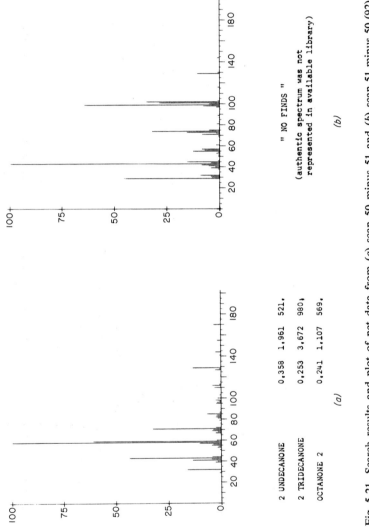

2 UNDECANONE   0.358   1.961   521.

2 TRIDECANONE   0.253   3.672   980,

OCTANONE 2   0.241   1.107   569.

(a)

" NO FINDS "

(authentic spectrum was not
represented in available library)

(b)

Fig. 5-21. Search results and plot of net data from (a) scan 50 minus 51 and (b) scan 51 minus 50 (92).

leading edge of the sixth GC peak. This net spectrum agrees reasonably well with the library spectrum of 2-undecanone. The net result of the alternate subtraction (Fig. 5-21b) is also more characteristic of the compound having the slightly greater retention time; however, the computer could not find a comparable spectrum in the available library. The spectrum was subsequently interpreted in the conventional manner (1); an ethyl ester of a methyl ketone was indicated, and the spectrum was confirmed as that of ethyl levulinate, which was not represented in the library.

The example discussed here required exclusive use of a medium-size on-line computer (IBM 1800). Time-sharing was not possible for the periods when the computer continually received and processed the consecutive GC–MS scans. However, this arrangement offers the most flexible, efficient, and economical method for recording and interpreting large numbers of spectra (92). In principle, a system of an A/D converter and tape recorder with off-line processing (30) can accomplish the same feats.

The systems described by Hites and Biemann (30,92,95) are not only revolutionary in the technique of conventional mass spectral data handling, but are also important in the production of standard libraries. Either the off-line (30) or on-line (92,95) system generates data directly in digital form required for ready-access libraries of mass spectra stored on magnetic tape. Again, the important features are convenience and elimination of human errors in the production of mass spectra. Furthermore, the comparative processing features of the program and the inherent advantages of the GC–MS system ensure that "purified" mass spectra, free from contamination or background, reach the standard library.

Even as the library search programs for the more common compounds are being developed and refined, it is recognized that programs for calculating types of spectra are also required. Biemann et al. (98) demonstrated the use of "ion types" to facilitate the interpretation of high-resolution mass spectra. In this way, the ions are listed according to relative abundance with the number of carbon atoms and other characteristic information such as the degree of saturation and number of heteroatoms. Computer programs for spectra interpretation of simple ketones, amides, and amines (99) have been designed to utilize the "ion-type" data which are derived from accurate mass measurements. Any comprehensive computer-based identification technique for conventional mass spectra cannot rely only on the sheer volume of its library, because the number of organic compounds is impossibly large. Computer-based methods of identifying fatty acids (94) and hydrocarbons (100) have been reported which supplement library retrieval programs. The general philosophy in the development of the combined calculation and search method suggests two broad classifications of unknowns. The calculation program should be able to process the large number of compound types which produce regular or fairly predictable mass spectra, e.g., straight-chain hydrocarbons, simple substituted open-chain aliphatics, etc. The library should then contain those spectra which are irregular or are otherwise difficult to distinguish with a systematic computer program.

The approach taken by Pettersson and Ryhage (100) incorporates a rectangular array (101) to effect a coarse screening of the unknown-compound for general classification. The rectangular array consists of lists of intensities of ions differing by 14 mass units, i.e., there are 14 columns of intensity entries, the first column contains the intensity of masses 14, 28, 42, 56, . . . , the second column that of masses 15, 29, 43, 57, . . . , etc., to the fourteenth column which contains the intensities of masses 27, 41, 55, ... The magnitudes of the 14 sums $(S_1, S_2, S_3, \ldots S_{14})$ can be used to characterize a class of compounds. Different compound types exhibit different sequences of decreasing sums; for example, the decreasing sequence in saturated hydrocarbons is $S_2 > S_{14} > S_1 > S_3, \ldots$, in unsaturated hydrocarbons it is $S_{14} > S_2 > S_1 > S_{12} \ldots$, while in fatty acids it is $S_4 \sim S_5 > S_{14} \sim S_2, \ldots$.

Further processing of the low to medium-resolution mass spectra by selected subroutines (94,100) is necessary to ascertain greater specificity of compound type; for example, the rectangular array cannot distinguish fatty acids from methyl esters of fatty acids. The development of such specific programs will supplement library searching techniques and broaden the general method for computer recognition of unknown mass spectra.

VIII. Applications

A. Routine Analyses

In most routine analyses, many of the components may be anticipated, but the GC–MS system confirms their presence and also ensures fraction homogeneity. In some laboratories it is not uncommon to find chromatograms which consists of nearly 100 peaks; in fact, there are examples in the literature (12,52,71,95,102) where all the components greater than 0.5% in a complex chromatogram have been identified by combined GC–MS analysis. For example, Figure 5-22 is the gas chromatogram of a mixture of methyl esters of fatty acids from buttermilk separated on an open tubular column (25 m × 0.25 mm) containing Castor wax (102). The gas

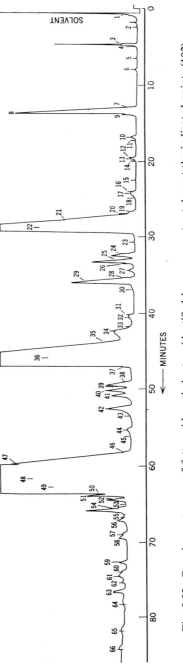

Fig. 5-22. Gas chromatogram of fatty acid methyl-esters identified by mass spectra taken at the indicated points (102).

chromatograph was temperature programmed at 1°C/min from 115 to 200°C with a helium flow rate of 5 ml/min. The components were identified by the mass spectra taken at the indicated points on the gas chromatogram (Fig. 5-22). It should be noted that even very small peaks—possibly less than 0.1% of the GC sample—can give interpretable mass spectra. However, in a single GC–MS run, the dynamic range of components for reasonable mass spectra is seldom greater than 100.

The GC–MS technique is finding increasing utilization in biochemical problems. Actually, it was the complexity of natural and biological mixtures that provided the incentive and necessity for GC–MS development. For example, considerable work has now been done on the characterization of sphingolipid bases of the general formula R—CH(OH)CH(NH$_2$)CH$_2$OH. Several long-chain bases were indicated by early gas–liquid chromatographic studies of sphingomyelin after various preparative oxidations (see reference 103 for a summary of this work). Later, separation of the trimethylsilyl (TMS) derivatives of these intact bases (104,105) facilitated structural determination (except for stereochemistry, of course) by mass spectrometry (105). Recently, GC–MS analysis was used to establish the structure of sphinga-4, 14-dienine and two lower homologs of sphingosine by analysis of the corresponding N-acetyl-O-TMS derivatives (103).

This method should be quite useful in the characterization of sphingomyelin fractions from human plasma, for instance, after preparative isolation and degradation (103). The N-acetyl derivatives (106,107) of the compounds are oxidized at the points of unsaturation with osmium tetroxide (108,109) before preparation of the TMS derivatives (103) in a mixture of pyridine, hexamethydisilazane, and trimethylchlorosilane (10:2:1). These derivatives have retention times ranging from 20 to 60 min on 3% OV-17 silicone oil at 240°C. The mass spectrum of such a derivative permits one to deduce the location of the original double bond(s), since fragmentation at these points is now enhanced due to the stabilizing effect of the O-TMS group. For example, fragmentation of the N-acetyl-poly-O-TMS derivative of 4,14-sphingadiene whose structure is illustrated on page 211, to yield ions of mass 145 and 694 confirms the double bond at Δ^4, while m/e 378 and 461 indicate Δ^{14}.

The osmium tetroxide step and conversion to the N-acetyl-poly-O-TMS derivative provides a compound whose mass spectrum indicates chain length, degree of unsaturation, and location of the double bonds in the original molecule. However, it is not possible to detect branching or ascertain stereochemical configurations by this technique (103). Characteristic mass spectral data from a series of these sphingolipid bases provide

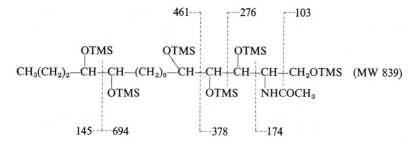

the basis for a computer program which will identify similar components in biological samples as the *N*-acetyl-*O*-TMS derivatives after osmium tetroxide oxidation (110).

Some diglyceride derivatives of phospholipids have been analyzed by GC–MS techniques (111). The phospholipids are degraded by enzymatic hydrolysis without rearrangement of the carbon skeleton. The resulting 1,2-diglycerides are then converted to the TMS derivatives which do not rearrange to the 1,3-diglyceride under GC–MS thermal conditions. These preliminary results (111) also demonstrate the utility of the GC–MS technique in biochemistry.

Many other applications of GC–MS analysis have been summarized by McFadden (12,31,34,112). Most analyses require a good deal of human attention, in selecting critical times for scanning the mass spectrum as well as reducing many mass spectra (e.g., establishing mass scale, subtracting background, measuring abundances, and tabulating or replotting the data). The semiautomatic (30) or "on-line" (92) data reduction system described earlier would be especially useful in those cases where hundreds of spectra are generated daily. These data-reduction techniques developed by Hites and Biemann, not only save time but acquire the raw data in digital form for immediate processing to the most representative net appearance of the mass spectra (see the example of low-resolution analysis in Section VII-B).

Generally, the GC–MS analyses are attempted with conventional mass spectrometry, but inevitably, there are some GC fractions which must be submitted for high-resolution mass analysis when this more expensive tool is available. The GC fraction in question can be (serially) collected at the GC exit and introduced through one of the conventional MS inlets or better, particularly if there are several dubious fractions in the chromatogram, a GC–MS combination which incorporates a high-resolution instrument can be used. Of course, it is also often necessary to obtain data from IR, NMR, UV, etc., analyses when a sufficient quantity of the GC fraction can be obtained.

There are several operational advantages to the GC–MS combination which utilizes a high-resolution mass analyzer. First, a calibration compound can be used simultaneously to define the mass scale at all times without interfering with the sample analysis. As explained earlier, the calibration compound is chosen for its unique elemental composition so that it may be easily resolved from the sample ions. Second, ions derived from the partitioning phase material (column bleed) are usually mass resolved from the sample ions due to their different characteristic elemental compositions. Third, it is possible to resolve impurities—even those comparable to the sample in molecular weight—if the elemental compositions differ. Finally, if the instrument employs photographic recording, the photoplate provides an integrating device for any variation in ion current as a result of sample pressure changes during the GC elution into the ion source. Some of these advantages are exemplified in Figure 5-23,

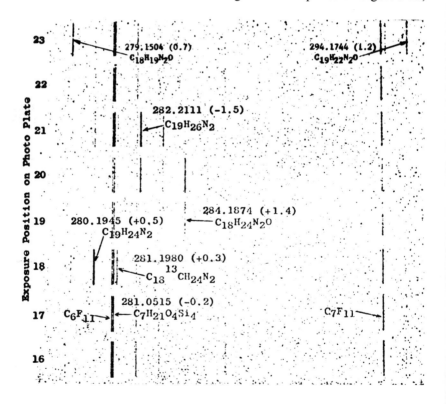

Fig. 5-23. Magnification of a portion of photographic plate containing spectra of alkaloids recorded during chromatogram represented by Figure 5-24 (50).

which is a magnification of a portion of a photographic plate in the region of m/e 280–295 (taken from a CEC-21-110, which is designed with Mattauch-Herzog geometry). The enlargement shows several exposures (16 through 23), each of which consists of a horizontal row of many well-defined vertical lines. The separation of these lines is related to the square root of the mass of the corresponding ions. Selected characteristic ions are indicated with elemental composition, mass measurement, and accompanying error in millimass units. Note that in every exposure there is at least a doublet at nominal m/e 281. As indicated in exposure 17, the line at the slightly lower m/e is due to the calibration compound fragment, C_6F_{11}, while the other line results from $C_7H_{21}O_4Si_4$ due to bleeding of silicone compounds from the gas chromatograph. Further to the right at m/e 293 is a second calibration line resulting from C_7F_{11}. The other lines in the enlargement resulted from the sequential exposure of the photoplate to the compounds that are indicated by the "beam monitor chromatogram" (monitogram) in Figure 5-24.

The numbers on the various sections of the monitogram (Fig. 5-24) correspond to the position of the photoplate exposures. The monitogram is a tracing of the beam monitor output which is interrupted when the total ion beam is deflected in changing the position of the photoplate. This interruption of the trace conveniently indicates the exact portions of the chromatograms exposed to the photoplate. The spectrum at position 18 was obtained during the emergence of the first chromatographic peak which contained a compound having a molecular ion of mass 280. Positions 19, 20, and 21 show an interesting sequence. First, a compound appears that has a fragment at m/e 284. In position 20, a mixture is apparent and finally in position 21 mass 284 virtually disappears and the compound with molecular ion at m/e 282 clearly dominates the spectrum. Henneberg (113) has suggested that moving the photoplate continuously throughout an exposure to a poorly resolved doublet (or an inhomogeneous fraction) more easily discloses the contributing species. This technique, however, rather inefficiently utilizes the photoplate and diminishes the sensitivity by spreading the ion current over a larger surface. Furthermore, it is necessary that the plate movement be extremely smooth or the inherent resolution of the instrument will be lost due to lateral excursions of the plate. While there may be cases where trends of appearance and disappearance are easier to detect with the moving plate, it has been demonstrated that discrete movements of the plate also reveal trends in composition of a GC fraction (20,50).

The monitogram shown in Figure 5-24 resulted from the injection of a portion of an extract from *Aspidosperma quebracho blanco* bark. The

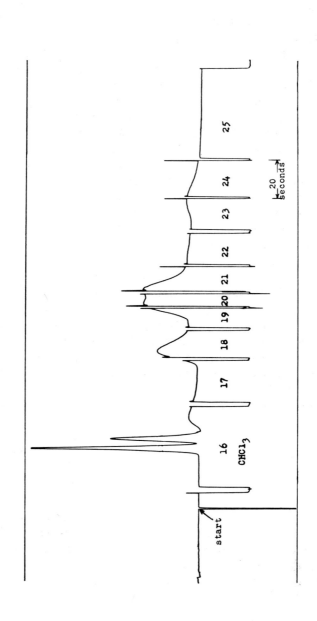

Fig. 5-24. Trace obtained by recording ion current after injection of a portion of *Aspidosperma quebracho blanco* alkaloids (50). (GC conditions: 6 ft, 1/8 in. column of 3% JXR on GasChrom P at 265°C with ca. 20 ml/min helium.)

major components were previously identified by low-resolution mass spectroscopy (114) as 1,2-dehydroaspidospermine, deacetylaspidospermine, and quebrachamine, respectively. The trace quantity shown near the end of the beam monitor chromatogram (exposures 23 and 24) was not identified in the earlier study of these alkaloids. This small GC peak was deliberately recorded as two separate exposures on the photographic plate in order to check its homogeneity. The fraction was, in fact, found to consist of two compounds.

Sufficient data were obtained from the compounds to suggest plausible structures (20,35). The magnified plate area (Fig. 5-23) shows exposure 23 and indicates a compound having ions at nominal m/e 279 and 294. Exposure 24 (not illustrated) had no line at m/e 294, but did contain lines at m/e 279 and 308. The mass spectrum at exposure 23 indicated a molecular ion of $C_{19}H_{22}N_2O$ and the fragmentation pattern suggested the structure of eburnamonine (**1** in Fig. 5-25). Most of the high-resolution mass spectral data from exposure 24 are consolidated in the element map discussed earlier (Fig. 5-5). The element map shows $C_{20}H_{24}N_2O$ as the molecular ion. The ions $C_{18}H_{19}N_2O$ and $C_{16}H_{16}NO$ indicate that the compound contains an ethyl group and easily loses the alicyclic nitrogen with 4 carbons and 8 hydrogens. Furthermore, the hydrogen-poor fragment, $C_{15}H_{13}NO$, indicates that the oxygen is bound to the aromatic part of the molecule and not to the alicyclic periphery. The elemental composition of the molecular ion as well as the fragments mentioned above preclude the possibility of aspidospermine or one of its derivatives (even though this class of compounds predominates in the mixture), but indicates a derivative of eburnamenine (**2**) previously extracted from this natural source (115). The compound of molecular weight 308 should be, therefore, eburnamenine with a methoxy group in the benzene ring (114). Since eburnamine (**4**) is converted to eburnamenine (**2**) under GC

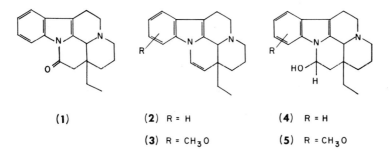

(**1**)

(**2**) R = H

(**3**) R = CH₃O

(**4**) R = H

(**5**) R = CH₃O

Fig. 5-25. Alkaloid structures: **1**, eburnamonine; **2**, eburnamenine; **3**, methoxy eburnamenine; **4**, eburnamine; **5**, methoxy eburnamine.

conditions at 250–300°C (114), it is not possible to determine whether structure **3** or **5** was present in the original extract.

The position of the aromatic methoxy group cannot be ascertained from a mass spectrum. Such a verification would require a good UV or NMR spectrum. However, in this particular case, there was not a sufficient quantity of material available for further analysis.

The entry at m/e 124 ($C_8H_{14}N$) on the element map is somewhat misleading, but since this line appeared in several consecutive spectra with significant intensity, it was concluded that the line was background from the large quantity of aspidosperma derivatives that preceded the compound under study by a few seconds (20). The intensity of m/e 124 is too low relative to the other fragments discussed to consider a structure analogous to aspidospermine.

Even though these data did not unequivocally prove the structure of this alkaloid, formula **3** is most likely. It is doubtful that any other analytical technique could derive as much unambiguous information (i.e., elemental composition) from such a small fraction of a complex mixture.

B. Structure Investigations involving Diagnostic Reactions and Kinetic Studies

In addition to the expediency it brings to routine analyses of complex mixtures, the GC–MS combination makes possible or at least simplifies other types of investigations. Some of these investigations of structure require diagnostic reactions in which a single compound may yield several derivatives; final elucidation requires identification of each of the derivatives. Ordinarily, this would require isolating the individual derivatives from the reaction mixture, or interpreting the mass spectrum of a mixture. The technique is further complicated, of course, if there are several compounds present in the initial sample.

In a study of the biosynthesis of cyclopropane compounds (116,117), it was necessary to determine the location of the cyclopropane ring in long-chain fatty acid esters. The mass spectra of intact cyclopropane acid esters reveal no characteristic peaks which permit location of the ring. Thus, in this case (117) the diagnostic reaction involved catalytic hydrogenation of the natural and synthetic chain and branched-chain derivatives:

$$CH_3(CH_2)_nCH\!\!-\!\!-\!\!CH(CH_2)_mCO_2CH_3 \xrightarrow[\text{Pt}]{H_2}$$
$$\diagdown \diagup$$
$$CH_2$$

$$CH_3(CH_2)_nCHCH_2(CH_2)_mCO_2CH_3 + CH_3(CH_2)_nCH_2CH(CH_2)_mCO_2CH_3$$
$$| \hspace{6.5cm} |$$
$$CH_3 \hspace{6cm} CH_3$$

$$+ CH_3(CH_2)_nCH_2CH_2CH_2(CH_2)_mCO_2CH_3$$

An aliquot (a few micrograms) of the reaction mixture was injected into the GC–MS system for analysis, and the mass spectra were recorded in 4–6 sec on the apex of the GC peak. The two branched-chain esters were not resolved by GC (1% SE-30, 100–200 mesh, 4 mm × 3 ft, 140–185°C), but this mixture—about 85% of the total reduction products—was completely separated from the straight-chain compound and unreacted cyclopropane ester. The different locations of the methyl branches, and hence the former cyclopropane ring, are reflected in mass spectral intensity differences associated with preferential bond cleavage at the points of branching.

In the spectrum shown in Figure 5-26, a peak appears at m/e 213. This results from the loss of mass 43 from the molecular ion, and is observed for most long-chain aliphatic esters (118). Since it is not related to the chain branching, care must be taken that this loss of mass 43 is not misinterpreted. Furthermore, the two expected branched methyl esters give rise to four simple "bond-break" peaks, each differing by 14 mass units from the preceding one. In the present case, these occur at masses 157, 171, 185, and 199; enhancement at these points permits location of the reduced cyclopropane ring.

Recognition of the branch points is further facilitated, and confirmed, by the fact that the primary ion, a, which results from alpha cleavage on the ester side of the branch point is always accompanied by two rearrangement ions, $a + 1$ and $a + 2$, due to abstraction of 1 and 2 hydrogens from the other part of the molecule (93). Thus:

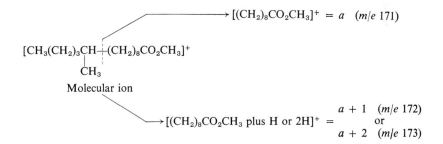

$$\longrightarrow [(CH_2)_8CO_2CH_3]^+ = a \quad (m/e\ 171)$$

$[CH_3(CH_2)_3CH\!\!-\!\!(CH_2)_8CO_2CH_3]^+$
$\overset{|}{C}H_3$
Molecular ion

$$\longrightarrow [(CH_2)_8CO_2CH_3 \text{ plus H or 2H}]^+ = \begin{array}{l} a + 1 \quad (m/e\ 172) \\ \text{or} \\ a + 2 \quad (m/e\ 173) \end{array}$$

These characteristic sets of three adjacent peaks (a, $a + 1$, $a + 2$) unambiguously designate the branch points in esters. The branch points can also be determined solely from intensity enhancement, but this procedure is more difficult and requires careful comparison with reference spectra.

Reductive ring opening of natural cyclopropane acid esters isolated from cultures of *L. arabinosus* and *Sterculia foetida* gave mixtures whose

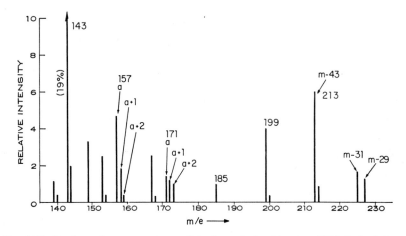

Fig. 5-26. Portion of mass spectrum of the branched acid esters (117) derived from methyl 9,10-methylenetetradecanoate (117).

mass spectra were identical with those of the synthetic derivatives, demonstrating that the technique is reliable and unambiguous. When used with the GC–MS system, this simple diagnostic means of ring location can be carried out on as little as a few hundred micrograms of starting material (117).

The GC–MS system is also well suited for problems where the partial labeling (e.g., deuterium) of compounds must be determined. In these cases the compounds are known, and it is often only necessary to examine the molecular ions to establish the relative abundance of the labeled and unlabeled species. However, in general, the two species are only slightly resolved on the GC column, and a single scan of the superimposed spectra is not adequate since the composition of the effluent entering the mass spectrometer changes during the emergence of the poorly resolved doublet. The MS scanning circuitry can be modified so that simultaneous recordings can be made of the changing intensities of the two selected m/e values during elution of the mixture from the gas chromatograph. Slightly displaced tracings for the two ions are observed with a single collector and one oscillographic recorder by maintaining a constant magnetic field while rapidly switching the accelerating voltage with a time-actuated relay and voltage-dividing circuit (119). The integrated areas under the superimposed, but distinguishable, tracings indicate the relative composition of the mixture. The ionization efficiency of each species must also be known or determined. This technique, particularly if the precision is improved, will be quite useful in quantitative determinations of the extent of labeling or the composition of heterogeneous GC fractions.

Investigations involving kinetic measurements are greatly facilitated by the GC–MS combination by virtue of the rapid termination of the reaction when an aliquot of the reaction mixture is injected onto the GC column, thereby separating the products and reactants. A good example is the kinetic study of ^{18}O incorporation into ketosteroid nuclei by exchange at the carbonyl functional group (38). The relative rates of ^{18}O incorporation by equilibration with $H_2^{18}O$ in the presence of acid may be used to uniquely characterize the stereochemical and electronic environment of the carbonyl function, while the total incorporation reveals the number of exchangeable groups.

In practice, several milligrams or less of sample are dissolved in an acidic (HCl) isopropanol solution, and excess $H_2^{18}O$ is added. Sample withdrawals are made at intervals and injected directly onto the GC column to stop the exchange reaction:

$$\text{>}=\!O + H_2{}^{18}O \rightleftharpoons \overset{{}^{18}OH}{\underset{OH}{\text{>}\!\!\!<}} \rightleftharpoons \text{>}=\!{}^{18}O + H_2O$$

The molecular ion region is rapidly scanned as the components of interest elute into the mass spectrometer, and the extent of ^{18}O incorporation during the given time interval is determined from measurement of the ion abundances on the MS record.

Absolute rate constants for the exchange reaction are obtained from the first-order kinetic plots of ^{18}O incorporation levels at various times. The rates are greatly influenced by (1) the steric environment of the carbonyl, partly because of the bulky intermediate diol, and (2) the presence of electron-donating or -withdrawing groups in the vicinity of the carbonyl, since the reaction proceeds by attack with the lone electron pair of oxygen on the electron-deficient carbonyl carbon. Thus, the rates in general decrease with increasing steric hindrance and/or conjugation and are essentially unaffected by structural differences in other parts of the molecule. This kinetic diagnosis of molecular structure enjoys a high degree of precision, since several compounds can be subjected to identical conditions simultaneously (i.e., same solution). An indication of the structural sensitivity of the technique is indicated by the fact that the exchange rate of the 3-one function group was found to be 11,600 times greater than that of the $\Delta^{1,4}$-3-one moiety. Furthermore, the method has sufficient precision to distinguish the 17-one carbonyl as being only 2.7 times more active in the exchange reaction than the $\Delta^{1,4}$-3-one functional group (38).

Without the GC–MS system the kinetic method of structure elucidation requires a much larger sample and a lengthy procedure for the isolation and quantitative determination of the products. Furthermore, most conventional approaches have poor precision due to difficulties in terminating the reaction and isolating the products from several different solutions. In this case, with a GC–MS combination, the reactions are conveniently terminated upon injection and within a few minutes the products are "isolated" and the quantitative information is recorded.

IX. Summary

Many of the developmental technicalities and difficulties have been simplified or eliminated in the GC–MS combination, and it is now a convenient and manageable routine technique.

The major problems of interfacing have been solved. "Separators" have been developed to effect an efficient combination of the instruments under mutually acceptable operating conditions. More explicit literature is now available on the fundamental and practical requirements of these separators, and this should stimulate further refinement in the technology of GC–MS interfacing.

The GC–MS combination provides a potent instrument for analysis whether on a routine identification basis or as a support technique for other studies. Furthermore, the gas chromatograph has distinguished itself as a convenient and facile inlet to the mass spectrometer. In utilizing a GC inlet to the mass spectrometer, it is possible to administer only enough sample to the ionization chamber to produce a usable mass spectrum instead of "flooding" the entire ion source and sample reservoir with sample vapor which, in some instances, may take hours to remove after recording the spectrum. Thus, in general, the gas chromatographic inlet probably provides the most convenient means of optimally utilizing the inherent sensitivity of the mass spectrometer.

Data acquisition techniques are becoming available which can reduce GC–MS data as they are produced. Furthermore, the data are available in digital form for immediate processing and comparison. For example, techniques have been reported recently which reduce the data to normalized plots of intensity vs. mass, subtract selected scans to produce net "purified" spectra, and compare the unknown spectrum with those available in a standard library. Such automatic techniques permit a considerable saving of time and produce readily interpretable data which are free from human fatigue error.

As mass spectrometry is applied more widely to molecular structure problems, systematic patterns of spectra interpretation are recognized which are amenable to computer programming. Even though the chemist's intuition may be required to complete an interpretation, a computer can accomplish the routine tasks faster and more thoroughly. Computer programs have been used for the complete interpretation and identification of the spectra of some simple compounds; however, these computer interpretations are more successful and reliable when based on high-resolution mass spectral data.

References

1. K. Biemann, *Mass Spectrometry*, McGraw-Hill, New York, 1962.
2. J. H. Beynon, *Mass Spectrometry and its Applications to Organic Chemistry*, Elsevier, Amsterdam, 1960.
3. J. B. Farmer, "Types of Mass Spectrometers," in *Mass Spectrometry*, C. A. McDowell, Ed., McGraw-Hill, New York, 1963, pp. 7–44.
4. W. H. McFadden, M. Lounsbury, and A. L. Wahrhaftig, *Can. J. Chem.*, **36**, 990 (1958).
5. F. W. McLafferty, Ed., *Mass Spectrometry of Organic Ions*, Academic Press, New York, 1963.
6. H. Budzikiewicz, C. Djerassi, and D. H. Williams, *Interpretation of Mass Spectra of Organic Compounds*, Holden-Day, San Francisco, 1964.
7. H. Budzikiewicz, C. Djerassi, and D. H. Williams, *Structure Elucidation of Natural Products by Mass Spectrometry*, Vols. I and II, Holden-Day, San Francisco, 1964.
8. H. A. Bondarovich and S. K. Freeman, "Mass Spectrometry," in *Interpretative Spectroscopy*, S. K. Freeman, Ed., Reinhold, New York, 1965, pp. 170–210.
9. F. W. McLafferty, *Interpretation of Mass Spectra*, Benjamin, New York, 1966.
10. H. C. Hill, *Introduction to Mass Spectrometry*, Heyden, London, 1966.
11. H. Budzikiewicz, C. Djerassi, and D. H. Williams, *Mass Spectrometry of Organic Compounds*, Holden-Day, pan Francisco, 1967.
12. W. H. McFadden, "Mass Spectrometric Analysis of Gas Chromatographic Effluents," in *Advances in Chromatography*, Vol. 4, J. C. Giddings and R. A. Keller, Eds., Dekker, New York, 1967, pp. 265–332.
13. *Index of Mass Spectral Data*, American Society for Testing and Materials, Philadelphia, Pa., 1963.
14. (a) A. Cornu and R. Massot, *Compilation of Mass Spectral Data*, Heyden, London, 1966. (b) A. Cornu and R. Massot, *Compilation of Mass Spectral Data —First Supplement*, Heyden, London, 1967.
15. E. Stenhagen, S. Abrahamsson, F. W. McLafferty, Eds., *Atlas of Mass Spectral Data*, Interscience, New York, 1968.
16. E. Stenhagen, S. Abrahamsson, F. W. McLafferty, Eds., *Archives of Mass Spectral Data*, Interscience, New York, in press.
17. F. E. Saalfeld, National Research Lab., 6525 (1967).
18. J. A. McCloskey, Ph.D. dissertation, MIT, Cambridge, Mass., May, 1962.
19. H. E. Duckworth and S. N. Ghoshal, in *Mass Spectrometry*, C. A. McDowell, Ed., McGraw-Hill, New York, 1963; pp. 201–274.
20. J. T. Watson, Ph.D. dissertation, MIT, Cambridge, Mass., June, 1965.
21. K. Biemann, *J. Pure Appl. Chem.*, **9**, 95 (1964).

22. K. Biemann, P. Bommer, and D. M. Desiderio, *Tetrahedron Letters*, **26**, 1725 (1964).

23. F. A. J. M. Leemans and J. A. McCloskey, *J. Am. Oil Chemists Soc.*, **44**, 11 (1967).

24. R. M. Teeter, C. F. Spencer, J. W. Green, and L. H. Smithson, *J. Am. Oil Chemists' Soc.*, **43**, 82 (1966).

25. J. M. Hayes, Ames Research Center, Moffett Field, private communication, December, 1967.

26. R. S. Gohlke, *Anal. Chem.*, **34**, 1332 (1962).

27. L. P. Lindeman and J. L. Annis, *Anal. Chem.*, **32**, 1742 (1960).

28. B. H. Kennett, *Anal. Chem.*, **39**, 1506 (1967).

29. R. Ryhage, *Arkiv Kemi*, **20**, 185 (1962).

30. R. A. Hites and K. Biemann, *Anal. Chem.*, **39**, 965 (1967).

31. R. Teranishi, R. E. Lundin, W. H. McFadden, and J. R. Scherer, "Ancillary Systems," in *The Practice of Gas Chromatography*, L. S. Ettre and A. Zlatkis, Eds., Interscience, New York, 1967, pp. 407–459.

32. J. T. Watson and K. Biemann, *Anal. Chem.*, **36**, 1135 (1964).

33. R. S. Gohlke, *Anal. Chem.*, **31**, 535 (1959).

34. W. H. McFadden, *Separation Sci.*, **1**, 723 (1966).

35. K. Biemann and J. T. Watson, *Monatsh. Chem.*, **96**, 305 (1965).

36. R. Ryhage, *Anal. Chem.*, **36**, 759 (1964).

37. S. R. Lipsky, C. G. Horvath, and W. J. McMurray, *Anal. Chem.*, **38**, 1585 (1966).

38. A. M. Lawson, F. A. J. M. Leemans, and J. A. McCloskey, *15th Ann. Conf. Mass Spectrometry and Allied Topics*, *ASTM E-14*, *Denver*, *May 1967*, pp. 386–388.

39. R. F. Cree, presented at the Pittsburgh Conference on Analytical Chemistry and Applied Spectroscopy, 18th March 1967.

40. P. M. Llewellyn and D. Littlejohn, presented at the Pittsburgh Conference on Analytical Chemistry and Applied Spectroscopy, 17th March 1966.

41. E. W. Becker, "The Separation Jet," in *Separation of Isotopes*, H. London, Ed., George Newnes, London, 1961, p. 360.

42. A. E. Barrington, *High Vacuum Engineering*, Prentice-Hall, Englewood Cliffs, N. J. 1963, pp. 47–62.

43. R. E. Halstead and A. O. Nier, *Rev. Sci. Instr.*, **21**, 1019 (1950).

44. M. Knudson, *Ann. Physik*, **28**, 75 (1909).

45. P. D. Zemany, *J. Appl. Phys.*, **23**, 924 (1952).

46. P. M. Krueger and J. A. McCloskey (to be published).

47. R. Ryhage, *Arkiv Kemi*, **26**, 305 (1967).

48. R. Ryhage, S. Wikstrom, and G. R. Waller, *Anal. Chem.*, **37**, 435 (1965).

49. R. D. Present, *Kinetic Theory of Gases*, McGraw-Hill, New York, 1958.

50. J. T. Watson and K. Biemann, *Anal. Chem.*, **37**, 844 (1965).

51. J. M. Hayes, Ph.D. Dissertation, MIT, Cambridge, Mass., Sept. 1966, p. 249.

52. M. C. ten Noever de Brauw and C. Brunnée, *Z. Anal. Chem.*, **229**, 321 (1967).

53. W. D. MacLeod, Jr., University of California at San Diego, private communication, November, 1967.

54. W. D. MacLeod, Jr. and B. Nagy, *Anal. Chem.*, **40**, 841 (1968).

55. S. R. Lipsky, W. J. McMurray, and C. G. Horvath, in *Gas Chromatography 1966*, A. B. Littlewood, Ed., The Institute of Petroleum, London, 1967, pp. 299–317.

56. M. A. Grayson and C. J. Wolf, *Anal. Chem.*, **39**, 1438 (1967).
57. R. M. Barrer, *Diffusion In and Through Solids*, Cambridge University Press, London, 1941.
58. A. Lebovits, *Mod. Plastics*, **43**, 139 (1966).
59. W. Jost, *Diffusion in Solids, Liquids, Gases*, Academic Press, New York, 1952.
60. J. Punderson (du Pont Experimental Station, Wilmington, Delaware), private communication, November 1967.
61. J. E. Arnold and H. M. Fales, *J. Gas Chromatog.*, **3**, 131 (1965).
62. R. L. Foltz, M. B. Neher, and E. R. Hinnenkamp, *Anal. Chem.*, **39**, 1338 (1967).
63. R. F. Cree (General Electric, Vacuum Products Section, Schenectady, N. Y.), private communication, October 1967.
63a. M. Blumer, *Anal Chem.* **40**, 1590 (1968).
64. J. D. Adams, J. P. Conkle, W. E. Mabson, J. T. Watson, P. H. Wolf, and B. E. Welch, *Aerospace Med.*, **37**, 555 (1966).
65. J. M. Hayes (Ames Research Center (NASA), Moffett Field, California), private communication, October 1967.
66. R. Teranishi, R. G. Buttery, W. H. McFadden, T. R. Mon, and J. Wasserman, *Anal. Chem.*, **36**, 1509 (1964).
67. J. T. Walsh and C. Merritt, Jr., *J. Gas Chromatog.*, **5**, 420 (1967).
68. W. D. MacLeod, Jr. and B. Nagy, to be published.
69. W. H. McFadden, R. Teranishi, D. R. Black and J. C. Day, *10th Ann. Conf. Mass Spectrometry*, ASTM E-14, New Orleans, 1962, pp. 142–146.
70. W. S. Updegrove, J. Oro', and A. Zlatkis, *J. Gas. Chromatog.*, **5**, 359 (1967).
71. J. A. Völlmin, I. Omura, J. Seibl, K. Grob, and W. Simon, *Helv. Chim Acta*, **49**, 1768 (1966).
72. J. R. Chapman, *15th Ann. Conf. Mass Spectrometry and Allied Topics*, ASTM E-14, Denver, 1967, pp. 29–34.
73. W. M. Brubaker, in *Advances in Mass Spectrometry*, Vol. 4, E. Kendrick, Ed., Institute of Petroleum, London, 1968 pp. 293–300.
74. K. S. Quisenberry, T. T. Scolman, and A. O. Nier, *Phys. Rev.*, **102**, 1071 (1956).
75. W. J. McMurray, B. N. Greene, and S. R. Lipsky, *Anal. Chem.*, **38**, 1194 (1966).
76. K. Van Canwenberghe, M. Vandewalle, and M. Verzele, *J. Gas Chromatog.*, **6**, 72 (1968).
77. P. Bommer, W. McMurray, and K. Biemann, *12th Ann. Conf. Mass Spectrometry and Allied Topics*, ASTM E-14, Montreal, 1964, pp. 428–432.
78. J. W. Amy, E. M. Chait, W. E. Baitinger, and F. W. McLafferty, *Anal. Chem.*, **37**, 1265 (1965).
79. J. N. Damico, N. P. Wong, and J. A. Sphon, *Anal. Chem.*, **39**, 1045 (1967).
80. J. H. Beynon, R. A. Saunders, and A. E. Williams, *J. Sci. Instr.*, **36**, 375 (1959).
81. R. P. W. Scott, I. A. Fowlis, D. Welti, and T. Wilkins, in *Gas Chromatography 1966*, A. B. Littlewood, Ed., Institute of Petroleum, London, 1967, pp. 317–336.
82. R. D. Grigsby, E. J. Eisenbraun, D. V. Hertzler, and K. M. Piel, *15th Ann. Conf. Mass Spectrometry and Allied Topics*, ASTM E-14, Denver, 1967, pp. 400–405.
83. R. M. Elliott and W. J. Richardson, *15th Ann. Conf. Mass Spectrometry and Allied Topics*, ASTM E-14, Denver, 1967, pp. 351–357.
84. D. M. Desiderio and K. Biemann, *12th Ann. Conf. Mass Spectrometry and Allied Topics, Montreal*, ASTM E-14, June 1964, pp. 433–437.
85. R. Venkataraghavan and F. W. McLafferty, *Anal. Chem.*, **39**, 278 (1967).
86. D. D. Tunnicliff and P. A. Wadsworth, *Anal. Chem.* **40**, 1826 (1968).

87. B. H. Johnson and T. Aczel, *Anal. Chem.*, **39**, 682 (1967).
88. D. M. Desiderio, Jr., and T. E. Mead, *Anal. Chem.* **40**, 2090 (1968).
89. K. Biemann and P. V. Fennessey, *Chimia*, **21**, 226 (1967).
90. C. Merritt, Jr., P. Issenberg, and M. L. Bazinet, in *Advances in Mass Spectrometry*, Vol. 4, E. Kendrick, Ed., Institute of Petroleum, London, 1968, pp. 55–64.
91. (*a*) D. H. Smith, R. W. Olsen, and A. L. Burlingame, *16th Ann. Conf. Mass Spectrometry and Allied Topics*, *ASTM E-14, Pittsburgh, 1968*, pp. 101–108; (*b*) A. L. Burlingame, D. H. Smith, R. W. Olsen, and T. O. Merren, *ibid.* pp. 109–113.
92. R. A. Hites and K. Biemann, in *Advances in Mass Spectrometry*, Vol 4, E. Kendrick, Ed., Institute of Petroleum, London, 1968, pp. 37–54.
93. S. Abrahamsson, S. Stenhagen-Stallberg, and E. Stenhagen, *Biochem. J.*, **92**, 2 (1964).
94. B. Pettersson and R. Ryhage, *Arkiv Kemi*, **26**, 293 (1967).
95. R. A. Hites and K. Biemann, *Anal. Chem.* **40**, 1217 (1968); see also: R. A. Hites, S. Markey, R. Murphy, and K. Biemann, *16th Ann. Conf. Mass Spectrometry and Allied Topics*, *ASTM E-14, Pittsburg, Pa., 1968*, pp. 302–305.
96. A. Jasson, S. Melkersson, R. Ryhage, and S. Wikstrom, *16th Ann. Conf. Mass Spectrometry and Allied Topics*, *ASTM E-14, Pittsburg, 1968*, pp. 306–312.
97. (a) W. E. Reynolds, J. C. Bridges, R. B. Tucker, and T. B. Coburn, *16th Ann. Conf. Mass Spectrometry and Allied Topics*, *ASTM E-14, Pittsburgh, 1968*, pp. 77–84; (b) Tech. Rept. No. IRL-1062, Instrumentation Research Laboratory, Stanford University School of Medicine, Palo Alto, California.
98. K. Biemann, W. J. McMurray, and P. V. Fennessey, *Tetrahedron Letters*, **1966** (33), 3997.
99. A. Mandelbaum, P. Fennessey, and K. Biemann, *15th Ann. Conf. Mass Spectrometry and Allied Topics*, *ASTM E-14, Denver, May 1967*, pp. 111–113.
100. B. Pettersson and R. Ryhage, *Anal. Chem.*, **39**, 790 (1967).
101. M. C. Hamming and R. D. Grigsby, *15th Ann. Conf. Mass Spectrometry and Allied Topics*, *ASTM E-14, Denver, May 1967*, pp. 107–110.
102. R. Ryhage, *J. Dairy Res.*, **34**, 115 (1967).
103. A. J. Polito, T. Akita, and C. C. Sweeley, *Biochemistry*, **7**, 2609 (1968).
104. R. C. Gaver and C. C. Sweeley, *J. Am. Oil Chemists Soc.*, **42**, 295 (1965).
105. K. A. Karlsson, *Acta Chem. Scand.*, **19**, 2425 (1965).
106. R. C. Gaver and C. C. Sweeley, *J. Am. Chem. Soc.*, **88**, 3643 (1966).
107. H. E. Carter and R. C. Gaver, *J. Lipid Res.*, **8**, 391 (1967).
108. J. A. McCloskey and M. J. McClelland, *J. Am. Chem. Soc.*, **87**, 5090 (1965).
109. W. G. Niehaus, Jr. and R. Ryhage, *Tetrahedron Letters*, **1967**, 5021.
110. C. C. Sweeley, A. J. Polito, and J. Naworal, *Biochemistry* (in press).
111. M. Barber, J. R. Chapman, W. A. Wolstenholme, *J. Mass Spectrometry Ion Phys.*, **1**, 98 (1968).
112. W. H. McFadden and R. G. Buttery, "Applications of Mass Spectrometry in Flavor and Aroma Chemistry," in *Topics in Organic Mass Spectrometry* (*Advan. Anal. Chem. Instr.*, Vol. 8), A. L. Burlingame, Ed., Interscience, New York, in press.
113. D. Henneberg, *Anal. Chem.*, **38**, 495 (1966).
114. K. Biemann, M. Friedman-Spiteller, and G. Spiteller, *J. Am. Chem. Soc.*, **85**, 631 (1963).

115. K. Biemann, M. Friedman-Spiteller, and G. Spiteller, *Tetrahedron Letters*, **1961** (14), 485.
116. J. W. Polacheck, B. E. Tropp, J. H. Law, and J. A. McCloskey, *J. Biol. Chem.*, **241**, 3362 (1966).
117. J. A. McCloskey and J. H. Law, *Lipids*, **2**, 225 (1967).
118. R. Ryhage and E. Stenhagen, in *Mass Spectrometry of Organic Ions*, F. W. McLafferty, Ed., Academic Press, New York, 1963, pp. 399–452.
119. C. C. Sweeley, W. H. Eliott, I. Fries, and R. Ryhage, *Anal. Chem.*, **38**, 1549 (1966).

CHAPTER 6

Gas Chromatography and Infrared and Raman Spectrometry

Stanley K. Freeman, *International Flavors & Fragrances Inc.,*
Research Center, Union Beach, New Jersey

I. Introduction

Prior to the advent of gas chromatography, spectral data obtained during the course of organic chemical studies were used primarily to confirm or support conclusions determined by chemical transformations or degradations. Today, it is not unusual to perform reactions on the milligram scale, to separate and isolate the products by gas chromatography, and to characterize them completely by spectral means. The great economy of time and materials has wide appeal for the organic chemist and permits many experiments which were previously impossible. However, abandoning the complicated and sometimes uncertain classical procedures of identification places the burden of structure elucidation upon the spectral disciplines.

In his quest to ascertain the nature of GC isolates via spectrometric methods, the investigator sometimes is confronted with a problem pertaining to the quantities obtainable by this elegant separation technique. Adequate amounts usually are gained with analytical packed columns, but as the complexity of a mixture increases, it is necessary to employ relatively low capacity wall-coated or support-coated open tubular columns to achieve satisfactory separations. Under the latter conditions, at best only the major constituents can be collected for infrared and Raman study. Approximate loading capacities of several column types appear in Table 6-1. It is apparent that the routine minimum sample requirements of these spectrometers (ca. 10^{-5} and 10^{-4} g, respectively) militate against their general use with open tubular column eluents. For example, it would be extremely difficult to generate a spectrum on a 10% constituent emerging from an 0.02 in. i.d. open tubular column.

A mass spectrometer will give useful data on nanogram quantities of materials eluted from a gas chromatograph by batch or "on-the-fly" methods. The present state of the art does not allow as much freedom of choice in the infrared and Raman fields, and the sensitivities of the flame ionization detector (ca. 10^{-11} g) and the mass spectrometer (ca. 10^{-10} g) exceed that of conventional infrared dispersion instruments by approximately four orders of magnitude. This circumstance means that mass spectra can be recorded on many GC effluents which cannot be examined

TABLE 6-1

Approximate Loading Capacities of Various Chromatographic Columns

Column	Loading capacity, mg/component
Packed column (¼ in. o.d.)	25
Packed column (⅛ in. o.d.)	5
Support-coated open tubular column (0.020 in. i.d.)	0.2
Wall-coated open tubular column (0.030 in. i.d.)	0.2
Wall-coated open tubular column (0.020 in. i.d.)	0.02
Wall-coated open tubular column (0.010 in. i.d.)	0.005

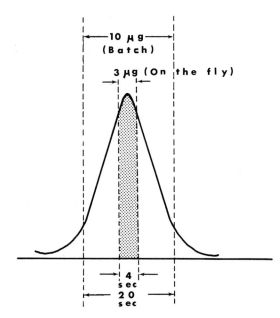

Fig. 6-1. Quantities of material available to a spectrometer by GC batch and "on-the-fly" methods.

by infrared or Raman. Furthermore, infrared and Raman instruments are static systems as opposed to mass spectrometers which consume materials, and hence collection is preferred when sample size is a factor. For small samples, the lower sensitivity of these spectroscopic instruments sometimes necessitates collecting the total eluted substance (batch) rather than recording only a portion of it during the course of its elution ("on-the-fly"). An example of the quantities of material that might be available to a spectrometer utilizing the two approaches is depicted in Figure 6-1. Batch and direct introduction systems will be discussed in greater detail in later sections.

An infrared spectrum often may serve as the sole datum required to identify a substance if its spectral mate can be found in a reference file. There are several commercially available infrared spectral compilations which range in quality from adequate to good and provide spectra for more than 60,000 compounds (1a–1d). The chemist and chemical spectroscopist have access to various retrieval systems (1b–1d,2) and an excellent computerized procedure for retrieving IR spectra has been reported recently (3).

Even in the absence of a reference spectrum, considerable information can be gleaned from an IR curve. In addition to ascertaining functionality, subtle structural features are evident to the scientist with some interpretive experience. Such information often plays an important role in conjunction with GC–MS and other spectral studies, and often supplies missing information such as the specific environment of a carbonyl group, the stereochemistry involving a hydroxyl moiety, the presence of a *gem*-dimethyl system, or the type of aromatic substitution.

II. The Combination of Gas Chromatography with Infrared Spectrometry

The usefulness of infrared in GC studies will depend on the ease with which small samples can be handled and measured. The literature is replete with reports on ancillary techniques and modified instrumentation whose aim is to obtain interpretable IR spectra on small samples (e.g., references 4–17). It is not practical to consider all the scientific papers and the spate of commercial pamphlets in this area, and the following selections appear to be of most value to the practicing chemist. Perhaps in the final analysis these choices were based predominantly upon the interests, preferences, and experience of the writer.

A. Batch Examination

There are several advantages to be gained by collecting GC fractions for presentation to the infrared spectrometer. It is usually more convenient to employ this approach when spectra are required on only a few trapped samples. Relatively inexpensive instrumentation (ca. $5,000) can be employed; in fact, most commercial devices equipped with a beam-condensing system can generate an interpretable spectrum on approximately 50 μg of material. Operating in the batch mode enables the investigator to repeat a run at his leisure and to readjust the instrumental parameters if necessary. This is particularly helpful where spectral features of marginal size samples can be improved by slow scanning rates, extreme scale expansion, etc.

Infrared spectra of materials may be recorded in the solid, liquid, and gaseous states of aggregation. A gaseous state spectrum corresponds to that of the free molecule, while a spectrum recorded on a solid substance corresponds to that of the unit cell. If intermolecular interactions are weak, as in the case of nonpolar compounds, the band positions and shapes will be approximately the same in all three states. However, for polar substances, where association or other strong forces are operative, the condensed state spectra will be considerably different. For example, a radical

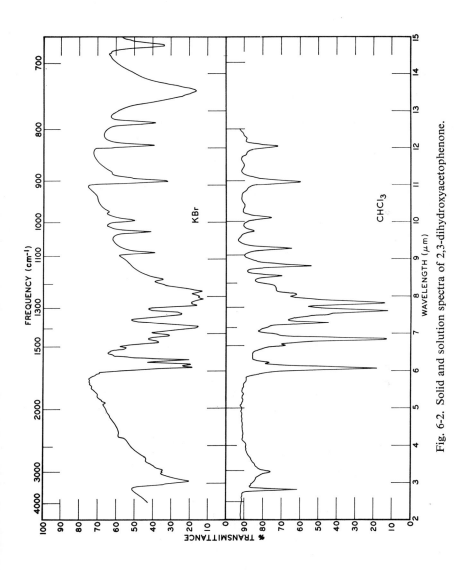

Fig. 6-2. Solid and solution spectra of 2,3-dihydroxyacetophenone.

231

difference is observed between the KBr pellet and CHCl₃ solution spectra of the crystalline solid 2,3-dihydroxyacetophenone (Fig. 6-2). Although it is sometimes stated that there is little difference between a vapor spectrum and a neat liquid spectrum (18), comparison of the curves of methanol makes apparent the significant spectral differences between vapor and liquid states (Fig. 6-3). As can be seen, the neat liquid displays a strong, broad absorption at 3.0 μm ascribable to the bonded hydroxyl group, while the vapor spectrum contains a moderate intensity, relatively narrow band at 2.75 μm, due to the unbonded hydroxyl moiety. In addition, the strongest absorption in the spectrum of liquid methanol appears at 9.7 μm, but the gaseous spectrum exhibits two intense, broad bands at 9.4 and 9.9 μm, and a sharp band at 9.7 μm.

Solute-solvent interactions in solution cause only minor spectral variations when compared with vapor spectra, and most molecular interactions are eliminated. The author of this chapter prefers to study solution spectra and accepts the fact that two solvents are required (CHCl₃ or CCl₄, and CS₂) to cover the region 2–15 μm (5000–667 cm^{-1}). In most instances, sufficient information for identification can be derived from a chloroform solution spectrum alone. Obviously, a mull or an alkali halide pellet must be used when a solubility problem exists.

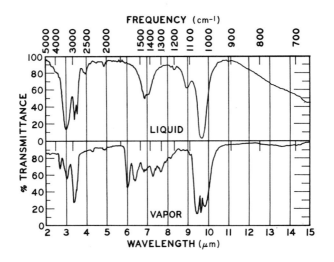

Fig. 6-3. Vapor and liquid spectra of methanol.

1. Solids

a. Pelletizing Technique. Although liquids comprise the majority of substances eluting from a gas chromatograph, materials which are solid at ambient temperature represent an important minority. For example, barbiturates, steroids, alkaloids, and aromatic acids are encountered in GC studies, and IR analysis is frequently used for confirmation or identification.

The technique of pelletizing samples for IR measurement, proposed about 20 years ago, involves mixing a powdered sample with pure, dry alkali halide matrix material and pressing the mixture into an IR-transparent disk (19,20). KBr is the most widely used matrix substance, although NaCl, CsBr, or KCl are employed for liquid samples in order to match the refractive index of the sample more closely and thus reduce light scattering. Many procedures are used to combine sample and alkali halide: grinding the mixture in the presence of a volatile solvent, dissolving the sample in a low boiling solvent and then mixing with KBr, dispersing a material in the solid matrix by means of ultrasonic vibration, lyophilization, or trapping a GC fraction directly onto powdered KBr. The spectral recording of such alkali halide disks occupied the attention of chemists and spectroscopists even prior to the commercial advent of gas chromatography (4,21), and continues to be widely used with collected GC fractions.

The technique undoubtedly is a valuable one, but some of the disadvantages should be borne in mind. In addition to association effects which occur in the solid state, physical and/or chemical changes may take place in a sample during the grinding process, and anomalous spectra result. Furthermore, band splitting due to intermolecular interactions in solids is observed with the pelletizing techniques, but is not observed in solution spectra.

The pellet procedure is superior in many ways to the mulling technique. A mull is prepared by grinding a solid sample in a mortar with mineral oil or hexachlorobutadiene to yield a paste which is then transferred to a salt plate. Alternatively, the sample may be mulled between halide plates. Interpretation of pellet spectra is simplified because of better resolution, less scattering, absence of interfering bands, and relative ease of small sample preparation. A recent study by Gore and Hannah on micro techniques indicated that the KBr disk methods offer the most sensitive means of analyzing very small samples with the presently available accessories (16). A digital readout device was used to collect IR data for computerization, and an interpretable spectrum of phenacetin (0.5 mm disk) was

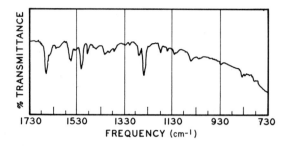

Fig. 6-4. 0.5 mm KBr pellet spectrum of 0.1 μg phenacetin in 1 mg KBr employing a 4× reflecting beam condenser (Perkin-Elmer Model 621 spectrometer equipped with digital encoders for frequency and transmission).

obtained on 0.1 μg with the aid of repetitive scanning (8 times) and curve smoothing (Fig. 6-4).

Good results with easily prepared KBr disks can be obtained routinely on 20–100-μg samples in the following way. The solid GC fraction, isolated in a cooled 200 mm × 1 mm i.d. glass capillary tube, is washed with 10–25 μl of a suitable solvent into a small mullite mortar containing ca. 15 mg of KBr. The mixture is ground lightly after solvent evaporation, transferred to a die (Fig. 6-5), and the threaded cap tightened to the die

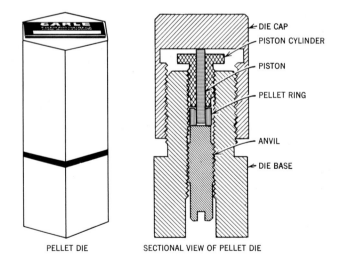

Fig. 6-5. Carle micropellet die.

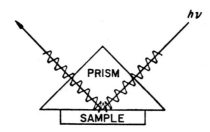

Fig. 6-6. Attenuated total reflection effect.

base with two end wrenches. Pressures exceeding 80,000 psi are achieved with moderate force. The press is disassembled, and the 3-mm pellet and ring unit inserted in the path of the spectrometer reference beam.

Leggon's method of trapping somewhat larger quantities of liquid and solid GC effluents directly on KBr powder (22) has been modified recently (23). A 50 mm × 1 mm glass tube containing 300 mg of this matrix is attached to the exit line of the chromatograph and a pellet pressed in the usual manner.

b. Internal Reflection Technique. Frustrated total reflection (FTR), or multiple internal reflection, offers a direct method for micro sampling (24). When a beam of radiation enters a prism, it will be reflected internally if the angle of incidence at the interface between sample and prism exceeds the critical angle. All the energy is reflected; however, the beam appears to penetrate beyond the reflecting surface (Fig. 6-6). The beam will lose energy at those wavelengths where the sample absorbs (attenuated total reflection—ATR). For most angles of incidence above the critical angle, the reflection spectrum (obtained by measuring and plotting the attenuated radiation as a function of wavelength) resembles the transmission spectrum. The absorption bands produced by the single-reflection ATR technique are relatively weak. Analogous to increasing the path length of transmission cells, the amount of absorption is enhanced by multiplying the number of reflections (FTR) (Fig. 6-7).

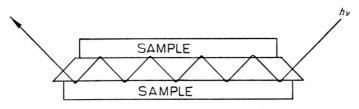

Fig. 6-7. Multiple internal reflection effect.

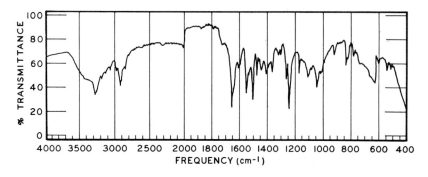

Fig. 6-8. Frustrated total reflection spectrum of 1 μg phenacetin (Perkin-Elmer Model 621 spectrometer, 4× beam condenser).

This approach is useful for the IR examination of solids which give anomalous spectra due to sample interaction with the pellet matrix, or if changes are induced while grinding. FTR cells can be used to analyze a thin layer of molecules deposited, condensed, or cast from solution onto the surface of an internal reflector plate. The spectra of very thin films are dependent on the film thickness. If the space between the plates (i.e., thickness of the sample) is less than 3 μm, energy transfer is effected between the plates, and the resultant spectrum is nearly identical to a transmission scan. Theoretically, the effective sample thickness of thin films can be several times the actual sample thickness, but in practice, experimental difficulties prevent taking full advantage of the nature of this effect. A spectrum of 1 μg of phenacetin deposited from chloroform solution on a limited portion of the crystal face is shown in Figure 6-8.

2. Neat Liquids

Sample size limitations encountered in IR spectrometry of undiluted samples are imposed mainly by handling techniques. The increase in sensitivity of pellet methods over neat liquid procedures lies in the fact that it is a relatively simple matter to place the entire disk in the condensed beam of the spectrometer. Fabrication of a cell designed to contain a few micrograms of liquid, all of which intercepts the IR radiation, is difficult. Even spectral examination of a liquid held between micro salt plates generally requires approximately 20 μg.

A convenient method for the isolation of a GC effluent utilizes a standard melting point tube. The fraction is trapped in a Dry Ice-cooled glass capillary tube (100–200 mm × 1 mm i.d.), one end of which is snugly

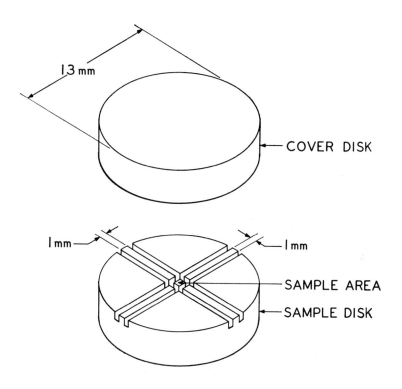

Fig. 6-9. Salt plates for small quantities of neat liquids.

inserted into the exit port of the chromatograph. Fogging, a phenomenon sometimes encountered in GC isolations, can be reduced or eliminated by touching a hot surface with the bottom of the tube while cooling the top section with Dry Ice. After collection, one end of the capillary is sealed and the material spun down. A micropipet, prepared by finely drawing a melting point capillary, is used to transfer the liquid onto a 13-mm diameter grooved KBr plate. A 1×1 mm or 1×4 mm area surface (scored with a jeweler's file) is utilized for a Perkin-Elmer Corp. Model 621 or a Beckman Instruments Model IR4 spectrometer, respectively (Fig. 6-9). Approximately 0.05-μl samples may be handled conveniently in this manner and give good spectra when used with appropriate beam-condensing accessories. Smaller GC fractions are transferred via solvent. After the sample has been centrifuged, the capillary tube is cut to a length of about 25 mm, and 1 μl of diethyl ether or another low-boiling solvent is added. The solution is placed on the KBr plate in small portions by means of a micropipet, and the solvent is allowed to evaporate between transfers. Finally, the sample is covered with a KBr disk and the plates are assembled in a demountable microcell holder.

Partial vaporization of a low-boiling substance from salt plates during the course of a scan may make interpretation or spectral matching difficult. Consequently, with this technique, it is good practice to remeasure the intensity of the absorption band first recorded immediately after running the total spectrum.

3. Gases

Many recent investigations on vapor spectra pertain to tandem GC–IR arrangements. However, there are occasions when batch examination of gases is required. Gas microcells for such studies are usually of small internal volume (ca. 3 ml volume and 5 cm long). They are equipped with self-contained heaters and heated transfer lines, allowing direct trapping of fractions boiling up to 300°C. Light-pipe cells, originally developed for direct GC–IR union, may be used for batch trapping (Fig. 6-10). Gas IR cells are used in GC work principally with eluents boiling below 100°C.

4. Solutions

In order to obtain maximum information from the IR spectrum of an unknown substance, molecular environmental effects should be mini-

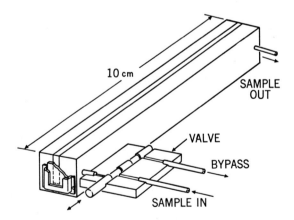

Fig. 6-10. "Light-pipe" cell, 3 ml volume (Beckman Instruments, Inc.). Cell efficiency (ratio of path length to volume): 3.33.

mized. Therefore, when possible, measurements should be conducted in nonpolar, dilute solutions. Some of the limitations of solution spectra are "dead" spectral regions due to solvent bands, occasional solubility problems, and poorer sensitivity compared with pellet methods. These are often balanced by the advantages. A GC trapping contained in a 200 mm × 1 mm i.d. glass capillary or other collection device can be washed into an IR cell with sufficient $CHCl_3$ to just fill it. ($CHCl_3$ is one of the best general solvents for the varied types of organic compounds commonly encountered.)

Subsequent to recording the spectrum of a medium or high boiling fraction dissolved in $CHCl_3$, it is a simple matter to withdraw the solution from the cell, allow the solvent to evaporate, and redissolve the sample in a liquid of different polarity. Comparison of the two spectra often yields additional information when dealing with polar substances. Effects of dilution also aid spectral interpretations. For example, bifurcated or multiple absorption bands observed in the 1650–1800 cm^{-1} region may be due to a variety of reasons such as two or more different carbonyls in a molecule, Fermi resonance, mechanical coupling, hydrogen bonding, or cis/gauche conformations. The task of determining the origin of these bands is best accomplished via solution spectra.

The two preferred rotamers that exist in compounds of the general type $RCOCH_2X$ and $RCOCHX_2$ offer an interesting example. Consideration of the cis and gauche conformers of methyl dichloracetate reveals that, in the case of the cis form, the partial negative charge on the starred Cl atom will tend to induce a partial positive charge on the carbonyl atom via a field effect. This causes a significant increase in the carbonyl stretching

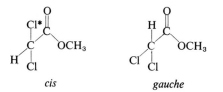

cis   gauche

frequency with the appearance of a second carbonyl band. That this higher frequency band pertains to the more polar cis form may be shown by comparing polar ($CHCl_3$) and nonpolar (CCl_4) solution spectra (Fig. 6-11). The solvent of higher dielectric constant favors the more polar conformer and consequently the intensity of the band corresponding to the cis form increases at the expense of the trans form.

The spectra of 2-pyrrolic esters in CCl_4 solution are very informative (25). In addition to a free carbonyl stretching frequency, many such

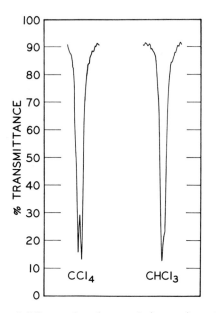

Fig. 6-11. The effect of difference in solvent polarity on the carbonyl absorption of methyldichloroacetate.

esters exhibit two other ester carbonyl bands. Dilution studies enable these to be assigned to an intermolecular and an intramolecular hydrogen-bonded ester group. The band of the latter occurs between the free carbonyl and the intermolecular hydrogen-bonded stretching frequencies. Only the intensity of the intermolecular bonded species is concentration dependent. Failure to recognize that the presence of two or more carbonyl

Free carbonyl
(1725 cm^{-1})

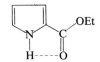

Intramolecular
hydrogen bonded
(1712 cm^{-1})

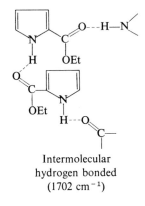

Intermolecular
hydrogen bonded
(1702 cm^{-1})

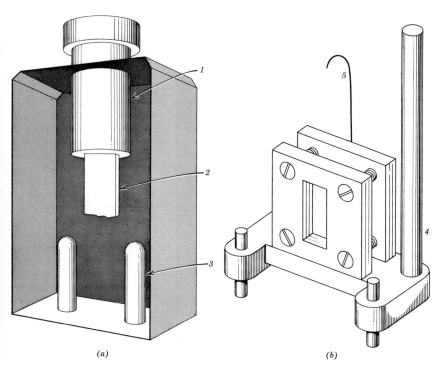

Fig. 6-12. Cells for small quantities of solutions. (*a*) Ultramicro cavity cell (1 μl) of Barnes Engineering Co.; (*b*) sealed micro-cell (3 μl) of Beckman Instruments, Inc. *1*, Stopper; *2*, sample volume; *3*, mounting holes; *4*, mounting base; *5*, filling device (gooseneck capillary).

absorption bands is due to one of the effects mentioned above could result in an extensive GC effort to purify a "clean" sample.

Two cells widely used for solution spectra are shown in Figure 6-12. The ultramicro cavity cell (Fig. 6-12*a*) has a total volume of about 1 μl, but, unfortunately, more than half is dead space. Good quality spectra are obtained at solution concentrations down to 1 μg/μl when this cell is used in a Perkin-Elmer Model 621 spectrometer equipped with a 4 × beam condenser. Other similar instruments will give equivalent performance. A spectrum of 1 μg of eugenol in 1 μl CCl$_4$ obtained in this manner is shown in Figure 6-13. At this dilution, it is difficult to correct completely for the solvent absorptions (1580, 1000, and 970 cm^{-1}). The sharp 3585 cm^{-1} band of unassociated hydroxyl accompanies the broad 3360 cm^{-1} bonded OH band.

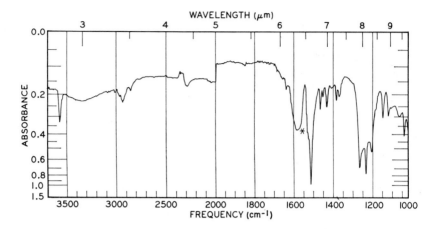

Fig. 6-13. Spectrum of 1 μg eugenol in 1 μl CCl$_4$ (Perkin-Elmer Model 621 spec-trometer, 4× beam condenser). The starred absorption band is due to partially uncompensated CCl$_4$.

Three microliters of sample is required to fill the cell shown in Figure 6-12b (0.1 mm sealed cell), and ca. 70% of this amount intercepts the light beam when a 4× beam condenser is used. Solvent band compensation is highly desirable when solution spectra are recorded.

Another practical cell for small liquid samples is the Carle liquid centrifuge microcell designed specifically for use with GC fractions (Fig. 6-14). Similar to other types of sealed cells, this cell may be employed for neat samples or solutions. One end of a glass capillary tube containing the chromatographic fraction is placed in the upper funneled opening, and the cell is filled by spinning the cell and tube in a 50-ml centrifuge cup.

B. Direct or Tandem Combination of GC and IR

There are many instances when direct or tandem IR spectral recording of GC fractions is preferable to batch methods. Sometimes the mixture under study may contain more than 100 components, requiring laborious and inconvenient collections. Poor trapping efficiencies are often en-countered with many low boiling materials and with compounds that tend to fog upon emerging from the chromatograph. In addition, some pure trapped samples will oxidize or polymerize after collection. A partial spectrum of benzaldehyde, isolated by gas chromatography, is shown in Figure 6-15. The presence of benzoic acid, arising from oxidation of the aldehyde subsequent to isolation, is clearly seen. Thus, it is evident that in

Fig. 6-14. Carle liquid centrifuge microcell for use with GC fractions.

many cases, direct IR spectral recording of GC effluents may be preferable to batch methods.

The use of an infrared spectrometer in conjunction with a gas chromatograph has not been without problems. At the present time, it can be fairly stated that no "on-the-fly" GC–IR combination is extant which approaches the utility of the many commercially available types of GC–MS unions. Some of the IR devices perform adequately for certain applications. Others augur well for future instrumentation which may permit small-sample IR spectra to be obtained with the facility now possible in the GC–MS field. Unambiguous identification of eluted components by their mass spectra alone is often difficult and sometimes impossible, and the complementary information derived from IR data is exceptionally helpful.

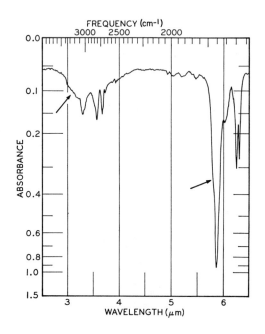

Fig. 6-15. Spectrum of benzaldehyde isolated by gas chromatography. The presence of benzoic acid is indicated by arrows.

1. Interrupted Elution Gas Chromatography

Interrupted elution chromatography actually represents a modified batch system. It has been suggested for mass spectrometric identification of GC effluents (26) and extensively used for this application by Ebert (27). In distinction to measuring spectra during or immediately subsequent to the emergence of a component, the "stop–go" method allows solute bands to be eluted intermittently from a gas chromatograph. In the Scott system (28), conventional infrared and/or mass spectrometers are linked to a packed column gas chromatograph. The design permits each peak to be examined automatically for a minimum of 10 min before the following peak is eluted. The practical application of this apparatus requires that no significant loss of GC column resolution results. Although seldom necessary in actual practice, holdups of even several days can be tolerated with little attendant decrease in resolution.

A diagram of part of Scott's apparatus (the glass manifold, infrared cell, and concentration trap) appears in Figure 6-16. The tapered 25-ml. 9-cm

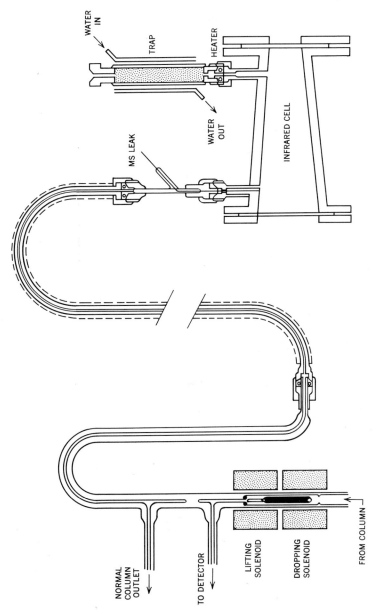

Fig. 6-16. Glass manifold, IR cell, and concentration trap of Scott's interrupted elution GC apparatus.

WATER IN

TRAP

HEATER

WATER OUT

MS LEAK

INFRARED CELL

NORMAL COLUMN OUTLET

TO DETECTOR

LIFTING SOLENOID

DROPPING SOLENOID

FROM COLUMN

single-path stainless steel cell is operated at 150°C. The cell's interior is gold plated to render it inactive to chemically labile substances. In order to avoid fogging due to water vapor, interaction with carboxylic acids, distortion on heating, brittleness, appreciable absorption regions, etc., silver chloride was selected as the most suitable window material. Concentration of the eluted materials is accomplished by the method of Desty (29). In operation, a substance leaving the chromatograph is trapped after passing through the IR cell. The trap and contents are heated swiftly to 180°C in 2 min and nitrogen (10 ml/min) is passed through the trap for an additional minute, forcing the concentrated substance back into the cell. For example, with a 40% Apiezon L–firebrick trap, the carrier gas is separated from pentane eluted from the gas chromatograph and a twelvefold concentration of the hydrocarbon is effected. Automatic control is provided for the carrier gas flow, IR spectral attentuation, temperature program, and flushing the cell after each fraction. Fowlis and Welti modified Scott's silver chloride cell so that samples as small as 50 μg can be spectrally examined (30). The device is useful for packed column studies but this sample size limits its application with open tubular columns.

Wilks Scientific Company introduced a modified low-cost commercial spectrometer employing a 25-ml light-pipe gas cell for moderate scan rate analysis of GC fractions (33). The range 2.5–15 μm (4000–667 cm^{-1}) was scanned in 45 sec. Owing primarily to relatively poor resolution and a recording speed too slow for many "on-the-fly" applications, the manufacturer has adopted Scott's approach. A new system has been developed which consists of a Perkin-Elmer Model 237 Infracord (3 min per complete spectrum), a 1-ml heated light-pipe cell, and a valving device which enables the operator to interrupt the gas flow of a packed column without affecting its resolution. The sensitivity of the spectrometer, approximately 20 μg for acetone with a scale-expansion accessory, makes the instrument unsuitable for gaining spectra on minor components (less than 1%) and the IR evaluation of open tubular column effluents is limited to major components only.

2. Moderate Scan Rate IR

a. Conventional IR Systems.   During the last several years, instruments have been developed for studying transient phenomena such as chemical kinetics and biological changes. These high-speed scanning IR spectrometers, at present not commercially available, generate spectra in 10^{-2} to 10^{-5} sec. Unfortunately, such instruments require specialized, expensive equipment and elaborate optical systems. Their resolution is inferior to grating spectrometers, since a compromise must be struck between scanning speed and resolution. Furthermore, they do not operate over a wide spectral range. The author of this chapter prefers the term "moderate scan speed" when considering instruments capable of producing spectra between 0.1 and 50 sec.

Bartz and Ruhl designed a spectrometer for components of 0.5 mg minimum with moderate resolution and a scan speed of 16 sec (31). However, a total of 32 sec is required for cell filling, scanning, resetting, and purging. Subsequent to passing through a 12-ml, 30 cm long light pipe, the source radiation is chopped and directed into two different single-beam grating spectrometers. One spectrometer covers the 2.5–7 μm (4000–1429 cm^{-1}) range while the other concurrently encompasses the 6.5–16 μm (1538–625 cm^{-1}) region. The instrument presently is utilized for monitoring reaction products of a small experimental chemical reactor and is not used with a gas chromatographic unit (32).

b. Filter-Wheel Spectrometer.   One commercial entry in the GC–IR field (Beckman Instruments Filter-Wheel spectrometer) employs a circular, variable filter monochromator (34). Its suitability for this application lies in the ease of coupling the rapidly rotating filter-wheel assembly with the high efficiency of transmitting energy through the sample to the detector.

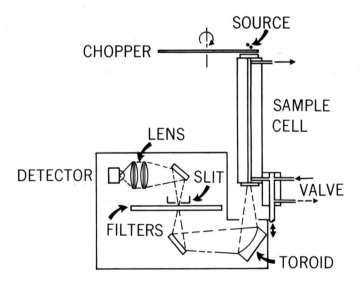

Fig. 6-17. Schematic of a filter-wheel spectrometer (Beckman Instruments Model 102).

The cell is constructed of optically polished gold-plated stainless steel for maximum reflectivity and resistance to oxidation at elevated temperatures. Both ends are sealed with KBr windows and cell heating is accomplished by means of a cartridge heater embedded in an aluminum housing. An optical schematic of this single-beam instrument is shown in Figure 6-17. Faster scanning speeds are possible with a filter-wheel monochromator than with conventional spectrometers. The spectral range 2.5–14.5 μm (4000–690 cm^{-1}) can be recorded and the instrument readied for another scan in 6 sec.

Limitations of the spectrometer are its relatively large sample requirement of ca. 0.5 mg and low resolution over most of its range (as compared with prism spectrometers), namely, 0.04 μm at 3 μm, 0.14 μm at 8 μm, and 0.26 μm at 14 μm. However, it has been employed for "on-the-fly" examination of commonly encountered aerosol propellants where sample size was not a limiting factor (Fig. 6-18) (35).

c. Interferometric IR Spectrometer.   It is well known that the typical IR dispersion spectrometer is energy limited. In order to increase resolution, the slits must be narrowed or the radiation spread widened. Thus, less energy is available to the detector and the signal-to-noise (S/N) ratio is lower. The energy restriction can be improved somewhat by opening the slits, but this results in a loss of resolution. In generating a spectrum, the detector measures the intensity of each segment of the radiation and only a limited amount of time can be spent on an individual measurement if a full spectrum is presented within a reasonable length of time. However, the S/N ratio is directly proportional to the measuring time. It is apparent, then, that the conventional dispersion spectrometer is limited in response speed, resolution, sensitivity, and S/N ratio, and any one characteristic can be materially improved only at the expense of others.

On the other hand, in interference spectroscopy, various performance characteristics can be improved simultaneously. A brief description of such an instrument is in order at this point. The source radiation of the spectrometer enters a Michelson interferometer, where it is reflected and transmitted by a semitransparent film (beam splitter) (Fig. 6-19). The reflected ray is again reflected at the stationary or fixed mirror, passes through the beam splitter, and reaches the detector. The transmitted ray is reflected at the movable mirror, returned to the beam splitter, and thence travels to the detector. For monochromatic radiation, if there is no difference in the length of the optical paths traversed by the two beams, they reach the detector simultaneously and the resultant intensity will be enhanced. Interference fringes are produced whenever two waves follow paths of

GLC–IR SCAN

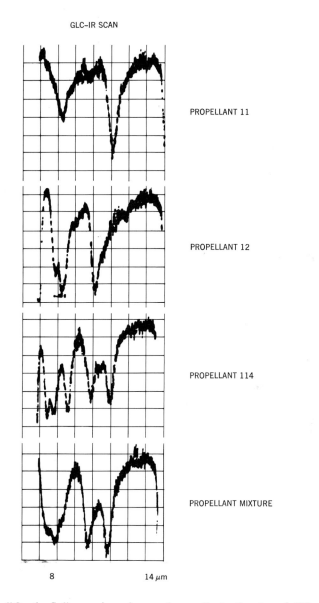

PROPELLANT 11

PROPELLANT 12

PROPELLANT 114

PROPELLANT MIXTURE

8   14 μm

Fig. 6-18. "On-the-fly" scanning of aerosol propellants. Spectra of GC eluents recorded in approximately 6 sec on a Beckman Instruments Model 102 spectrometer.

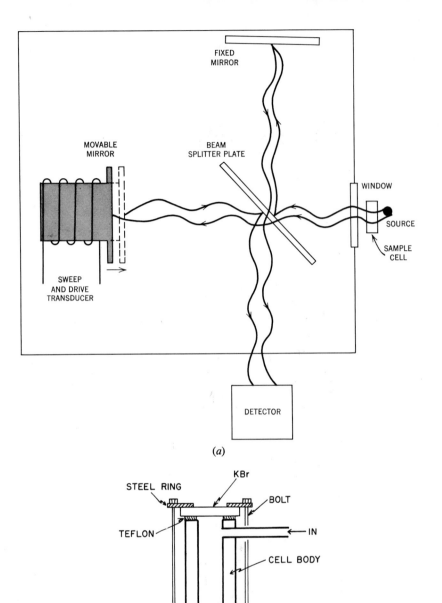

FIXED
MIRROR

MOVABLE
MIRROR

BEAM
SPLITTER PLATE

WINDOW

SOURCE

SAMPLE
CELL

SWEEP
AND DRIVE
TRANSDUCER

DETECTOR

(a)

KBr

STEEL RING

BOLT

TEFLON

IN

CELL BODY

(b)

Fig. 6-19. (a) Schematic of a Michelson interferometer and detector. (b) Flow-through IR cell for the interferometer. The cell is positioned between the optical system and the IR radiation source (36).

250

different lengths between the source and the detector so that they arrive at the detector out of phase. This is brought about by linearly moving the movable mirror a short distance and abruptly returning it to its original position in a total time of approximately 1 sec, which is equivalent to one scan. The polychromatic light, or composite radiation, falling on the detector represents the sum of all interferences and enhancements, and the detector's output rises and falls as the fringes move across it. The detector produces a complex audiofrequency signal (an interferogram, the analog of the original optical signal) which contains the frequency–intensity information present in the untransformed optical signal. This is analogous to the way in which a heterodyne radio receiver causes a complex radio-frequency wave to transfer its complexities to an audiofrequency signal. For any specific rate of mirror motion, a specific relation exists between the frequency of the optical signal and the detector output. The resultant interferogram is converted into a "normal" spectrum by a wave analyzer, which is actually a low-frequency spectrometer.

Low and Freeman explored the feasibility of applying this technique to obtain spectra of GC fractions (36). It appeared that an interference spectrometer would be superior to conventional dispersion instruments for GC applications because it had an inherently higher S/N ratio and could scan more rapidly and repetitively. The device employed in the first phase of their study was a commercial one consisting essentially of a repetitively scanning Michelson interferometer, a time-averaging computer for digitization, and an analog data reduction system which produced the spectra. This instrument scanned once per second, in the range 2500–250 cm^{-1} (4–40 μm) at a resolution of about 18 cm^{-1}. Gas sample spectra of completely resolved as well as unresolved GC peaks were obtained (Fig. 6-20). A more recent report indicated that it would be feasible to gain spectra of 1 μg material in 15 sec (37).

At present, a serious drawback of this technique is poor resolution (about 18 cm^{-1}). In addition, a somewhat unconventional spectral presentation is obtained in that the relative band intensities differ from those obtained by normal dispersion instruments and a large computer system is required to reduce the data to a conventional analog form (Fig. 6-22). The pseudo-double-beam spectra shown in Figures 6-20 and 6-21 were obtained by instrumentally subtracting background scans from scans of samples. This is the normal procedure, and a curve gained in such a manner appears in Figure 6-22 (curve C). Spectra gathered by these means are not entirely satisfactory for rapid comparison with ordinary double-beam spectrometric data because of differences in relative band intensities. These

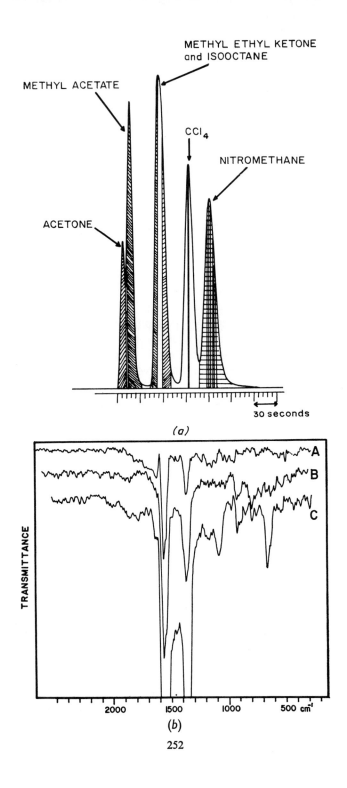

METHYL ETHYL KETONE
and ISOOCTANE

METHYL ACETATE

CCl$_4$

NITROMETHANE

ACETONE

30 seconds

(a)

TRANSMITTANCE

A

B

C

2000   1500   1000   500 cm^{-1}

(b)

252

variations arise because the intensity of a band in a subtraction spectrum is a function of the energy distribution of the source itself. A correction can be made by digital computation methods by calculating the ratios of the absorptions. Using an IBM Model 301 digital computer, the ratio of background (*A*, Fig. 22) to difference spectra of eugenol (*E*, Fig. 22) may be obtained to yield a conventional spectrum.

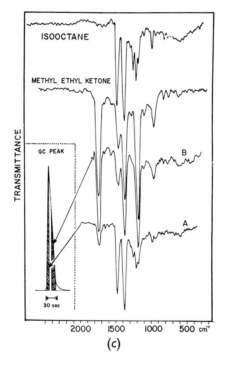

(*c*)

Fig. 6-20. Examination of a synthetic mixture by combined GC–IR. (*a*) Chromatogram of the mixture on an 8 ft × ¼ in. o.d. Carbowax 20M column. (*b*) Interferometric IR spectra of the nitromethane peak (*A*, 43 scans of flowing gas stream, taped. Spectrum resulted from examination of a sample approximated by the shaded area in the chromatogram of Figure 6-20*a*. Sample size: 0.3 μl. *B*, 200 scans of trapped nitromethane peak stored directly in the computer. Sample size, 0.3 μl. *C*, 200 scans of trapped 1 μl sample of nitromethane stored directly in the computer). (*c*) Interferometric IR spectra of the unresolved peak (methyl ethyl ketone + isooctane) shown in Figure 6-20*a*.

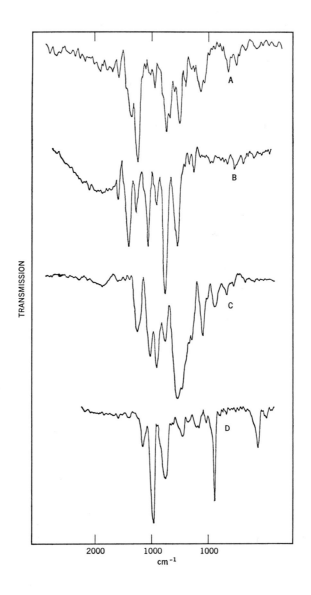

Fig. 6-21. Interferometric IR spectra of microgram quantities. *A*, 0.5 μg chalcone on KBr plate, 261 scans; *B*, 1 μg *m*-nitroacetophenone on KBr plate, 200 scans; *C*, 1 μg longifolene in solution, 250 scans; *D*, 2 μg eugenol in solution, 300 scans.

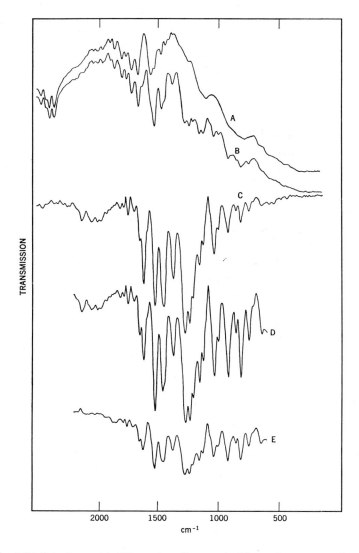

Fig. 6-22. Interferometric IR spectra of eugenol. All spectra were produced by digital computation. *A*, Background (globar radiation passing through two KBr plates), 100 scans; *B*, 1 μl of eugenol placed between KBr plates, 100 scans; *C*, 100 scans of background as in *A* subtracted from 100 scans of sample plus background as in *B*, resulting in the difference spectrum of 1 μl of eugenol; *D*, ratio of spectra *A* and *B*; *E*, ratio of spectra *A* and *C*.

d. Mirror Scan Wheel Spectrometer. A dispersion instrument capable of generating an infrared spectrum as fast as 1 msec has been developed by the Warner & Swasey Co. This spectrometer, originally designed to study transient phenomena, was modified recently for recording spectra of GC effluents in 200 and 500 msec (38). In contrast to high-speed recording instruments (millisecond scan rates) which are based on the rapid movement of an essential element of the dispersing train, the Warner & Swasey device employs a fixed dispersing system. Wavelength scanning is accomplished by sweeping a sequence of mirrors through an intermediate focal plane of the spectrometer. This single-beam, grating dispersion instrument is made possible by the use of two fast, sensitive detectors which simultaneously cover the range 2130–3700 cm^{-1} (2.7–4.7 μm) and 1110–2170 cm^{-1} (4.6–9.0 μm).

Nitrogen-cooled indium antimonide and helium-cooled copper-doped germanium are used for the low- and high-wavelength sections, respectively. The spectral region encompassing 714–1820 cm^{-1} (5.5–14.0 μm) is accessible by replacing the indium antimonide detector with a helium-cooled copper-doped germanium detector and changing the grating and filters.

A very brief discussion pertaining to performance figures of fast detectors is helpful for better understanding. A performance parameter which is useful for general measurement of detectors is the noise equivalent power (NEP), i.e., the smallest power detectable with an S/N ratio of unity. The NEP for any detector depends on all the characteristics of the incident radiation. Symbolically, NEP(500, 900, 1) denotes the fact that the detector has been evaluated at a blackbody temperature of 500°K, 900 Hz chopping frequency, and unit noise bandwidth. As a figure of merit, the NEP suffers from the disadvantage that large values correspond to low performance. For this reason, the term "detectivity" was introduced, which is defined as the reciprocal of the NEP. The symbol D^*, referring to detectivity per unit area, generally is used in detector specifications. For the Warner & Swasey detectors, D^*(500, 1800, 1) values are on the order of 10^9–10^{10} cm Hz$^{1/2}$ W^{-1} at 5 μm. The response of a detector also varies with wavelength (Fig. 6-23).

In the operation of the spectrometer, a component eluted from the gas chromatograph flows through a 10-cm optical path light-pipe cell (1 ml volume) located inside the instrument. The thermally insulated heated cell is internally gold plated and equipped with removable salt windows. The source radiation is modulated at 10,000 Hz before entering the light pipe in order to discriminate against emission from the heated light pipe. Subsequent to leaving the cell, the energy is directed to an entrance slit,

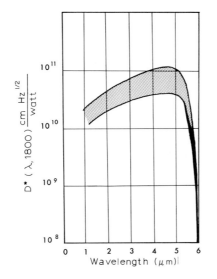

Fig. 6-23. Detectivity range of InSb (photovoltaic) detector, 180° field of view.

collimated, and dispersed by a grating (Fig. 6-24). The light is brought to focus on the mirror scan wheel subsequent to reflections from two mirrors, and thus focuses a spectrum in an intermediate focal plane. The scan wheel consists of 24 peripherally mounted, matched corner mirrors. Each one laterally displaces an incident ray and sends it back parallel to itself, thus causing a left-to-right reversal of the spectrum. The second pass grating dispersion thus adds to the first pass instead of canceling it, as would be the case if flat mirrors were used. The result is a double-pass spectrum. The radiations of the two separated spectral regions are thus directed toward the exit slits and then to the detectors. The signal can be recorded by photographing an oscilloscope display, by magnetic tape, or by a direct-writing oscillograph.

A spectrum of 25 μg acetone, recorded in 0.5 sec, is shown in Figure 6-25. Note that the absorption bands are superimposed upon the background globar source emission curve in a typical single-beam spectrometric presentation. Narrow spectral ranges can be studied utilizing higher resolution, interchangeable gratings.

e. I.&C.I.–Esso Double-Beam Spectrometer. Instruments & Communications, Inc. and Esso Engineering Corp. recently reported the initial results of a study on a modified conventional double-beam spectrometer

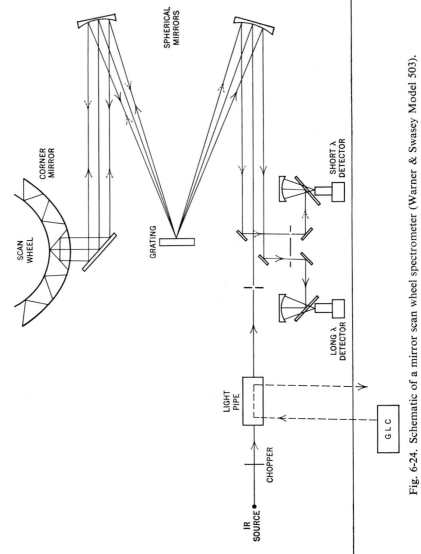

Fig. 6-24. Schematic of a mirror scan wheel spectrometer (Warner & Swasey Model 503).

258

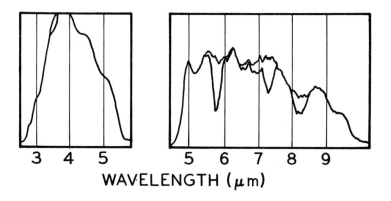

WAVELENGTH (μm)

Fig. 6-25. IR spectrum of 25 μg acetone vapor recorded in 0.5 sec (Warner & Swasey Model 503 spectrometer).

capable of covering 4000–4500 cm^{-1} (2.5–15 μm) in 10 sec (39). Nernst source radiation is chopped at 400 Hz, passed through a 0.5 ml × 10 cm windowless, double sample cell, and enters a Czerny-Turner design monochromator. Upon leaving the monochromator, the radiation is focused on a liquid helium-cooled, mercury-doped germanium detector.

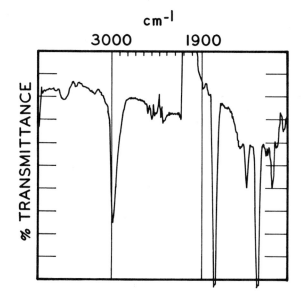

Fig. 6-26. IR spectrum of 40 μg ethyl acetate vapor recorded in 10 sec (I.&C.I.—Esso spectrometer).

A spectrum of 40 μg ethyl acetate vapor is shown in Figure 6-26. This prototype instrument holds promise of generating interpretable spectra in 10 sec on 1 μg of material. However, similar to other IR spectrometers designed for direct recording of GC eluents, conveniently obtaining data on a few micrograms to submicrogram quantities must await an interface which can handle the vast differences in concentration gradients, various flow rates, and wide range of band widths encountered.

III. The Combination of Gas Chromatography with Raman Spectrometry

The complementary and confirmatory information provided by use of both IR and Raman spectrometry has been long recognized. Vibrations which give rise to weak IR bands often display intense Raman bands (e.g., sulfur–sulfur, tri- and tetrasubstituted carbon–carbon double bond, and ring breathing vibrations). On the other hand, the intense IR signals attributed to esters, ketones, acids, alcohols, etc. are weak in Raman spectra.

Absorption of IR radiation occurs only when there is a change in dipole moment of the molecule during a normal vibration. The intensity of the resultant absorption band is proportional to the square of the change in dipole moment (transition moment). Thus, for an asymmetrical molecule (which must have a dipole moment greater than zero), all the normal vibrations will be IR active. A molecule with symmetry properties may or may not have a dipole moment. If the dipole moment differs from zero, all the normal vibrations should be IR active, but this is not necessarily so because of the molecule's symmetry. Thus, while a band for the C=C stretching vibration appears in the spectrum of *cis*-dichloroethylene, it is absent in the case of the *trans* isomer.

Raman spectrometry is valuable where the symmetry and dipole moment restrictions of a molecule do not permit the infrared spectrum to represent the complete vibrational spectrum. Raman scattering, or the Raman effect, occurs only when there is a change in the polarizability of a molecule during a vibration.

The interaction of a photon with a molecule in the ground state ($V = 0$) may momentarily raise the molecule to a higher energy level which is not stable at room temperature (Fig. 6-27). If the molecule leaves this unstable level, it can scatter a photon and return to the ground state. Consequently, the scattered photon has the same energy content as the existing photon and Rayleigh scattering occurs. On the other hand, if the molecule falls to an excited vibrational state, such as $V = 1$, the scattered photon's energy

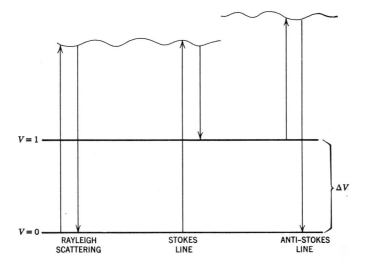

$V = 1$

$V = 0$

RAYLEIGH
SCATTERING

STOKES
LINE

ANTI-STOKES
LINE

ΔV

Fig. 6-27. Energy level diagram for the Raman effect.

now is equal to the energy of the exciting photon minus the difference in energy between the $V = 1$ and $V = 0$ levels. The frequency of the scattered photon thus is shorter than that of the incident light and this type of Raman line is called a Stokes line. An anti-Stokes line arises when a molecule in an excited level ($V = 1$) is raised to a higher, unstable level by reacting with an incident photon and then returns to the ground state upon scattering a photon. In this case, the energy of the scattered photon equals the energy of the exciting photon plus the energy difference between $V = 1$ and $V = 0$. At room temperature most molecules are in the ground state, therefore anti-Stokes lines are considerably weaker than Stokes lines.

The energies of the scattered photons are either increased or decreased relative to the exciting radiation, and this results in the appearance of bands shifted from the frequency of the incident light. These quantized energy increments correspond to the differences in rotational and vibrational levels of the molecule. Raman bands represent frequency differences, hence the term "Raman shifts." They are cited as v and have the dimension of cm^{-1}.

Until the recent advent of laser technology, the organic chemist paid scant attention to Raman spectroscopy because of the high instrumental cost, the painstaking care demanded in experimental technique, and particularly, the large sample requirements. Unlike absorption spectra—which represent strong first-order effects—Raman spectra are exceedingly weak, depending on light scattering, a second-order effect. Some advantages of obtaining Raman spectra by lasers rather than by their predecessor, the Toronto mercury arc, are:

(*1*) Red lasers, such as the He–Ne laser (6328 Å) avoid many fluorescence problems.

(*2*) The collimated nature of laser energy allows focusing for excitation of extremely small sample volumes.

(*3*) Multiple passing techniques permit increasing the signal-to-noise ratios for liquids of weak scattering power.

(*4*) The laser's brightness and nearly complete linear polarization simplify the measurement of depolarization factors.

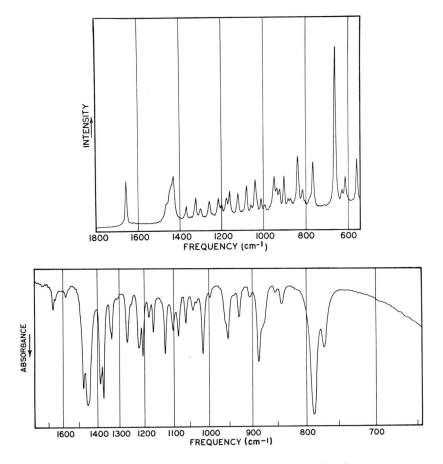

Fig. 6-28. Raman (upper) and IR (lower) spectra of α-pinene.

(5) The simplification of the excitation optics makes it easier to correct for reflection losses, which are dependent on refractive index, and therefore allows more accurate intensity measurements.

The Raman effect offers several advantages over infrared. For one thing, the Raman spectrum of water does not interfere with most spectral measurements. In addition, the intensity of Raman bands varies almost linearly with concentration. There is no need for several sets of prisms or gratings to conveniently record a continuous Raman spectrum covering the range from 0 to 4000 cm^{-1}. Selection rules are less restrictive than those for infrared so that a rotational–vibrational band may offer direct information relevant to molecular parameters, while the same transition, appearing as an IR band, permits only the determination of combinations of these parameters. A Raman and an IR spectrum of α-pinene appear in Figure 6-28.

Most present-day commercial instruments are equipped with 75 mW He–Ne or 250 mW argon ion lasers which are satisfactory for gaining spectra on liquids and solids. Organic chemical vapors have insufficient molecular density to yield good spectra with these low-energy excitation levels, but the 1-W argon ion and krypton ion lasers may be suitable for gas studies. Advantages of the argon (4880 Å) and krypton (5682 Å) lasers compared with He–Ne (6328 Å) are greater exciting power, more Raman scattering due to the shorter wavelengths of radiation, and increased quantum efficiency of photomultiplier tubes for the higher energy protons.

Fig. 6-29. 0.16 μl capillary cell (aluminizing removed).

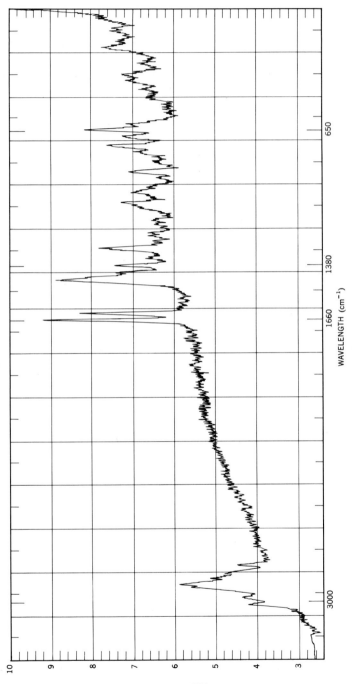

WAVELENGTH (cm^{-1})

Fig. 6-30. Raman spectrum of 0.15 μl linalyl methyl ether recorded in 7 min employing a 0.5 mm i.d. Pyrex capillary tube.

264

Bailey, Kint, and Scherer (40) reported a simple Raman microsampling technique which involved the modification of a Perkin-Elmer Model LR-1 single monochromator spectrometer with a 10 mW He–Ne laser. "Throw-away" glass capillary cells were made by blowing a small spherical lens on Pyrex capillary tubes (Fig. 6-29). The incoming laser beam thus was focused to a point above the capillary tip corresponding to the focal length of the capillary lens. The back side of the capillary was aluminized in order to utilize the back-scattering. Fractions isolated by gas chromatography can be transferred to the cells by means of smaller, open-end capillaries. The sample is emptied by gravity or air pressure as required by its viscosity. Viscous samples are transferred by evacuating the capillary cell, dropping the open end into the material, and restoring the system to atmospheric pressure. Solid spectra are obtained by filling the capillary with molten sample employing the same technique. Good Raman spectra have been obtained on samples as low as 0.04 μl, a sensitivity approaching that of commercial IR instrumentation.

A laser Raman system, equipped with microsampling capabilities, is available from Spex Industries, Inc. The author of this chapter has gained good spectra (Fig. 6-30) in 15 min (0–4000 cm^{-1}) with this double monochromator instrument on approximately 0.1 μl of material employing Pyrex microcapillary tubes. In actual practice, a 6-in. length of 0.5 mm i.d. glass tubing is employed in the usual manner to trap a component eluting from a gas-liquid chromatograph. The capillary section containing the

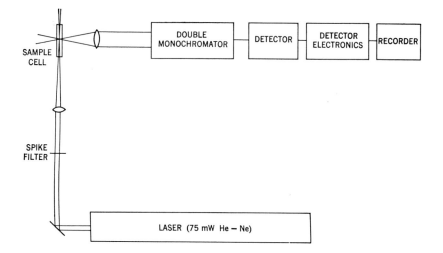

Fig. 6-31. Schematic of a laser–Raman spectrometer (Spex Instruments Co.).

material is cut, sealed with a microburner, and centrifuged. A Raman spectrum is recorded directly, but care is required to assure that the sample capillary is precisely aligned with the beam. If this is not well accomplished, Raman scattering intensity is reduced, and furthermore, the fluorescent scattering from the glass gives excessive background radiation. Very recently, Freeman and Landon (41,42) have reported on their studies pertaining to small sample handling. They obtained good spectra on 8 nl (0.008 μl) of CCl_4 and linalool in 0.1 mm i.d. capillaries. The schematic of the spectrometer's salient features is shown in Figure 6-31.

Although the increased interest in Raman spectrometry has already found applications on collected GC fractions, it is too early to assess possible fast-scan potentialities for directly coupled GC–Raman studies. The current Raman systems are at least an order of magnitude less sensitive than the best IR instruments. This will certainly limit near-future developments to more abundant fractions of the GC samples.

References

1. (a) Thermodynamics Research Center *Catalog of Infrared Spectral Data*, American Petroleum Inst., Research Project 44, Texas A&M University, College Station, Texas.
   (b) *Documentation of Molecular Spectroscopy*, Verlag Chemie, Berlin, and Butterworths, London.
   (c) *Standard Infrared Spectra*, Sadtler Research Laboratories, Inc., Philadelphia, Pa.
   (d) *Infrared Data Cards (IRDC)*, The Infrared Data Committee of Japan.
2. *Manual on Recommended Practices in Spectrophotometry*, ASTM, Philadelphia, Pa., 1966, p. 29.
3. D. S. Erley, *Anal. Chem.*, **40**, 894 (1968).
4. D. H. Anderson and N. B. Woodall, *Anal. Chem.*, **25**, 1906 (1953).
5. R. A. Pickering, *Anal. Chem.*, **28**, 168 (1956).
6. E. D. Black, J. D. Margerum, and G. M. Wyman, *Anal. Chem.*, **29**, 169 (1957).
7. F. Bissett, A. L. Bluhm, and L. Long, Jr., *Anal. Chem.*, **31**, 1927 (1959).
8. E. D. Blac, *Anal. Chem.*, **32**, 735 (1960).
9. J. E. Stewart, R. O. Brace, T. Jones, and W. F. Ulrich, *Nature*, **186**, 628 (1960).
10. J. Haslam, A. R. Jeffs, and H. A. Willis, *Analyst*, **86**, 44 (1961).
11. M. Sparagana and W. B. Mason, *Anal. Chem.*, **34**, 242 (1962).
12. A. C. Maehly, *Analyst*, **87**, 116 (1962).
13. R. W. Hannah and J. L. Dwyer, *Anal. Chem.*, **36**, 2341 (1964).
14. Anon., *Instrument News* (Perkin-Elmer), **15** (4), 7 (1965).
15. H. J. Sloane, T. Johns, W. F. Ulrich, and W. J. Cadman *Appl. Spectry.*, **19**, 130 (1965).
16. R. C. Gore and R. W. Hannah, *Annual Meeting, Society for Applied Spectroscopy, Chicago, Ill., June, 1966*.
17. J. T. Chen and J. H. Gould, *Appl. Spectry.*, **22**, 5 (1968).

18. G. D. Propster, *Pittsburgh Conf. Anal. Chem. Appl. Spectry.*, *March*, *1965.*
19. M. Stimson and M. J. O'Donnell, *J. Am. Chem. Soc.*, **74**, 1805 (1952).
20. U. Schiedt and H. Renwein, *Z. Naturforsch.*, **1**, 270 (1952).
21. E. R. Blout, "Microspectroscopy," in *Techniques of Organic Chemistry*, Vol. I, Part III, 2nd ed., A. Weissberger, Ed., Interscience, New York, 1954, p. 2179.
22. H. W. Leggon, *Anal. Chem.*, **33**, 1295 (1961).
23. M. K. Snavely and J. G. Grasselli, *Instr. News* (Perkin-Elmer), **17** (2), 6 (1966).
24. N. J. Harrick, *Internal Reflection Spectroscopy*, Interscience, New York, 1967, pp. 217–295.
25. R. Grigg, *J. Chem. Soc.*, **1965**, 5149.
26. J. H. Beynon, *Mass Spectrometry and its Applications to Organic Chemistry*, American Elsevier, New York, 1960, p. 190.
27. A. A. Ebert, Jr., *Anal. Chem.*, **33**, 1865 (1961).
28. R. P. W. Scott, I. A. Fowlis, D. Welti, and T. Wilkens, in *Gas Chromatography 1966*, A. B. Littlewood, Ed., Inst. of Petroleum, London, 1967, pp. 318–336.
29. D. H. Desty, T. J. Warham, and B. H. G. Whyman, in *Vapour Phase Chromatography*, D. H. Desty, Ed., Butterworths, London, 1957, p. 346.
30. I. A. Fowlis and D. Welti, *Analyst*, **92**, 630 (1967).
31. A. M. Bartz and H. D. Ruhl, *Anal. Chem.*, **36**, 1892 (1964).
32. A. M. Bartz, Dow Chemical Company, Midland, Mich., private communication. (1967).
33. P. A. Wilks, Jr. and R. A. Brown, *Anal. Chem.*, **36**, 18 (1964).
34. G. T. Keahl, *Pittsburgh Conf. Anal. Chem. Appl. Spectry.*, *February*, *1966.*
35. S. K. Freeman and C. T. Malone, *Chem. Spec. Mfgers. Assoc. Meeting, Florida, December, 1966.*
36. M. J. D. Low and S. K. Freeman, *Anal. Chem.*, **39**, 194 (1967).
37. M. J. D. Low and S. K. Freeman, *Ag. & Food Chem.* **16**, 525 (1968).
38. S. A. Dolin, H. A. Kruegle, and B. Krakow, *Pittsburgh Conf. Anal. Chem. Appl. Spectry., March 1968.*
39. R. A. Brown and J. C. Kelleher, *Pittsburgh Conf. Anal. Chem. Appl. Spectry., March 1968.*
40. G. F. Bailey, S. Kint, and J. R. Scherer, *Anal. Chem.*, **39**, 1040 (1967).
41. S. K. Freeman and D. O. Landon, *Anal. Chem.* **41**, 398 (1969) Feb. (1969).
42. S. K. Freeman and D. O. Landon, Spex Speaker (Spex Industries) **4**, No. 4. 1968.

CHAPTER 7

Gas Chromatography and
Nuclear Magnetic Resonance Spectroscopy

Gordon E. Hall, *Unilever Research Laboratory, Colworth House,*
Sharnbrook, Bedford, England

I. Introduction

Of the spectroscopic methods most commonly used in conjunction with
gas chromatography, nuclear magnetic resonance (NMR) spectroscopy is
the least sensitive. This is unfortunate since it yields a great deal of specific
information which is often necessary for elucidating the structure of a
compound isolated by gas chromatography. This information can include
the chemical nature of the groupings present, their structural relationship
with each other, and their spatial (i.e., stereochemical) relationship. The
magnetic resonance spectra of the hydrogen nuclei (protons) in organic

compounds are especially informative. The present article sketches the physical basis for NMR spectra, shows how the information obtained is related to structures, outlines the associated difficulties, and describes the apparatus required.

II. The Magnetic Resonance Phenomenon

When an elemental particle such as an electron, neutron, or nucleus with a nonzero spin quantum number is placed in a magnetic field, certain energy levels become nondegenerate by virtue of the particle's interaction with the magnetic field. An electron has a spin quantum number of $\frac{1}{2}$, and thus has two such distinct energy levels which form the basis for electron spin resonance (ESR) spectroscopy. Certain nuclei (e.g., ^1H, ^{13}C, and ^{31}P) also have spin quantum numbers $I = \frac{1}{2}$, and they similarly have two distinguishable energy levels due to interaction with a magnetic field. These energy levels are the basis for NMR spectroscopy. Some nuclei have higher spin quantum numbers, which result in more than two energy levels, but in the present article only particles with spin number $\frac{1}{2}$ will be considered.

The normal thermal processes ensure a Boltzmann distribution of particles between the two energy levels, with a greater number in the lower energy state than in the higher. A transition of particles from the lower to the higher level can be induced by electromagnetic radiation of the correct (resonance) frequency. This frequency, ν, is given by the usual quantum relation;

$$\Delta E(\propto H_0) = h\nu \qquad (7\text{-}1)$$

ΔE is the difference in energy between the two energy levels, and for a particular particle is directly related to the magnetic field H_0. For ordinary magnetic fields ($H_0 \sim 10{,}000$ G), ν is in the microwave region for electron spin resonance, and in the radiofrequency region (~ 1–100 MHz) for nuclear magnetic resonance. A resonance signal may be located either by holding ν constant and varying H_0, or by holding H_0 constant and altering ν. Not only may nuclei go from the lower to the higher energy levels by absorption of radiation, but the reverse process can occur with emission of radiation. The energy difference, ΔE, is actually so small that, in ordinary circumstances, the populations of the two energy levels are almost identical and the absorption and emission of radiation almost cancel each other. Hence the low sensitivity of NMR spectroscopy. Nuclei with a zero nuclear spin quantum number (e.g., ^{12}C and ^{16}O) have single energy levels in a magnetic field and thus cannot exhibit the NMR phenomenon.

Now where does all this help the elucidation of structure? Different elements (and different isotopes of the same element) have different (see Table 7-1) characteristic frequencies v, at any given magnetic field strength. In principle this allows an elemental analysis of an unknown sample. Far more importantly, however, individual nuclei of any one isotope show slight variations around their characteristic frequency. These variations arise from the fact that the individual nuclei find themselves in differing environments. The applied magnetic field is modified by the nuclei constituting the environment, so that neighboring nuclei experience different magnetic fields. There are corresponding differences in the resonance frequency and these differences are referred to as *chemical shifts*.

For high-resolution NMR spectroscopy, magnetic environmental effects arising from intermolecular interactions generally average so that the observed chemical shifts reflect only the intramolecular interactions. Hence the value of chemical shifts for structure elucidation. A simple example that can be used to illustrate this point is the proton spectrum of methyl acetate. It consists of two peaks, the chemical shift of one being similar to that found with other methyl esters, and the other chemical shift similar to that observed for other acetates. The different chemical shifts reflect the fact that the two methyl groups within the molecule are different, each of the two types of proton experiencing different magnetic fields owing to the environmental differences.

Since chemical shifts correspond to very small changes in the magnetic field strength or radiofrequency, it is not yet possible to measure them absolutely, and a standard substance is used as a reference; for proton spectroscopy this is usually tetramethylsilane (TMS). This compound is highly symmetric and contains only one type of proton which principally gives a single sharp resonance line. The compound is also chemically inert, which means that intermolecular effects are kept to a minimum. It has the added advantage of high volatility, so that it may readily be removed from a solution. When using a reference compound the chemical shifts become the difference in resonance frequency between the observed and the standard nuclei, and are expressed either in dimensionless units or in frequency units. The units employed are, for convenience, the same whether the spectrum is obtained by varying the magnetic field strength or by varying the radiofrequency. In frequency units the chemical shift is linearly related to the applied magnetic field. Thus, a shift of 100 Hz at 10,000 G changes to one of 200 Hz at 20,000 G. If the resonance frequency of the nucleus were 100 MHz, a chemical shift of 100 Hz would correspond in dimensionless units to $100/(100 \times 10^6) = 1$ ppm. Such a value is independent of magnetic field strength.

For comparing data from different instruments it is advantageous to have the chemical shifts independent of field strength, and so they are generally recorded in the dimensionless ppm. Proton shifts are normally referred to either the delta (δ) or tau (τ) scale. In the former units the shift is simply in ppm downfield from the TMS line; the τ-value is given by the relationship $\tau = 10.00 - \delta$. This relationship shows that the two scales are not different in kind, but differ in their zero position and in the direction in which the numbers increase.

Another important parameter obtained from an NMR spectrum is the relative amount of each type of nucleus present in the molecule. Under certain conditions the *area* of any resonance absorption peak is directly proportional to the number of nuclei responsible for that peak. Integration across the spectrum thus allows an estimate of the relative proportions of nuclei in the different environments represented by the various chemical shifts. As mentioned earlier, the proton magnetic resonance spectrum of methyl acetate consists of two peaks. The areas of these two peaks are the same since each corresponds to three protons. Such information clearly is of great value.

The third important spectral feature is the fine-structure caused by spin–spin splitting. In any collection of nuclei, a particular nucleus not only gets a general view of its environment, which determines the chemical shift, but its spin also interacts specifically with the spins of certain nuclei among its neighbors. The extent of this interaction is represented by the so-called *spin–spin coupling constant J*, usually recorded in frequency units, Hz, in which its value is independent of the magnetic field. The proton spectrum of ethyl acetate, like methyl acetate, has a singlet (relative area 3) for the acetate methyl group. However, the absorptions corresponding to the methylene (relative area 2) and the other methyl group (area 3) are multiplets owing to spin–spin coupling. All three parameters—chemical shift, area, and spin–spin splitting—are discussed further in Section IV.

Standard texts (1,2) should be consulted for a more complete description of the NMR phenomenon.

III. Sample Size

Before proceeding further, it is important to consider the sample size required to obtain a useful spectrum (3). An indication can be obtained from Table 7-1 which shows the approximate sample sizes required to obtain spectra using a typical piece of high-resolution NMR equipment working at maximum sensitivity. The significance of the experimental times referred to in Table 7-1 is made clear later, when discussing the enhancement of signal to noise where the technique of averaging spectra by computer methods is discussed. It must be emphasized that the exact sample size required varies with the problem so that Table 7-1 is only a guide.

Inspection of Table 7-1 shows immediately that investigation of certain nuclei in samples derived by gas chromatography borders on the impossible. There are two factors which influence this: the inherent sensitivity of NMR for the particular isotope, and the abundance of the element as that isotope. Both factors arbitrate against carbon NMR spectroscopy using the natural abundance of ^{13}C. (The abundant isotope ^{12}C has a zero spin quantum number.) The most readily observed nucleus occurring commonly in organic compounds is the proton (^1H). Only PMR spectroscopy will be considered in the remainder of the discussion, but it is important to realize that the principles covered in general are applicable to the study of other nuclei (4). Several isotopes shown in Table 7-1 could clearly be investigated in conjunction with gas chromatography.

TABLE 7-1

Approximate Minimum Sample Sizes for Various Nuclei (1,3)

Nucleus	Resonance frequency (MHz) at 10,000 G	Sample weight (mg) required for indicated experimental times		
		10 min	1 hr	16 hr
^1H	42.577	5	1.2	0.3
^{19}F	40.055	8	2.0	0.5
^{31}P	17.235	35	1.8	0.5
^{11}B	13.660	10	0.5	0.1
^{13}C (at natural abundance)	10.705	6500	1000	250
^{14}N	3.076	1200	60	15

The sample weight data apply to a Perkin-Elmer Ltd. R-10 (14,092 G) spectrometer with an NS-544 accessory using a conventional cylindrical cell and a sample of molecular weight 250.

All figures relate to a single sharp resonance from a single nucleus, with a final signal-to-noise ratio 10.

The 10 min experiment refers to a single scan, while the 1 and 16 hr experiments involve computer averaging the spectra continuously obtained during those periods (see Section VI-B-2).

No doubt the steady improvement in sensitivity achieved over the last few years will continue. In the meantime it is not a bad idea to try to provide a minimum of either 1–2 mg or 20–40 mg samples for routine PMR spectra, the actual size depending on whether or not the microcells and averaging methods described later in the experimental procedure, are available.

IV. Extraction of Information from a Spectrum

The interpretation of a spectrum such as that shown in Figure 7-1 involves two main steps: (*1*) the measurement of the analog traces and conversion into a convenient digital form, and (*2*) interpretation of that

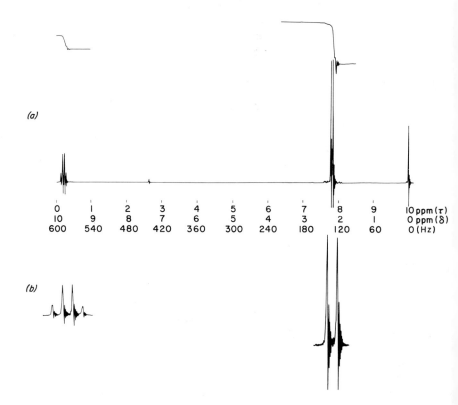

Fig. 7-1. The PMR spectrum of a 6% solution of acetaldehyde in deuterochloroform at 33.4°C and 60.004 MHz.

data in terms of structure. Figure 7-1 follows the modern convention in that the magnetic field strength increases from left to right. As is usual, the scale is given in both the dimensionless units and the frequency units used for chemical shifts. The single sharp line at the high-field (right-hand) end of the spectrum is due to the reference compound TMS, and the acetaldehyde spectrum consists of the doublet and quartet shown on an expanded scale in Figure 7-1b. This spectrum is used below as an example of a so-called first-order spectrum, the definition of which is given later.

A. Interpretation of a First-Order Spectrum

The chemical shifts for the different types of hydrogen can be obtained by measuring the center of gravity of the multiplets. In the present case, these are 2.17 and 9.78 ppm downfield from TMS. Consequently, on the δ scale they are at 2.17δ and 9.78δ, and on the τ scale the absorptions are at 7.83τ and 0.22τ. Standard chemical shift tables (2,5,6), which have been compiled by the examination of many known compounds, show that these observed shifts are in the ranges to be expected for the protons in a methyl group α to a carbonyl group and an aldehyde proton, respectively. The integral shown in Figure 7-1 is consistent with such groupings in that it gives an area ratio of 3:1 for the two multiplets. Further structural evidence is obtained from the fine structure due to spin–spin coupling. In the case of acetaldehyde there is little choice, but protons in general only couple significantly with each other if they are close to each other in the molecular structure, and particularly if they are on vicinal carbon atoms. (There are countless exceptions (7) to this statement, but this is not the place for detailed discussion.) A proton coupled equally to, and in a first-order manner with, n other protons splits into $n + 1$ lines; in the present example the aldehydic proton is coupled with three protons and so its resonance is observed as a quartet. The relative intensities of the four lines (1:3:3:1) can be predicted by binomial theory (1,2,5). The spacings between any adjacent pair of lines in either of the multiplets must all be the same, and this spacing (in Hz) is equal to the magnitude of the coupling constant. In the present case the coupling constant J has the value 2.9 Hz.

In general, one set of protons may be coupled to more than one other set, each coupling possibly involving different constants; due allowance must be made for this possibility. Double resonance methods such as spin-decoupling (2) can establish which multiplets are coupled. By consideration of the character of each multiplet, the order in which particular groupings are put together in the molecule can then be deduced. Although the magnitude of the coupling constant is not in itself of direct importance in this case, it should always be checked (2,5) against those recorded for similar situations. On the other hand, the magnitude of the coupling constant is of great importance in stereochemical studies allowing, for example, *cis* and *trans* isomers to be distinguished and the conformation of certain cyclohexane derivatives to be deduced (8).

Only one line of the spectrum shown in Figure 7-1 is left unaccounted for by the above discussion: the singlet 7.35 ppm downfield from TMS. This line derives from the chloroform present as an impurity in the deutero-chloroform used as the solvent. The possibility of impurities must always be considered. Stationary phases from the chromatography in particular may confuse the spectral integral, since they frequently have absorptions in the same region as the sample.

B. Interpretation of a Complex Spectrum

When the chemical shift between two multiplets is much greater than the coupling constant involved, the PMR parameters can be deduced by the simple methods outlined in the previous section. For the historical reason that early attempts to explain spectra used perturbation theory, such a situation is called first-order. When the shift and coupling constant are comparable, the situation is more complicated, as can be seen in Figure 7-2, which shows the PMR spectrum of methanol in acetonitrile. The appearance of this spectrum depends on both temperature and concentration, changes in either of which alter the extent of hydrogen bonding. At high methanol concentrations a first-order spectrum results, being a quartet and doublet similar in appearance to Figure 7-1, the spectrum of acetaldehyde. As the concentration decreases, the lower field quartet (due to the hydroxyl proton) moves upfield; a spectrum not far removed from first-order is shown in Figure 7-2a. As the concentration is decreased further, the resonance due to the hydroxyl proton moves further upfield, finally passing upfield of the methyl resonance (c). While the chemical shift is small, great distortion of both multiplets is observed, as can be seen in both (b) and (c). Since, in frequency units, any coupling constant is independent of the magnetic field strength while the shift is proportional to it, a spectrum which appears complex at one magnetic field strength may be first-order at a higher field strength.* If a spectrum is complex and the requisite parameters cannot be deduced by inspection, two possibilities remain. Either the parameters may be guessed with sufficient accuracy for

---

* If there is no chemical shift between the protons involved, no splitting due to their spin–spin coupling is observed. In this case the coupling constant cannot be directly measured.

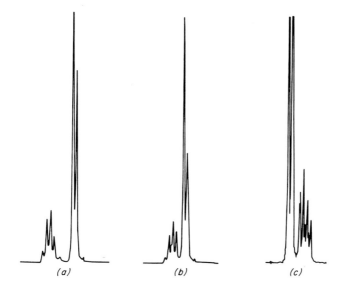

Fig. 7-2. The PMR spectrum of methanol at various concentrations in acetonitrile. (a) High concentration, (b) moderate concentration, (c) dilute concentration.

the problem in hand, or they may be accurately obtained by calculation. In all but the simplest cases (1,2,8), digital computer techniques are required and various programs are available (9).

Electronic computers have been used for several years to solve NMR problems. Many calculations are completely impractical by manual computations because they could take several years. Although there is no exact basis for theoretically calculating chemical shifts and coupling constants, it is possible to calculate the spectrum to be expected on the basis of a set of postulated parameters. The basic mathematical methods for doing this were worked out by analogy with the wave mechanics previously applied to electrons. The mathematical principles are not in themselves difficult, but with a several-spin system the algebra and arithmetic required for the solution of the many equations become long and tedious. Hence the need for computers. Unfortunately, a spectroscopist is normally in the position that he has a spectrum from which he wishes to deduce the chemical shifts and coupling constants, and is not in the position of knowing the parameters with which to calculate the spectrum. Although direct methods for doing the former have been tried, at present the only generally applicable methods involve iterative techniques. Trial parameters, which are either guessed by inspection of the spectrum or postulated by analogy

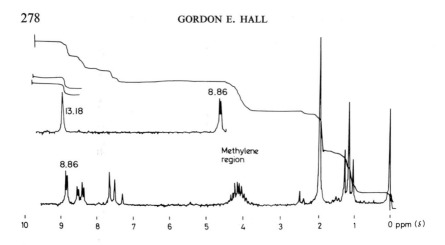

Fig. 7-3. The PMR spectrum of ethyl 2-(2,4-dinitrophenyl) acetoacetate in
deuterochloroform at 33.4°C and 60.004 MHz.

with other compounds, are first used. The computer calculates the expected spectrum which is compared with the experimental spectrum. If
there is a reasonable correspondence between these spectra, the trial
parameters may be adjusted by the computer in some logical manner. A
new expected spectrum is calculated and this in turn is compared with the
experimental spectrum. The procedure is repeated until a good fit (mathematically defined in the computer program) between the calculated and
experimental spectra is obtained.

Two pitfalls in interpretation must be briefly mentioned. First, it sometimes happens (10) that protons which at first glance might be expected
to be identical are in fact magnetically distinguishable. For example, as
the spectrum in Figure 7-3 shows, the two protons of the methylene group
in the enolic form of ethyl 2-(2,4-dinitrophenyl) acetoacetate (1), have
different chemical shifts (11). In consequence, the methylene region is
more complex than the quartet which would otherwise be expected. So it

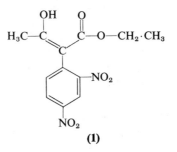

(1)

must not be assumed that identical groups attached to the same atom are
magnetically equivalent.

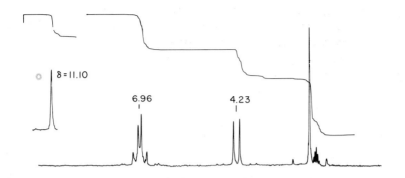

8 = 11.10

6.96

4.23

Fig. 7-4. The PMR spectrum of *p*-methylphenoxymethyl phosphonic acid in deuteroacetone at 33.4°C and 60.004 MHz.

The second pitfall is that protons may be spin-coupled with other nuclei (such as ^2H, ^{13}C, ^{31}P) in the molecules. In such a case, those particular protons show a first-order splitting which is not reproduced anywhere else in the proton spectrum. An example is shown in Figure 7-4, which is the spectrum of *p*-methylphenoxymethyl phosphonic acid (**2**). The methylene

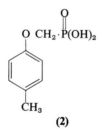

(2)

protons occur as a doublet at 4.23δ ($J = 10.2$ Hz) by virtue of the coupling with the ^{31}P atom. The same splitting should be observed in the phosphorus spectrum, which should be a triplet, but as seen in Figure 7-4, the phosphorus spectrum does not appear in the PMR spectrum. In such a case, one neither should assume the two proton lines to arise from chemically shifted protons, nor search for the corresponding, but nonexistent, splitting elsewhere in the proton spectrum.

V. Advantages and Disadvantages of the Utilization of PMR in Combination with Gas Chromatography

Undoubtedly no one spectroscopic method holds such preeminence that it can be used to the exclusion of the others. For use in conjunction with gas chromatography, mass spectroscopy comes nearest to this ideal, largely because the small sample requirement is most compatible. In general, however, a combination of spectroscopic methods is far better than the use of any one method individually. If the compound under investigation is a known one with well-documented spectra, the comparison of its IR and/or mass spectra with those of the authentic sample frequently suffices for its identification. These methods generally give specific fingerprint spectra. To a lesser extent this is also true of PMR spectra, but these are not commonly recommended for this task for the following reasons:

1 The effort required to obtain PMR spectra of gas chromatographic fractions is greater than for infrared and mass spectra.

2 Relatively few standard PMR spectra are available (12,13)—a few thousand compared with the tens of thousand infrared spectra available.

3 The exact appearance of PMR spectra can vary considerably with the solvent, with the operating magnetic field strength, and sometimes with the operating temperature. However, these are not serious difficulties for an experienced spectroscopist and infrared and mass spectra also vary in appearance depending on the instrument with which they have been obtained.

4 The presence of non-proton-containing groups is only evidenced indirectly, and a change in such groups may not make significant changes in the spectrum. In comparison, infrared spectra sometimes only identify functional groups and mass spectra can be misleading owing to structural rearrangements. In such cases the PMR spectrum is an essential complementary fingerprint.

The above comments summarize the disadvantages of PMR compared with infrared and mass spectroscopy when allied to gas chromatography.

The *advantages* of PMR spectroscopy are considerable when the structure elucidation of a new compound or the identification of one with no recorded spectra are involved. If the molecular weight of the compound is known, either by another method or by internal standardization of the PMR spectrum itself (14), then the spectral integral is very valuable. For example, a degradation product of the antibiotic *Micrococcin P* was for many years believed to have the formula $C_{24}H_{23}N_5O_5S_4$. On the basis of a PMR spectrum, this was revised (15) to $C_{24}H_{17}N_5O_5S_4$, and this molecular formula was confirmed by X-ray crystallography (16). Next, the chemical shift evidence, considered in combination with the infrared spectrum, indicates which groupings are present in the molecule. Finally, the splitting patterns and coupling constants provide evidence for the order in which these groupings are joined together and for their stereo-chemical relationships. Even the presence of many non-proton-containing groupings (e.g., carbonyl, nitro, nitrile) can be inferred by their effects on neighboring protons. In this respect, ^{13}C NMR data (17) would be valuable complementary information, but, as can be seen from Table 7-1, the available sensitivity is normally too low.

The advantages of PMR spectroscopy may be summarized as follows.

1. By employing modern techniques it is usable in combination with gas chromatography.

2. It affords evidence regarding the presence of particular groupings.

3. It indicates the relative numbers of the different protons.

4. If furnishes information regarding the structural relationship of the different groupings.

5. It yields information regarding the stereochemistry and the con-formation of the molecule.

Much of the information outlined above cannot be obtained from the IR and mass spectra.

In addition, the great specificity of PMR spectroscopy to look at certain parts of a molecule should be noted. For example, many carotenoid end groups nowadays can be identified in this way (18). The older chemical methods for the detection of the β-ionone (**3**), α-ionone (**4**), and iso-propylidene end groups have, as a result, been largely abandoned.

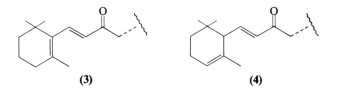

(3)   (4)

VI. Experimental Procedures

A. General Comments

High-resolution spectra must be obtained under conditions in which the molecules are tumbling rapidly, in order to remove effects due to inter-molecular interactions. Although moderately high-resolution spectra have been obtained from liquids adsorbed onto pyrogenic silica (19), spectra from solids and extremely viscous liquids or solutions are not ordinarily suitable. This is because the intermolecular effects cause line broadening and the fine spectral details, so important for structural assignment, are lost. For combination with gas chromatography, vapor-phase spectra would seem on first consideration to be the most suitable, but no systematic study of such spectra yet appears to have been made. Perhaps the most important consideration is the usual one of sensitivity—in this case, of obtaining sufficiently high concentrations of vapor at relatively low operating temperatures. Another problem concerns the line broadening frequently observed in a vapor spectrum. Some peaks in a PMR vapor-phase spectrum may be narrow while others in the same spectrum may be quite broad (20). These lines may be narrowed by collision processes induced either by working at high sample pressures or by the addition of a foreign gas such as argon (21). Some chemical information has been obtained from vapor-phase spectra, an example being (22) the determination of the barrier to rotation about a nitrogen–nitrogen bond. However, at the present time, the liquid phase is the preferred state for structural studies.

If the sample is a plentiful liquid of low viscosity, its spectrum may be obtained directly. Otherwise, it is dissolved in a solvent which is preferably proton-free in order to avoid confusion in the spectrum. This solvent must be carefully chosen to be compatible with any other experiment to be done. Carbon tetrachloride is commonly employed for gas chromatographic fractions because it is proton-free, it is practicable for IR spectra, and it is a highly symmetric and chemically inert molecule. The symmetry means that it is magnetically isotropic, which is important because anisotropic solvents such as benzene and carbon disulfide can induce specific chemical shifts. Indeed, these shifts can yield further structural information and they have been widely used (8), particularly for the study of steroids.

The reference compound may either be dissolved in the solution (internal method) or placed in a separate tube (external method). The internal reference method using TMS is normal for PMR spectra. Cyclopropane derivatives absorb in the same region as the TMS band, and so if such compounds are suspected it is good practice first to obtain the spectrum with no added reference compound.

The conventional PMR tube is a glass cylinder of about 5 mm o.d. After adding the sample, the tube should be sealed either by heating the end of the glass tube or by means of a convenient plastic cap, and then inserted in the rf coil(s) between the magnet pole faces of the spectrometer and spun about its long axis. The spinning ($\sim$ 2000 rpm) can be simply accomplished by blowing air onto the vanes of a turbine placed over the tube. In this case a friction bearing is required to locate and steady the tube. Alternatively, an air bearing which both locates and spins the tube may be used. The spinning effectively improves the homogeneity of the magnet, since the molecules then all experience the same average field strength. The effective homogeneity in turn defines the spectrometer's resolution. This is commonly referred to in terms of the line width, which is ordinarily constant through a high-resolution spectrum.

The spectrometer should, of course, be operated under optimum conditions (3). These need not concern us here, except to point out that the experimental temperature can be an important variable. Subject to two conditions, the height of a PMR peak obtained from a given sample increases as the temperature is decreased because the Boltzmann distribution increases the difference in nuclear population of the two energy levels. The first condition is that the line width should not alter (owing to viscosity changes, for example) over the temperature range considered. Second, the temperature change should not have caused a chemical change in the sample. For most gas chromatographic fractions in carbon tetrachloride, operation at about room temperature is generally satisfactory.

B. Application in Conjunction with Gas Chromatography

To obtain a PMR spectrum using a conventional system as outlined above requires several milligrams of sample. The exact weight requirement depends on the complexity of the compound and on the operating magnetic field strength, but using present-day spectrometers it is about 5–40 mg. The sample size required can be reduced in two ways: first, by more efficient use of the sample and, second, by the enhancement of signal-to-noise (S/N).* These two approaches are discussed in turn, followed by a section dealing with handling the samples.

1. Microcells

For good resolution to be maintained using a cylindrical PMR cell, the ratio of the length divided by the diameter for the column of liquid should (23) be at least about 5–10. In consequence, the sample length is large in comparison with that of the detector coil, so not all of the sample is contributing to the signal. Several studies have been made to improve this situation by means of microcells and the subject has been extensively reviewed (3,24). All microcells involve dissolving the sample in a smaller

---

* The meaning of S/N need not be considered, except to recognize that it is a quantitative measure of the fact that random fluctuations are superimposed on any signal. The commonest definition of S/N of a PMR peak is:

$$S/N = \frac{2.5 \times \text{peak height}}{\text{peak-to-peak value of noise}}$$

volume of solvent, so that if a low S/N is due to poor solubility of the sample in the solvent(s) available, then the use of microcells actually worsens the situation. Fortunately, samples isolated via gas chromatography are usually readily soluble in suitable solvents, so the S/N ratio of the spectrum is limited by the sample size. This is the situation assumed in the ensuing discussion, in which the enhancement factor is taken to indicate the relative improvement in S/N ratio observed using a given weight of sample. Two general types of microcell are discussed.

The first type is spherical, or nearly spherical. A sphere is a special case of an ellipsoid and deviations from an ellipsoidal shape lead to loss of resolution. Such a cell (most often made in glass) commonly has a sample volume 0.03–0.05 ml rather than the approximately 0.4 ml recommended for ordinary cylindrical cells. Some microcells are commercially available (3,24) but they are difficult to clean. The assembly illustrated in Figure 7–5 works successfully in at least one type of spectrometer (25). The cell is spun by means of the air-bearing shown in the figure.

The spheres are blown on the end of small-diameter glass ("melting

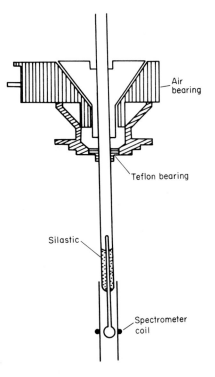

Fig. 7-5. The microcell and spinner assembly as used by Frost et al. (25).

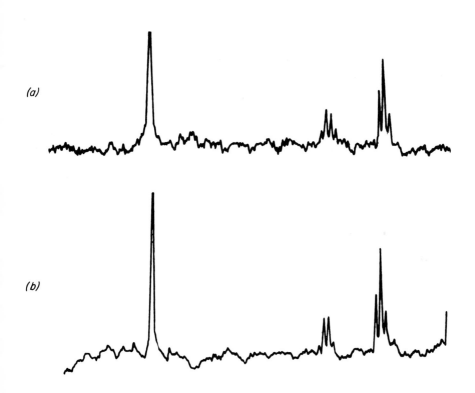

Fig. 7-6. The PMR spectrum of ethylbenzene in carbon tetrachloride. (*a*) 610 μg in 45 μl, single-scan, spherical microcell. (*b*) 3.5 mg in 0.4 ml, single-scan, cylindrical cell, using a Perkin-Elmer Ltd. R-10 spectrometer at 33.4°C and 60.004 MHz. After Frost et al. (25).

point'') tubes, using as a glass lathe an electric motor with a hollow drive shaft through which air pressure may be applied to the developing sphere. The glass capillary is sealed at one end, the open end is placed in the chuck, and the whole rotated while the sealed end is gently heated and blown into a sphere. These microcells are disposable and so present no cleaning problems.

A spectrum obtained using such a microcell is shown in Figure 7-6*a*. Comparison with the spectrum taken with a standard cell (Fig. 7.6*b*) indicates an enhancement factor of about five.

With the assembly described, great care must be taken to ensure that the tube spins accurately about its long axis. The silicone tubing holding the microcell is embedded in Silastic, which has sufficient flexibility to allow adjustment of the cell alignment. Other workers (26) have recommended that to avoid alignment difficulties, a slightly smaller spherical cell should be used inside a conventional cylindrical tube which then more accurately locates the cell along the long axis. To further minimize wobbling, the space around the sphere is filled with a liquid such as carbon tetrachloride. Whichever method is used, the optimum position for the sphere in the spectrometer probe is determined by adjusting it for the maximum observed S/N ratio of a suitable signal.

The second type of microcell involves a cylindrical sample of smaller dimensions than ordinarily employed. The diameter (and hence the length) may be considerably reduced (23,27) by containing the sample within a small tube inserted inside, and along the central axis of, the normal one. The solution is more concentrated, but the filling factor in the probe coil is worse and the overall enhancement factor is 2–3. Alternatively the column length in a conventional cylindrical cell may be reduced a certain amount, using plugs if necessary. Care must be taken to ensure that the vortex formed in the liquid by the spinning (28) does not extend into the coil region. In practice the volume can be halved, to about 0.2 ml (24). This means that the enhancement factor achievable with a sphere is only about two compared with the reduced volume cylindrical cell. In general when dealing with small samples it is advisable to use a microcell first, then, if desired, to dilute the solution into a cylinder. This is easier than attempting to concentrate a solution previously run unsatisfactorily in a cylindrical cell.

2. Enhancement of Signal-to-Noise (S/N)

In an NMR spectrometer, much of the noise arising from the instrument is introduced at the first-stage radiofrequency amplifier. A few attempts have been made to reduce this noise in particular spectrometers, but this is really a problem for the instrument manufacturers' design engineers. The present section is only concerned with what can be done using an instrument in good working condition. The S/N ratio of any system can be increased either by enhancing the signal or by decreasing the noise, or by a combination of these two approaches.

The simplest method of reducing noise is to filter the high frequency components electrically from the spectrometer output by means of a resistance–capacitance circuit. All commercial spectrometers have such filters, and the optimum time constants to use for given spectrometer sweep speeds can be obtained from the manufacturers. If the rate of spectrometer sweep is lowered, a higher time constant can be used without causing unacceptable peak distortion, so more noise is removed and the S/N ratio increases. This is so if the effect known as radiofrequency saturation is not encountered. Saturation is a phenomenon resulting when the population of the nuclear energy levels become more or less equal, in which case the resonance signal disappears. The very act of observing the resonance signal causes the population to tend toward that state, but if this disturbance is insignificant then the signal obtained is proportional to the radiofrequency power. Consequently, to observe maximum S/N ratio, the greatest possible power must be used. In this case, if the rate of sweep is decreased, the power must also be decreased to avoid saturation. As always, it is a question of balancing the various factors. The method of enhancing S/N by means of a low sweep rate and a high time constant is a particular example of the time-averaging methods which have become very popular in recent years.

Time-averaging methods can be used to produce enhancement factors in excess of an order of magnitude. The methods all involve increasing S/N by using longer experimental times. By taking more time for the measurements, the noise is allowed to cancel out to a certain extent. There have been various methods proposed (cf. refs. 3, 24, and 29 for reviews) but at present the most satisfactory, albeit expensive, method is that of summing spectra digitally. The analog spectrum is obtained under optimum conditions and digitized at regular, predetermined intervals along the magnetic field frequency axis. Additional spectra are produced independently, digitized at exactly the same points, and the corresponding points of each spectrum added together and stored in a small computer. Noncoherent noise tends to average out while the signal, which is coherent, increases in size. The theoretical enhancement factor obtainable by summing n spectra is $\sqrt{n}$. For PMR spectra a factor of 15 is readily achieved by adding 250 spectra obtained in an overnight experiment. An example is shown in Figure 7-7a, the spectrum of 26 μg of ethylbenzene.

To obtain the best spectrum by this averaging process it may not be necessary to operate the spectrometer at the conditions producing the maximum S/N in a single scan. Some S/N can be sacrificed for speed, if in so doing a sufficiently greater number of spectra may be accumulated (in a given time) to give a greater final S/N. For nuclei such as phosphorus-31 for which fast scan rates may be used, several thousand spectra may be readily accumulated (25) in an overnight experiment.

Slow magnetic field drifts are most usually compensated by triggering the computer/digitizer sweep from a prominent spectral peak, such as that of TMS. Alternatively drift can be compensated retrospectively by programming the computer to ensure that the major peaks in all the spectra are overlapped correctly before summing. The inherent stability of internally locked instruments is such that these devices may not be necessary, and a mechanically reproducible spectrometer sweep position may then be used to trigger the computer/digitizer sweep.

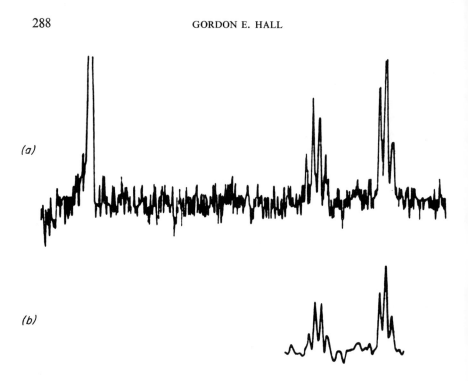

Fig. 7-7. The PMR spectrum of ethylbenzene (26 μg) in carbon tetrachloride (45 μl) after 344 accumulations with a NS-544 computer. (a) The direct output; (b) the result of a nine-point least mean square, cubic smoothing of part of (a) with measured points every 0.75 Hz.

General purpose electronic digital computers are ordinarily too expensive to be used on-line with the spectrometer, although such a computer could be accepting information from several sources in a sequential switching process. Off-line operation may proceed by storing the analog spectra on magnetic tape from which the data may be fed at a fast rate through the digitizer into the computer. Alternatively, digitized spectra can be stored in a suitable computer input medium (e.g., paper tape) and this read in at a later time. A special purpose computer and digitizing system operating on-line is, however, normally used, and such equipment is frequently termed a CAT ("computer of average transients").

The spectrum obtained in this way is, of course, in digital form. For display purposes this is converted into an analog trace (cf. Fig. 7-7). The digital spectrum is in a form ideally suited for further calculations, such as those involved in spectral analysis, or for further smoothing of the spectrum. Mathematical smoothing has the advantage of not introducing the peak distortion caused by a time constant, which is the analog equivalent. The result of smoothing the data in part of the spectrum in Figure 7-7a is shown in 7-7b. The method seems to be of limited value when used subsequently to spectral summing (25,30).

The dangers to be faced when using computer-enhanced spectra will not be discussed in detail but are pointed out by the following summary:

1. The sample must be stable at the experimental temperature for the period of the experiment, usually several hours.

2. Incorrect overlapping of spectra may occur and result in spectral artifacts.

3. A large peak, may, after multiple addition in a computer, digitally exceed the highest possible word length in the memory. In this case the word effectively resets to zero and the peak may appear split on display.

4. The solvent or the internal reference compound may be impure. Previously invisible side bands (e.g., the so-called ^{13}C satellites) may in any case then be observed. So, if in doubt, always run a blank.

On the other hand, the advantages of using computer-enhanced spectra are obvious from Figure 7-7 if this is compared with Figure 7-6. The necessary sample size has been lowered by more than an order of magnitude.

3. Sample Handling

There has been a general attempt in the last few years to combine gas chromatography directly with spectroscopic methods, so that the spectra are obtained at the time of elution of any chromatographic peak. So far, such a direct combination has not looked practicable in the case of NMR spectroscopy. The reasons can be summarized as the lack of sensitivity of NMR and its need for long experimental times. A higher S/N is obtained from a flowing sample compared with a stationary one of the same concentration because a higher radiofrequency power can be used without danger of saturation, owing to the fact that fresh nuclei are continuously entering the spectrometer. However, even so, if one considers that concentrations of tens of milligrams per milliliter are required for proton spectra, the sample concentrations in the gas eluting from a chromatograph are far too low. This concentration must also be maintained for several minutes, the time required to scan the spectrum. If a lower concentration is used, a longer experimental time is required, to allow enhancement of S/N. At the present, the only significant possibility in the future is the use of Fourier transform spectroscopy (see Section VIII.) which may allow a considerable reduction in the necessary experimental time.

There are two other less important problems associated with the direct combination of gas chromatography with NMR spectrometry. First, the effluent gas will frequently be at a temperature which is considerably higher than that at which the NMR spectrometer would ordinarily be operated. Such disparity would result in magnetic field gradients which would lead to a loss in resolution. In any case, as explained earlier (Section VI-A), a rise in temperature results in a decreased signal. However, it should be possible to overcome this problem with a suitable variable temperature probe. Second, some of the effluent will normally be diverted through the chromatograph detector, thereby reducing the sample available to the spectrometer and aggravating the already difficult sensitivity problem. Clearly, the use of either a high sensitivity detector or one in series with the spectrometer would obviate this difficulty.

For the present, then, a batch method of obtaining PMR spectra of gas chromatographic fractions is necessary and the fractions must be trapped in some way. Any method available for trapping is suitable providing it is convenient, gives high sample recoveries at the sample size involved, and allows ready transfer to the spectrometer. Relatively large amounts of sample are required even when using the microcells and computer techniques described in the previous sections. For this reason, long packed gas chromatography columns are generally preferred for the analysis of complex mixtures such as essential oils and aroma constituents. Packed columns are used to get a high loading, and long columns for sufficient resolution. Even then, a fraction must frequently be trapped several times. If the sample is available in large quantities and the compound to be investigated is clearly separated from any others, then normal preparative gas chromatography may be used. Indeed, in these circumstances it may be possible to trap 50–100 mg and obtain a spectrum in the conventional way.

Three trapping methods suitable for use with the smaller samples are outlined below. A large number of other methods or modifications are available (24,32) varying in their recovery efficiencies and in the sample sizes at which they are sufficiently efficient.

Sample handling is cut to a minimum in the first method, in which the effluent is trapped directly into a glass microcell (31). The column effluent is led directly into the cell by means of a syringe needle which reaches right down to the sphere. The sample is condensed from the gas stream by means of a Triclene and solid carbon dioxide cooling bath, and then washed down with solvent injected into the needle. Two objections can be raised to the method as described. First, the sample has to pass through metal tubes, which might catalyze rearrangements. The use of all-glass lines would obviate this difficulty. Second, the resolution obtained in the spectrometer is reported to be poorer than usual. This deterioration in resolution is probably due to the fact that special tubes were used to allow the use of a wide-bore syringe needle in order to avoid a large pressure drop at the chromatograph exit. The problem might also arise from insoluble impurities. No claim is made regarding the recovery yields, although the method presumably must suffer the misting problems experienced by related trapping techniques. It is similar to the method commonly use for milligram quantities, involving condensation of the sample into a tube inserted into the outlet of the gas chromatograph. The sample is then transferred by means of solvent.

The second method is a variant of the technique of bubbling effluent through liquid carbon tetrachloride. A small volume of a solvent such as carbon tetrachloride is placed in a capillary U-tube. Immediately before the peak emerges from the gas chromatograph, the tube is dipped into liquid argon or nitrogen and then placed at the exit of the gas chromatograph. The solvent forms a solid lattice which is apparently sufficiently open to avoid a significant back pressure, but which traps the sample. The method is claimed (32) to give good recoveries; in experiments in our laboratory, the recoveries have been about 50–80%. Two difficulties have sometimes been experienced using this method. First, the hot effluent gases may melt the crystalline solvent which must then be resolidified by a second dipping while the peak is emerging. Second, higher boiling compounds tend to condense at the beginning of the tube rather than filter through the solid matrix. The latter then becomes redundant.

The third method involves trapping the fraction onto a column packing material, e.g., Apiezon L on Celite, in a tube which is connected to the column exit and cooled with solid carbon dioxide. The open end of the trap should be protected by a calcium chloride drying tube. The tube (B in Fig. 7-8) has ground glass cones at each end and so may be stoppered and stored in the refrigerator after collection (33). It would appear that the rationale for using solid carbon dioxide to cool the trap is that this process is convenient and that it appears to work. It is quite probable that the Apiezon L, if not actually a solid at those temperatures, might well be a mush. In this case, true gas–liquid partition will not occur. However, this does not mean that the trapping process is not efficient. Experiments using Carbowax 20M poly(ethylene glycol), squalane, and SE-30 methylsilicone gum indicate (34) that in many cases at certain temperatures adsorption by the solid stationary phase is very efficient. The optimum trapping temperature range for the stationary phase used should

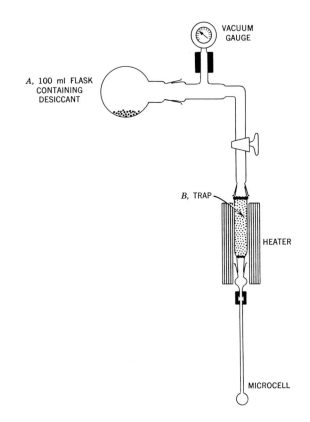

Fig. 7-8. Schematic of the apparatus used for recovery of a sample directly from a packed trap into a spherical microcell.

therefore be determined experimentally. When the trap is warmed for recovery of the sample, the system presumably reequilibrates to gas–liquid partition.

The sample may be recovered in several ways, but the apparatus shown in Figure 7-8 has proved (35) convenient. The microcell touches the underside of a ground glass socket which has been drawn out into a tube of the same diameter as the stem of the microcell, and held in place by silicone rubber tubing. The flask *A* is first evacuated and filled with argon, and the stopcock turned off. The rest of the apparatus is then assembled in such a way that the gas flow through the trap will be in the opposite direction to that when the sample was trapped, and the heater turned on. When the trap temperature reaches 200°C, the stopcock is opened and liquid nitrogen is brought up until the microcell is almost completely immersed. Condensation of the argon (and air) produces a flow of gas down through the trap, thus eluting the sample which is recondensed in the cell. The heater is then switched off and the liquid nitrogen allowed to evaporate, whereby the liquid argon in the cell slowly evaporates. The remaining sample is then washed down into the sphere with carbon tetrachloride containing a reference compound. Virtually quantitative recoveries are claimed for this method.

When using microcells, great care must be taken to ensure the total absence of any insoluble material. Particulate matter of any kind in the sample cannot be tolerated. If the sample is to be transferred from another vessel, such as an infrared cell, it is good practice to filter the solution. A method for doing this has been described (24). It consists of pumping the solution through a filter from a 50- or 100-μl syringe fitted with a Luer lock, a 13 mm Swinny-type Polythene filter holder, and a long needle which fits into the microcell. The holdup volume of the filter system is, however, excessive. In our laboratory the simple procedure of passing the solution through lens tissue placed in a piece of glass tubing has proved satisfactory. The tubing is drawn out at one end to fit into the microcell. All the apparatus must be dust and fiber free.

The stages involved in obtaining a PMR spectrum from gas chromatographic effluents may be summarized as follows:

(i) Trap the fraction from the chromatograph.

(ii) Dissolve the recovered fraction in a solvent compatible with any other technique which is to be subsequently used.

(iii) Add the reference compound, avoiding TMS if a cyclopropane derivative is suspected.

(iv) If necessary, transfer the solution to the microcell, filtering in the process.

(v) Seal the microcell, in order to prevent contamination or loss of sample.

(vi) Place the cell in the spectrometer. Check that the spinning is steady.

(vii) While the sample and cell holder is coming to temperature equilibrium, set up the spectrometer and computer.

(viii) Start the repetitive scans and leave for the required length of time.

(ix) Read out the final averaged spectrum.

VII. Available Instruments

Spectroscopists are frequently asked: "Which instrument should I buy?" The answer given depends on a variety of factors which can be classified as (a) the requirements for the envisaged work, (b) the capital and running costs that can be afforded, and (c) the quality of the personnel available to operate the instrument.

For the purposes of the present discussion, a few commercially available instruments are listed in Table 7-2, along with a few salient facts. For the convenience of this discussion, only spectrometers manufactured by Perkin-Elmer Ltd. (in England) and Varian Associates (in the United States) are recorded. Spectrometers are also available from other manufacturers such as Japan Electron Optics Ltd. (J.E.O.L), Trüb-Täuber & Co. A.G. (Switzerland), NMR Specialties (U.S.A.), and Bruker-Physik A.G. (Germany).

From the point of view of using the spectrometer in conjunction with gas chromatography, two important considerations are the sensitivity of the instrument and whether it can be used in conjunction with a computer. Table 7-2 lists such data regarding the instruments listed. In each case the S/N refers to the middle two peaks of the methylene PMR quartet observed with a 1% ethylbenzene solution using a conventional cylindrical (~ 5 mm o.d.) cell. The prospective customer must consider not only the scale at which he wishes to work but whether the resonance of nuclei other than protons is to be observed. Some spectrometers are able only to observe protons. Some instruments are able only to operate at one fixed temperature, and of those for which variable temperature probes are available, the cost and specification of this accessory vary considerably.

Since prices vary from country to country it is not practicable to give the cost of any instrument in Table 7-2. The terminology low, medium, and high has been used, but at the moment the cheapest high-resolution spectrometer costs about $16–20,000, while a more flexible instrument with a wider range of accessories can cost in excess of $70,000. With regard to running costs, all instruments operate better in a room with a certain measure of thermostatic control (say $\pm 2°C$). Perkin-Elmer Ltd. have used permanent magnets for all their instruments, but Varian have relied mainly upon electromagnets. The latter require water cooling and their electricity consumption raises the running costs, but they can be operated at different field strengths and are not so sensitive to temperature changes as are permanent magnets. All spectrometers require regular servicing, the costs for which can be ascertained by reference to the manufacturers.

VIII. Trends in NMR Spectroscopy

Large reductions in the necessary sample sizes may be anticipated for the future since NMR spectroscopy is developing rapidly. The important

TABLE 7-2

Some Data Concerning Six NMR Spectrometers[a]

Manufacturer	Type	Cost	Specification S/N	Frequency for protons, MHz	Is it compatible with a computer?	Can nuclei other than ^1H be observed?	Is variable temperature probe available?	Does the instrument include an integrator?	Is a spin decoupler available?
Varian	HA-100	high	40	100	Yes	Yes	Yes	Yes	Yes
Perkin-Elmer Ltd.	R-14	high	25	100	Yes	Yes	No	Yes	Yes
Varian	A-60D	medium	18	60	Yes	No	Yes	Yes	Yes
Perkin-Elmer Ltd.	R-10	medium	18	60	Yes	Yes	Yes	Yes	Yes
Varian	T-60	low	12	60	Yes	No	No	Yes	Yes
Perkin-Elmer Ltd.	R-12	low	12	60	Yes	No	No	Yes	Yes

[a] This information has been obtained from the manufacturer's brochures. This table is meant only as a guide to the instruments available and is no substitute for full discussion with spectroscopists and the manufacturers. Those instruments capable of observing the resonance of nuclei other than protons require accessories to do so.

developments cannot be prophesied with certainty, but three particular trends are briefly discussed below (cf. refs. 3, 24, and 29 for other reviews).

1. High magnetic field strengths increase the chemical shift separations (in Hz) and thus are more likely to produce first-order spectra. The resulting simplification in interpretation alone is desirable, but higher field strengths also should lead to smaller sample sizes. Not only is the signal area then distributed through fewer lines of the simplified spectrum, but the inherent sensitivity for a given nucleus theoretically increases with the magnetic field strength, although various practical difficulties often limit the increase. The fields obtainable with iron-core magnets are limited in any case both by magnetic saturation of the metal and by the greatly increased power consumption of high field strength electromagnets. The operating and capital costs rise steeply. However, the advent of superconducting magnets has given hope that practical high-field, high-sensitivity spectrometers will become available. At the present, both the capital and running costs (mainly in connection with the liquid helium system) are high.

2. Double resonance methods, especially those involving the Overhauser effect, lead to very high sensitivity enhancement factors. If one magnetic particle (e.g., an electron) is undergoing a transition from one energy level to another, the populations of the energy levels of a second magnetic particle (e.g., a nucleus) in the same sample are altered. This population alteration is the basis for the Overhauser effect. Even an inversion of the energy populations can occur, resulting in a higher population in the upper energy states than in the lower. If this does occur, the resonance signal of the second magnetic particle is inverted. Enhancement factors exceeding two orders of magnitude have been observed for nuclei in the presence of a free radical whose electron spin resonance is being stimulated extensively. The method so far has not found wide applicability owing to two main considerations: (i) very high microwave power is required to stimulate the electron transitions and (ii) the free radical causes line broadening of the nuclear resonances and, in consequence, loss of spectral detail.

3. Finally, pulse methods (29) promise a large decrease in the experimental times recorded in Table 7-1. The operation of a spectrometer in the conventional way is extremely wasteful of time. Consider a spectrometer sweeping the spectrum by altering the frequency. At any given moment, information is obtained from only one point in the spectrum because the frequency corresponds only to that point. If the entire spectrum could be studied at one time (e.g., by using several frequencies at once in a "multichannel" spectrometer), the data clearly could be acquired at a very much faster rate. It turns out that this can in effect be done using particular pulses of radiofrequency radiation. The resulting spectrum is the Fourier transform of the conventional absorption spectrum which then must be regenerated by suitable computations. For this reason the technique is commonly referred to as Fourier transform spectroscopy. At present the method requires sophisticated equipment, but the large saving in experimental times may stimulate work to simplify the procedure.

IX. Summary

Magnetic resonance spectroscopy is a valuable tool for detecting chemical groups and elucidating their molecular relationships in a sample of which the structure is required. Proton magnetic resonance spectroscopy is feasible in combination with gas chromatography, at the expense of a certain degree of care and, usually, a considerable amount of experimental time. Present developments, however, promise a significant reduction in this time and in the requisite sample size. In exceptional circumstances, even carbon-13 spectroscopy may be possible using isotopically normal gas chromatographic samples.

References

1. J. A. Pople, W. G. Schneider, and H. J. Bernstein, *High Resolution Nuclear Magnetic Resonance*, McGraw-Hill, New York, 1959.
2. J. W. Emsley, J. Feeney, and L. M. Stucliffe, *High Resolution Nuclear Magnetic Resonance Spectroscopy*, Pergamon Press, Oxford, 1965.
3. G. E. Hall, in *Annual Review of N.M.R. Spectroscopy*, E. F. Mooney, Ed., Academic Press, London, 1968, p. 227.
4. P. C. Lauterbur, in *Determination of Organic Structures by Physical Methods*, Vol. II, F. C. Nachod and W. D. Phillips, Eds., Academic Press, New York, 1962, p. 465.
5. (a) R. H. Bible, *Interpretation of N.M.R. Spectra. An Empirical Approach*, Plenum Press, New York, 1965. (b) R. M. Silverstein and G. C. Bassler, *Spectrometric Identification of Organic Compounds*, Wiley, New York, 1963, p. 71.
6. H. A. Szymanski and R. E. Yelin, *N.M.R. Band Handbook*, Plenum Press, New York, 1967.
7. S. Sternhell, *Rev. Pure Appl. Chem.*, **14**, 15 (1964).
8. N. S. Bhacca and D. H. Williams, *Applications of N.M.R. Spectroscopy in Organic Chemistry*, Holden-Day, San Francisco, 1964.
9. (a) J. D. Swalen, in *Progress in N.M.R. Spectroscopy*, Vol. I, J. W. Emsley, J. Feeney, and L. H. Stucliffe, Eds., Pergamon Press, Oxford, 1966, p. 205. (b) R. A. Hoffman, *Pure Appl. Chem.*, **11**, 543 (1965).
10. M. van Gorkom and G. E. Hall, *Quart. Rev.*, **22**, 14 (1968).

11. G. E. Hall, D. Hughes, D. Rae, and A. P. Rhodes, *Tetrahedron Letters*, **1967**, 241.

12. (a) N. S. Bhacca, L. F. Johnson, and J. N. Shoolery, *N.M.R. Spectra Catalog*, National Press, New York; Vol. I, 1962; Vol. II, 1963. (b) *The Sadtler Standard Spectra*, Sadtler Research Laboratories, Philadelphia, Pa.

13. M. G. Howell, A. S. Kende, and J. S. Webb, Eds., *Formula Index to N.M.R. Literature Data*, Plenum Press, New York, Vol. I, 1965; Vol. II, 1966.

14. S. Barcza, *J. Org. Chem.*, **28**, 1914 (1963).

15. G. E. Hall, N. Sheppard, and J. Walker, *J. Chem. Soc.*, C, **1966**, 1371.

16. M. N. G. James and K. J. Watson, *J. Chem. Soc.*, C, **1966**, 1360.

17. J. B. Stothers, *Quart. Rev.*, **19**, 144 (1965).

18. See e.g. B. C. L. Weedon, in *Terpenoids in Plants*, J. B. Pridham, Ed., Academic Press, New York, 1967, p. 119.

19. J. H. Pickett and L. B. Rogers, *Anal. Chem.*, **39**, 1873 (1967).

20. G. W. Flynn and J. D. Baldeschwieler, *J. Chem. Phys.*, **37**, 2907 (1962).

21. L. G. Alexakos and C. D. Cornwell, *J. Chem. Phys.*, **39**, 1616 (1963).

22. R. K. Harris and R. A. Spragg, *Chem. Commun.*, **1967**, 362.

23. S. C. Slaymaker, *Appl. Spectry.*, **21**, 42 (1967).

24. R. E. Lundin, R. H. Elsken, R. A. Flath, and R. Teranishi, *Appl. Spectry. Rev.*, **1**, 131 (1967).

25. D. J. Frost, G. E. Hall, M. J. Green, and J. B. Leane, *Chem. Ind. (London)*, **1967**, 116.

26. R. A. Flath, N. Henderson, R. E. Lundin, and R. Teranishi, *Appl. Spectry.*, **21**, 183 (1967).

27. S. Ito and I. Miura, *Bull. Chem. Soc. Japan*, **38**, 2197 (1965).

28. R. C. Hewitt, *Rev. Sci. Instr.*, **38**, 831 (1967).

29. R. R. Ernst, in *Advances in Magnetic Resonance*, Vol. 2, J. S. Waugh, Ed., Academic Press, New York, 1966, p. 1.

30. J. M. Read and J. H. Goldsteain, *Anal. Chem.*, **37**, 1609 (1965).

31. E. G. Brame, *Anal. Chem.*, **37**, 1183 (1965).

32. K. Witte and O. Dissinger, *J. Chromatog.*, **28**, 413 (1967).

33. M. D. D. Howlett and D. Welti, *Analyst*, **91**, 291 (1966).

34. A. G. Altenau, R. E. Kramer, D. J. McAdoo, and C. Merritt, *J. Gas Chromatog.*, **4**, 96 (1966).

35. I. A. Fowlis and D. Welti, *Analyst*, **92**, 639 (1967).

CHAPTER 8

Gas Chromatography and Thin-Layer Chromatography

Rudolf Kaiser, *Badische Anilin- & Soda-Fabrik A.G.,*
Ludwigshafen am Rhein, Germany

Translated by L. S. Ettre

I. Introduction

A. Principles

Thin-layer chromatography (TLC) is a separation technique as well as an analytical method which is particularly useful in qualitative analysis. It can often supplement gas chromatography, which is generally weak in respect to qualitative analysis. In thin-layer chromatography, the mobile phase consists of volatile liquids and the stationary phase of a thin layer of very fine solid particles (particle diameter: 1–50 μm). The stationary phase may be an adsorbent, a mixture of adsorbents and partition liquids insoluble in the moving phase, or a mixture of inert solid supports and partition liquids. The mobile phase travels through the stationary phase due to capillary forces.

The technique, potentialities, limitations, and conditions of TLC are discussed in detail in a number of excellent monographs (1–4), and for detailed information on the various aspects of the technique, we refer to them.

The requirements of TLC do not differ significantly from those of GC:

Required minimum sample amount:	About 10^{-6} g per component
Analysis time:	About 10–30 min
Separation efficiency:	Comparable to an about 1-m long packed GC column

The basic difference between TLC and gas chromatography is in the mode of detection. In gas chromatography, separation and detection follow each other in a dynamic system. On the other hand, in TLC, the dynamic separation process is stopped and the result of separation is "frozen" prior to detection. Thus, the individual separated compounds are "stored" and available for further investigations.

B. Modes of Detection in TLC

In TLC, the detection is performed almost always with various optical methods such as

(1) Fluorescence excitation

(2) Fluorescence emission

(3) Evaporation of the solvent and spraying the "frozen" fraction with reagents which

(a) dissolve in the fraction but not in the thin layer and therefore lift the fraction spot from the pure layer (e.g., iodine vapor);

(*b*) make the fraction observable by changing it chemically (e.g., destruction with conc. H_2SO_4);

(*c*) react with the fraction, forming compounds which can be detected in UV or visible light; or

(*d*) result in color changes.

It is evident from this short summary that by the proper selection of the specific chemical reactions, whole molecules, certain molecular structures, or constituents (e.g., functional groups, elements, specific structures) can be identified with high reliability.

The disadvantage of such a detection is similar to the disadvantage of the application of substance-specific detectors in GC: sometimes, only unimportant impurities are observed while the main components remain undetected.

C. Warning

The direct coupling of gas chromatography and TLC is very simple and cheap. Practically every commercial gas chromatograph can be modified for combination with TLC in a simple way. The necessary parts of a thin-layer chromatograph are all low-cost items. For this reason, it can be expected that the coupling of the two techniques will find an increasing application in the future.

However, the users of the combined GC–TLC systems should be warned: by their use, results may be obtained which were not expected; similarly, results which have been observed with the existing methods when used alone may be found incorrect. These facts may be very irritating and may require additional work. The author of this chapter, having gone through this experience, wants to caution the readers not to be discouraged.

II. GC–TLC Combinations

There are two possible types of GC–TLC combinations. In the first, the thin layer serves only as an ancillary detector while in the second, the separation power of both chromatographic methods is skilfully utilized.

A. The Thin-Layer Plate as a GC Detector

Due to the possibility of specific detection methods in TLC, the thin-layer plate itself may serve as an additional substance-specific detector for GC. Its utilization as an ancillary detector has the advantage that it is highly specific, but it has the basic shortcoming that it can be used only for fractions which are retarded by the sorbents used in the plate.

In order to utilize a thin-layer plate as an ancillary detector for gas chromatography, the column effluent has to be split and one part of it conducted to the plate. Assuming that there is no condensation in the connecting line and that the sorption power of the thin layer is sufficient, substance i will be sorbed by the layer with sorption efficiency up to 80%.

B. Coupling of the Two Chromatographic Techniques

A much more important GC–TLC combination is represented by the system where the separation efficiency of both techniques is utilized. In this way, for example, one can achieve by TLC an additional separation of the fractions preseparated in the gas chromatograph. This complex system supplies information of much higher degree than could be obtained by either method alone.

III. Principles of the Dynamic Coupling of Gas Chromatography and Thin-Layer Chromatography

Fractions separated in the gas chromatograph are dosed on the "start" line of the thin-layer plate. As shown in Figure 8-1, the column effluent is split into two fractions: a small stream is conducted to the flame ionization detector and the main flow is directed onto the plate. Depending on the analytical problem, one may dose only one peak, or even only part of a peak, or every peak from the column effluent. Correspondingly, the dosing can be performed either stepwise ("point" dosing) or continuously ("line" dosing).

In stepwise dosing, the TLC plate is moved step by step with the help of an appropriate mechanism directed by the gas chromatographic detector. In continuous dosing, the speed of the plate can be selected according to the requirements. Below, we list the advantages and disadvantages of three possible arrangements.

(a) The TLC plate is moving with a constant speed (5).

Advantages: It is easy to compare the GC chromatogram and the TLC chromatogram (both having a linear movement) since the individual peaks can be correlated directly.

Disadvantages: As the chromatogram progresses, the later peaks become broader, particularly in isothermal gas chromatography. This results in a decreasing amount of fraction per unit length of TLC plate start line, and significantly reduces the sensitivity of TLC in this system.

Possible improvement: The GC column is temperature programmed so that the later peaks become sharper, while the TLC plate is still

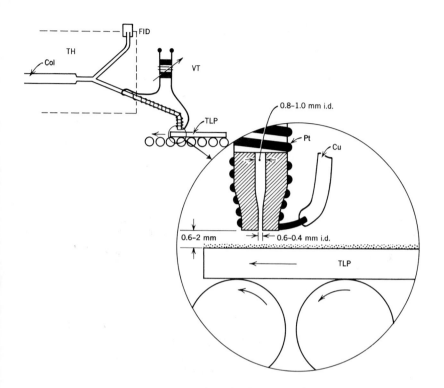

Fig. 8-1. Coupling of the end of the gas chromatograph with the thin-layer plate. If the dosing line is made of a glass capillary (6 mm o.d.; 0.7 mm i.d.), a platinum wire (0.5 mm o.d.) heated by 50 W is sufficient to heat the line up to 200°C. *Col*, separation column; *TH*, GC column thermostat; *FID*, flame ionization detector of the gas chromatograph; *VT*, variable transformer, e.g., 6 A, 20 V; *TLP*, thin layer plate; *Pt*, platinum wire; *Cu*, copper connecting wire.

moved linearly. In this way, the concentration per unit length of plate for these peaks is increased. At the same time (if the temperature program started low enough and proceeded through the whole analysis), spots for members of a homologous series will be evenly spaced on the plate.

(*b*) The gas chromatograph is operated isothermally while the TLC plate is moved logarithmically with time.

Advantages: The spots on the TLC plate will be spaced linearly with the retention index; the molecular weight of members of a homologous series increases linearly (6).

Disadvantages: The instrumentation is more complex.

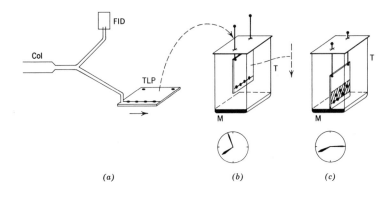

Fig. 8-2. (*a*) Dosing on the thin-layer plate. (*b*) Conditioning of the plate in the vapor chamber. (*c*) TLC separation. *Col*, GC separation column; *FID*, flame ionization detector of the gas chromatograph; *TLP*, thin layer plate; *T*, tank (vapor chamber); *M*, mobile phase.

(*c*) The plate is moved in steps.

Advantages: Trace fractions are collected as dots on the plate, thus increasing their amount per unit length. In this way, one can detect traces which remain undetected by even a very sensitive GC detector.

Disadvantages: The real advantages of direct correlation of the GC chromatogram with the TLC chromatogram are lost. Thus, for example, one can no longer see the substances which might be hidden in the front, middle, or tailing of a GC peak.

When the GC analysis is completed, the TLC plate is removed from the outlet line of the gas chromatograph and placed in the usual saturation chamber where, first, it is conditioned with the vapor of the future mobile phase. Subsequently, the plate is brought into contact with the mobile phase and the TLC separation process is initiated. When finished, the plate is dried, developed, and evaluated in the usual ways (Fig. 8-2).

IV. Instrumental Requirements of GC–TLC Coupling

A. Gas Switching System

Kaiser developed a simple gas switching system which directs the entire gas flow either to the flame ionization detector or to the TLC plate, or divides it in any desired ratio between the detector and the plate. The system is similar to the column switching system of Deans (7).

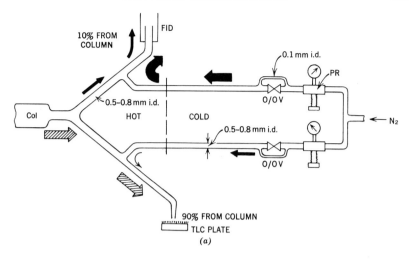

Fig. 8-3. Principles of the gas switching system. (a) 90% of the column effluent is conducted to the TLC plate and 10% to the flame ionization detector (*FID*).

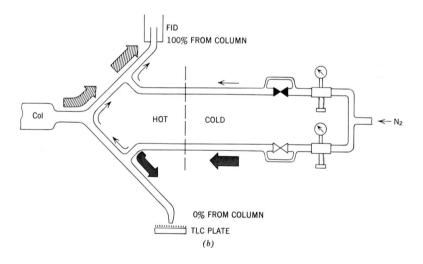

Fig. 8-3. *Continued.* (b) 100% of the column effluent is conducted to the flame ionization detector. Here, the top on–off valve is closed. *Col*, GC separation column, *O/O V* = on–off valve; *PR* = pressure regulator.

The system consists of three capillary tubes and the switching is performed with the help of an ancillary gas flow. Figures 8-3a and b illustrate the operation of this system; its principles can be explained as follows.

If an additional gas flow F is added to an existing flow in a narrow tube, then the pressure drop in that part of the tube will be increased. As a result of this, the pressure drop, and therefore also the gas flow, will decrease in that part of the tube which is before the point where the new flow is added. By proper selection of F the flow upstream of the mixing point can be stopped completely or even reversed. In this way, the added adjustable gas flow behaves like a continuously regulated valve resulting in changes in the pneumatic resistance. With proper adjustment of the auxiliary gas flow, one can divide the original (main) gas flow into any desired ratio or even stop it completely.

B. Requirements Concerning the Connecting Tube ("Dosing Tube")

The connecting tube between the gas chromatograph and the thin-layer plate should be as short as possible and has to be properly heated. There should be no overheated parts ("hot spots") and the temperature should

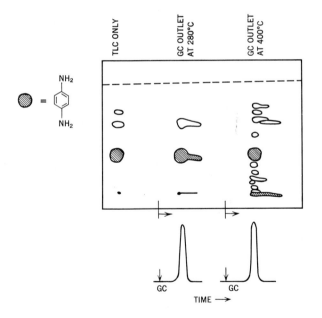

Fig. 8-4. Decomposition in an overheated connecting tube. Substance: p-phenylenediamine. After Kaiser (12).

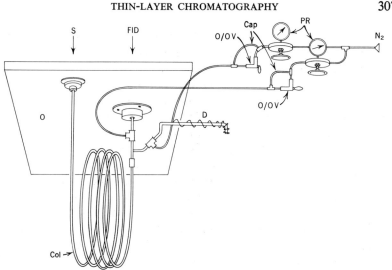

Fig. 8-5. Schematic of a system where the chromatographic separation column as well as the gas switching valve and the dosing tube are mounted on the door of the column oven. *O*, oven door; *S*, sample inlet system; *Col*, separation column; *FID*, flame ionization detector; *D*, dosing tube; *O/O V*, on–off valve; *Cap*, capillary tubing; *PR*, pressure regulator. For more detailed schematic of the gas switching system, see Figure 8-3.

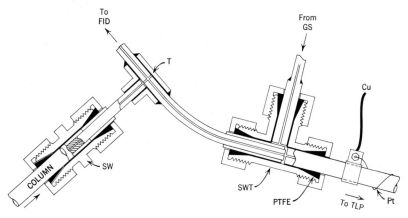

Fig. 8-6. Schematic of the connection of a glass dosing tube with metal components by the use of Swagelok fittings. Teflon, soft copper, or soft aluminum may be used as the material of the ferrule. *T*, no-dead-volume connection tee with inside connecting tubes (0.5 mm i.d.); *GS*, gas switching system; *SW*, Swagelok fitting; *SWT*, Swagelok tee piece; *PTFE*, polytetrafluorethylene; *Pt*, platinum heating wire; *Cu*, copper connecting electrical line; *FID*, flame ionization detector; *TLP*, thin-layer plate.

308 RUDOLF KAISER

be a few degrees higher than that of the GC separation column. If even a short part of the connecting tube is overheated, significant decomposition of certain sample components may be observed (see Fig. 8-4).

It is also important to select the material of the connecting tube with care. According to our experience, the best material is zirconium with a zirconium dioxide layer on the inside surface of the tube. Glass is also suitable, and it is fairly easy to construct the gas switching valve and the dosing tube of glass. However, it is more complicated to connect glass components to the metal parts of the gas chromatograph. The best solution is to mount the sample introduction system, the detector, and the dosing tube on the door or wall of the column oven. Figure 8-5 illustrates such a system. Another possibility is to connect the glass dosing tube to the standard metal tubes of the gas chromatograph with Swagelok fittings (Fig. 8-6). All-glass systems are preferred when the sample is so unstable that it requires the use of glass columns in the gas chromatograph. Figure 8-7 is a schematic of an all-glass system.

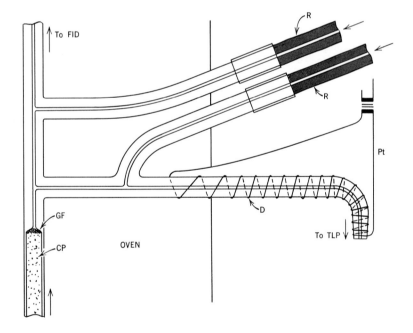

Fig. 8-7. Schematic of an all-glass gas switching and connecting system. *CP*, column packing; *GF*, glass frit; *D*, dosing tube; *TLP*, thin layer plate; *R*, rubber heavy wall tubing connecting the all-glass system to the gas switching system; *Pt*, platinum heating wire.

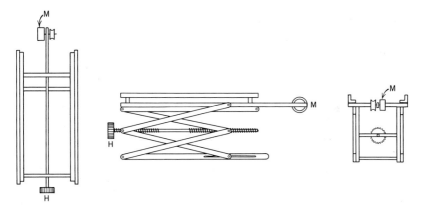

Fig. 8-8. A simple system for the transfer of the thin-layer plate. The plate is moved in horizontal direction by motor *M*. The height of the plate can be adjusted with help of screw *H*. After Kaiser (12).

In most cases, stainless steel capillaries of 0.35–0.50 mm i.d. are suitable for the dosing tube. This tube has to be heated with appropriate methods, e.g., by direct electrical heating. Another very good material for the construction of the dosing tube is a thick-wall gold capillary.

C. Requirements Concerning the Transport Mechanism of the Thin-Layer Plate

Figure 8-8 shows the simplest possible solution; similar systems are also described in the literature (5). In this system, one end of a fine wire is attached to the thin-layer plate and the other end to the circumference of a wheel rotated by a small motor. If the radius of the wheel is *r* (mm) and

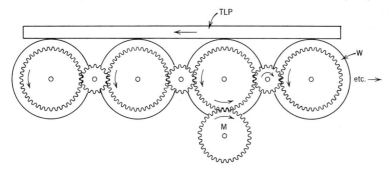

Fig. 8-9. Thin-layer plate (*TLP*) transfer system consisting of cylinders *W* moved by gears driven by motor *M*. After Willhalm (8).

it rotates U times per minute, then the thin-layer plate will travel with a speed of $2\pi rU$ mm/min. A more complicated but better solution, according to Willhalm (8), is shown in Figure 8-9.

V. Special Possibilities of a Combined GC–TLC System

A. Detection of Impurities Originating outside the Sample

It is very important to evaluate whether the spots on the thin-layer plate are related only to the sample or also represent substances not present in the sample. A few examples for such possibilities are stationary phase vapor (bleeding), decomposition products of the stationary phase, or residues from previous analyses which remained in the gas chromatographic system. In temperature programming, one should always realize that the amount of stationary phase bleeding will increase with time (temperature). In order to be able to evaluate whether a spot is related to stationary phase bleeding, one should also apply a small amount of stationary phase solution to the starting point of the thin-layer plate. Figure 8-10 shows some typical results which are related to sources other than the sample.

Generally, if the thin-layer plate indicates a substance immediately at the start of the analysis, it must be related to sources other than the sample. If it came from the sample, the first spot would be observed only after the gas holdup time of the GC system (+ dosing tube) had elapsed.

The information obtained in this way in a GC–TLC system concerning the impurities coming from sources other than the sample itself is also of general importance for other ancillary techniques. Impurities which can be detected here would generally result in a high background in the combination of gas chromatography and mass spectrometry. In preparative chromatography, they would reduce the purity of the separated fractions and make the additional investigations more difficult.

B. Special Information which can be Obtained from the Combination of Gas Chromatography and TLC

Besides improved component identification, the combination of the two chromatographic methods also gives additional information concerning the efficiency of the individual methods. Here, we briefly summarize a few such possibilities and illustrate them with the help of the gas chromatogram as well as the picture of the TLC plate. On the plate, the longitudinal axis

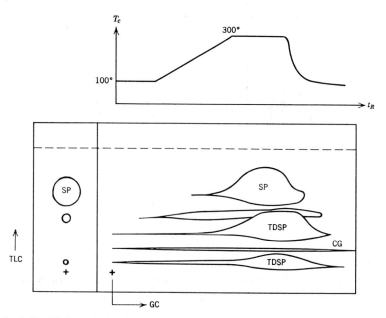

Fig. 8-10. TLC analysis obtained in the investigation of the bleeding and de-composition of the stationary phase of the gas chromatographic column. *SP*, volatile stationary phase; *TDSP*, thermal oxidation decomposition product of the stationary phase; *CG*, impurities present in the carrier gas; T_c, column temperature; t_R, retention time.

corresponds to the movement of the plate while the vertical axis represents the movement of the individual components on the plate during development.

1. Was the GC Separation Complete?

If the GC separation was incomplete, a specific GC peak will give more than one spot. Figure 8-11 illustrates such a case; here, the second and third peaks contained four and two components, respectively. If such a result is obtained, the gas chromatographic separation column should be modified, if possible, until it separates all the components shown on the original TLC plate. Figure 8-12, for example, shows such a case. Figure 8-12*a* is the result of the first, insufficient analysis: the plate shows four spots while there were only three peaks in the gas chromatogram. By proper selection of the GC separation column, the result shown in Figure 8-12*b* was obtained. As seen, it was established that the third dot on the first plate actually consisted of two substances, while the black dot on the first plate had another small impurity in it.

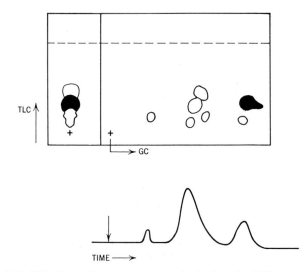

Fig. 8-11. Thin-layer chromatogram obtained when the GC separations are insufficient.

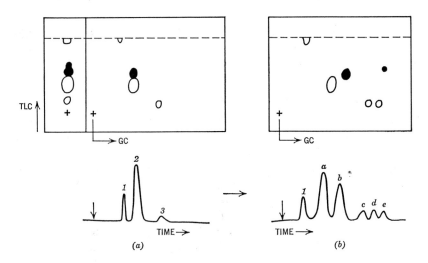

Fig. 8-12. Insufficient and sufficient separation. (a) Gas chromatographic column with poor separation, (b) Gas chromatographic column with the proper separation efficiency. Peak 2 actually consisted of peaks a, b, and e, and peak 3 of peaks c and d.

2. Is Every Original Sample Component Present in the Column Effluent?

The combination of gas chromatography and TLC provides an easy way to investigate whether some of the sample components are lost somewhere in the gas chromatograph. Independent of the combined GC–TLC analysis, one should also apply the original (unseparated) sample to the starting point of the plate. If more spots are obtained here than in the analysis of the individual fractions, then some of the components are lost. Figure 8-13 illustrates this case; the original sample was applied to the left side of the plate.

3. Are the Sample Components Changed Chemically in the System?

The GC–TLC combination is convenient to determine whether the fractions in the column effluent are identical to the original sample components or are the result of chemical changes in the system. Here, again, we compare the thin-layer chromatogram of the original sample with the sum of the individual TLC spots obtained from the column effluent. If the original sample showed spots different from that obtained in the combined GC–TLC system, then the sample has undergone chemical changes somewhere in the system. We may have two cases:

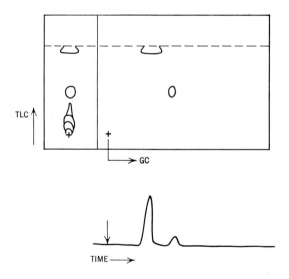

Fig. 8-13. Thin-layer chromatogram showing that part of the sample is lost in the gas chromatograph.

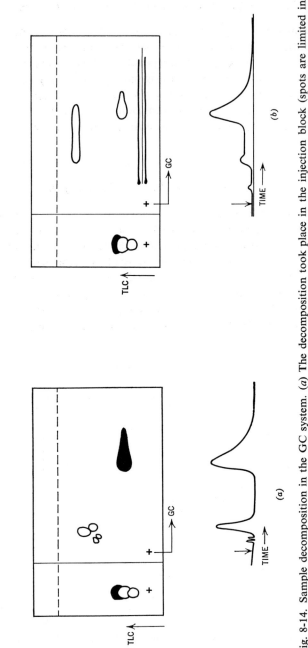

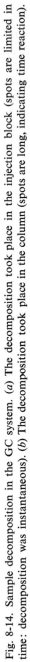

Fig. 8-14. Sample decomposition in the GC system. (a) The decomposition took place in the injection block (spots are limited in time: decomposition was instantaneous). (b) The decomposition took place in the column (spots are long, indicating time reaction).

(*a*) The sample decomposed in the injection block. As a result, the plate shows spots not present in the original sample. These spots are sharp since the decomposition took place instantaneously (see Fig. 8-14*a*).

(*b*) The sample decomposed in the column. As a result, the plate shows spots not present in the original sample, but now the spots are spread because the decomposition is not instantaneous (see Fig. 8-14*b*).

4. Does the TLC Plate Show All the Components Present in the Column Effluent?

If for any reason, some of the fractions in the column effluent do not get to the starting line of the TLC plate, then at certain time intervals which can be established easily from the GC chromatogram, spots will be missing. This is indicated in Figure 8-15, where there is no spot which would correspond to the second and fourth peaks.

There are two reasons why a sample component might not give a spot on the plate even though it resulted in a peak in the GC chromatogram: either the sorbent in the plate is not adequate or the fraction is decomcomposed before reaching the plate, through reactions with air, light, or moisture present in the air.

This method then establishes whether the result of the TLC plate corresponds to the gas chromatogram. Another possibility is to extract the

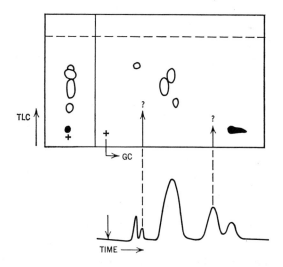

Fig. 8-15. Thin-layer chromatogram obtained when certain fractions do not get onto the plate.

sample components from the plate (one can expect yield up to 70%) and analyze their mixture again in the gas chromatograph. If the gas chromatogram differs from the one obtained originally, the sample had to suffer changes either between the gas chromatograph and the TLC plate, or on the plate.

C. Enrichment of Trace Components on the TLC Plate

In the complex GC–TLC system, trace components of the sample can be enriched so that they can be distinguished better on the plate than in the GC chromatogram. There are two ways to achieve this:

(*1*) The motion of the plate is stopped during the time the fraction elutes from the column. In this case, one should be careful to identify the spots which precede or follow the main peak because the GC chromatogram may not indicate their presence. Figure 8-16 illustrates such a case.

(*2*) One may analyze repetitively the same sample in the gas chromatograph and adjust the step-by-step movement of the TLC plate so that the components always reach the plate at the same place. In some cases, one may dose two closely spaced trace impurities at the same starting point of the plate; this may improve the enrichment of the two components on the plate (particularly if they are closely spaced in the gas chromatogram) but,

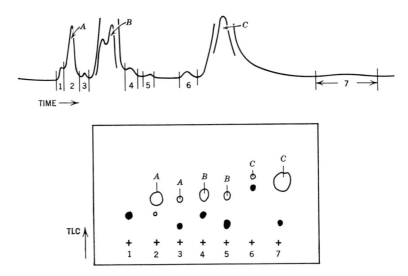

Fig. 8-16. Trace analysis in a GC–TLC system. The movement of the plate is stopped during the emergence of a fraction from the column. Dark spots on the plate represent the trace impurities. After Kaiser (12).

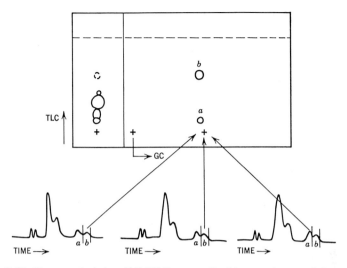

Fig. 8-17. Trace analysis in a GC–TLC system. In this case, the sample is analyzed repetitively in the system and peaks *a* and *b* are condensed at the same place of the plate; the plate does not move during the emergence of these fractions. The left side of the plate shows the direct analysis of the original sample on the TLC plate. As seen, impurity *b* cannot be seen well until after repetitive analyses.

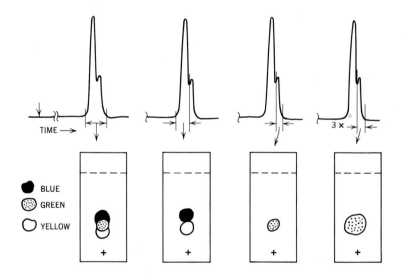

Fig. 8-18. Trace analysis with the cut technique. Various parts of the column effluent are condensed on a nonmoving plate in order to establish the presence of a trace component hidden in the tail of the GC peak.

of course, makes the relative identification of the two closely spaced GC peaks difficult or impossible. Figure 8-17 illustrates such a case.

A very special method in trace analysis is illustrated in Figure 8-18. Here, the first analysis of the sample in the GC–TLC system resulted in three spots, one blue, one green, and one yellow. It could not be established with certainty whether the middle (green) spot corresponded to a third component which could not be seen in the GC chromatogram, or simply resulted from mixing the colors blue and yellow. Therefore, various cuts from the original GC fraction were condensed on the plate in repetitive analyses. These investigations clearly revealed that the green spot indeed corresponded to a trace impurity hidden in the tail of the composite peak.

VI. Indirect Coupling of Gas Chromatography and TLC

In the previous discussions, we assumed that the two chromatographic techniques are coupled directly in a continuous system where the column effluent flow is continuously conducted to the plate and the individual fractions present in the effluent are sorbed by the plate. There is also a possibility for a discontinuous system, where the individual GC fractions are condensed in some way and are analyzed subsequently by TLC, usually while in the form of a solution. Of course, this method depends on the efficiency of the collection system. Assuming a good collection yield, this method might have certain advantages, particularly in the analysis of trace components which may be condensed from repetitive analyses in the same trap.

A recent method of Witte and Dissinger (9,10) represents a definite improvement in fraction collection for subsequent TLC analysis. A drop (up to 2 µl) of solvent, usually carbon tetrachloride, is brought into a glass capillary U-tube (5–7 mm o.d., 1–3 mm i.d.) and the capillary is immersed in liquid nitrogen. In this way, the original liquid droplet will become crystalline and porous. For convenience, a small metal capillary tube is sealed into one end of the glass capillary. When a peak starts to emerge from the column (as indicated by the GC detector connected parallel to the collection outlet), the metal capillary is inserted into the heated collection outlet of the gas chromatograph. A silicone rubber or Teflon tube permits a gastight seal at the outlet during condensation. During this operation, the lower part of the glass capillary U-tube where the solvent droplet is located is immersed in liquid nitrogen. The porous crystalline solvent droplet represents a pressure drop equivalent to about a 0.5–1-m long standard packed column. The fraction eluting from the column is

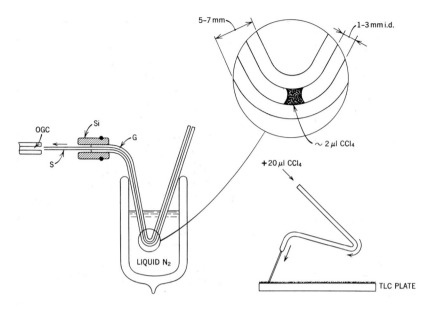

Fig. 8-19. Micropreparative collection system and its application, according to Witte and Dissinger (9,10). *OGC*, outlet of the gas chromatograph; *S*, metal capillary sealed into the glass capillary *G; Si*, silicone rubber tube.

generally completely adsorbed on the surface of the porous crystalline solvent droplet. After the peak has emerged, the capillary tube is removed from the outlet of the gas chromatograph and stored in liquid nitrogen. After the gas chromatographic analysis is completed, each of the individual fractions can be analyzed separately on thin-layer plates. The tube is removed from the liquid nitrogen bath, warmed up, and washed with about 20 μl of solvent onto the starting point of the plate. Figure 8-19 illustrates this type of fraction collector and its application.

The basic disadvantages of the discontinuous system are that the direct relationship between the GC chromatogram and the TLC plate cannot be established and that trace compounds undetected by the GC detector may also be undetected by TLC and thus escape the attention of the analyst. Also, a discontinuous method requires at least one preliminary GC analysis of the sample in order that the collection program may be established for the desired GC peaks.

VII. Typical Examples for the Application of GC–TLC Systems

Finally, we give two examples for the practical application of combined GC–TLC systems (with direct coupling).

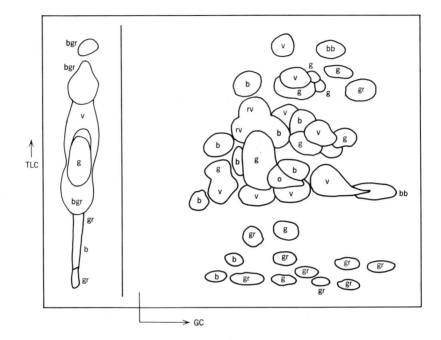

Fig. 8-20. Analysis of a xylenol fraction in a combined GC–TLC system. The spots on the thin-layer plate, after visual development with dibromoquinonechloroimide, were exposed first to ammonia and then to HCl vapors. After Kaiser (11,12). *v*, violet; *rv*, red violet; *bb*, bright blue; *b*, blue; *g*, yellow; *gr*, grey; *o*, orange. Silica gel GF with chloroform.

Figure 8-20 is from the work of Kaiser (11,12). The sample analyzed was a very close distillation fraction of xylenols in which the sulfur-containing components plus the other nonphenolic sample components had to be identified. It is evident that neither gas chromatography nor the use of TLC alone (left side of the plate) could permit the proper solution of this problem. On the other hand, the combined use of GC and TLC permitted the identification of over 42 components. This greatly facilitates further analysis by IR, GC–MS, or other available ancillary techniques.

Figure 8-21 is an example from a recent publication of Winter et al. (13) and illustrates a portion of the analysis of a coffee aroma sample in a combined GC–TLC system. The GC separation column contained a nonpolar silicone oil liquid phase; thus, in temperature-programmed analysis, the retention times are linear with the retention indices and therefore the abscissa of the chromatogram can be divided into retention index units. This is a part of the chromatogram obtained illustrating the retention

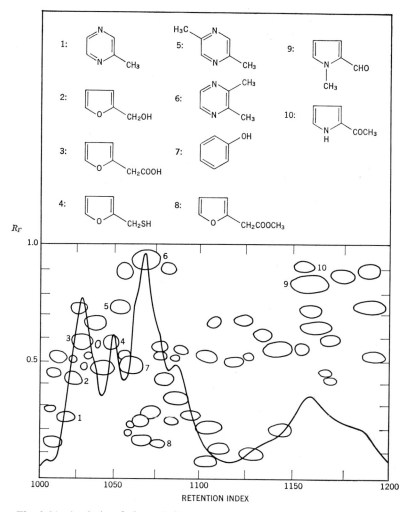

Fig. 8-21. Analysis of the volatile aroma substances of coffee in a combined GC–TLC system, according to Winter et al. (13). The GC separation column contained a nonpolar liquid phase and was temperature programmed. The figure shows the 1000–1200 retention index region and the picture of the TLC plate is superimposed on the gas chromatogram. Ten spots are identified.

index range between 1000 and 1200; the picture of the TLC plate is superimposed on the gas chromatogram. The TLC plate consisted of kieselgel and the moving phase was a 5:1 mixture of benzene and diethyl ether. The spots were visualized with anisaldehyde and sulfuric acid by heating the plate for 10 min at 100°C. The GC chromatogram shows in this range about 12 peaks and shoulders; on the other hand, the TLC plate reveals the presence of over 60 different substances, 10 of which are identified on the top of the figure.

VIII. Conclusions

The combined application of GC and TLC results, in many cases, in additional information not available from either technique used alone. The coupling of the two techniques is fairly simple.

The advantages of the direct combination of gas chromatography and thin-layer chromatography can be summarized as follows:

(1) By comparing the results of separate TLC and GC analyses with the results obtained in a combined GC–TLC system, the completeness of the individual analyses can easily be evaluated.

(2) Differences in the three results point to inadequacy in the analysis. The comparative results permit an easy correction of the inadequacy.

(3) The substances corresponding to the individual gas chromatographic peaks are automatically collected, in this way permitting subsequent, particularly chemical, investigations.

(4) The thin-layer plate coupled with the gas chromatograph can be considered as a special qualitative detector which in many cases can be made highly specific.

(5) The instrumental setup and its application are very simple and result in very important information.

In certain cases, the combined GC–TLC technique provides decisive information for the qualitative identification of the individual sample components. However, it should always be kept in mind that although the TLC analysis might show whether or not the GC separation was complete, it is still not an absolute method and inadequacies of the GC analysis might still remain undetected.

In the interpretation of the results with a combined GC–TLC system, one should remember that (1) the TLC analysis is very specific and trace impurities which are unimportant might give more dominant spots than important sample components; (2) it is quite possible that the TLC method will be "blind" to the important components; (3) it is possible that somewhere in the system, the components of the sample undergo chemical changes; (4) it is more difficult to reproduce retention times in TLC and they are much more influenced by the presence of other substances than in gas chromatography; and (5) if the number of spots on the plate is identical with the number of peaks in the GC chromatogram, it still does not prove that no additional substances are present.

References

1. E. Stahl, *Thin-Layer Chromatography*, Academic Press, New York, 1965, 553 pp; 2nd ed. (presently available only in German), Springer, Berlin, 1967, 979 pp.
2. K. Randerath, *Thin-Layer Chromatography*, 2nd ed., Academic Press, New York, 1966, 285 pp.
3. J. G. Kirchner, *Thin-Layer Chromatography* (*Technique of Organic Analysis*, Vol 12), Interscience, New York, 1967, 788 pp.
4. A. A. Akhem and A. I. Kuznetsova, *Thin-Layer Chromatography*, Davey, New York, 1965 (translation from Russian).
5. J. Janák, *J. Chromatog.*, **15**, 15 (1964).
6. J. Janák, I. Klimes, and K. Hana, *J. Chromatog.*, **18**, 270 (1965).
7. D. R. Deans, *Chromatographia*, **1**, 18 (1968).
8. B. W. Willhalm, *Symposium on the Combination of Gas Chromatography and Mass Spectrometry*, Zürich, February 24, 1968.
9. K. Witte and O. Dissinger, *J. Chromatog.*, **28**, 413 (1967).
10. K. Witte and O. Dissinger, *Z. Anal. Chem.* **236**, 119 (1968).
11. R. Kaiser, *Z. Anal. Chem.*, **205**, 284 (1964).
12. R. Kaiser, *Chromatographie in der Gasphase, IV. Quantitative Auswertung*, Bibliographisches Institut, Mannheim, 1965, 278 pp.
13. M. Winter, F. Gautschi, I. Flament, B. Willhalm, and M. Stoll, in *Aroma und Geschmacksstoffe in Lebensmitteln,* Forster Verlag, Zürich, 1967, pp. 165–198.

CHAPTER 9

Chemical Identification of Gas Chromatographic Fractions

Charles Merritt, Jr., *Pioneering Research Laboratories, U.S. Army Natick Laboratories, Natick, Massachusetts*

I. Introduction

In the preceding chapters the use of a number of elegant ancillary instruments for qualitative analysis of gas chromatographic eluates has been described. The success of many of these methods, especially those in which the gas chromatograph is directly coupled to the ancillary device, is spectacular. Nevertheless, the cost of mass spectrometers, infrared spectrophotometers, hydrogenation reactors, etc., is high, and often specialized knowledge is required to be able to interpret the data obtained. In appropriate circumstances it is also possible to perform qualitative analysis of the gas chromatographic eluates by using the information obtained from simple chemical tests and the retention volume data provided by the gas chromatogram. This chapter is concerned with the use of functional group classification reagents to provide a direct, rapid, and inexpensive method of qualitative gas chromatographic analysis.

The inability of gas chromatography to provide complete qualitative analysis from retention volume data alone is due primarily to the inability of the gas chromatograph to determine or distinguish organic functionality (1). The fact that certain members of different homologous series of compounds may have identical retention characteristics thus prevents the optimum use of the gas chromatograph for qualitative analysis of heterologous mixtures. Retention volume data alone from a single column without a knowledge of compound functionality, do little more than indicate several possibilities for the identity of an individual compound of given retention volume. If the functionality can be established, however, the identity of a particular compound can then be determined from its retention volume.

The first application of classification reagents in gas chromatography appears to have been by Dubois and Monkman (2), who used them primarily as confirmatory tests to establish the presence of various types of compounds found in the analysis of solvents or solvent mixtures. Other workers (3–7) have from time to time also used reagents in a specific way to identify or classify components of column effluents, but the most efficacious use of reagents is that which couples the retention volume data obtained from the chromatogram with the classification information obtained by means of the reagents (8). This procedure provides the analyst with a means for systematic identification of column eluates.

II. Compound Identification

A. The Use of Retention Volume vs. Carbon Number Plots

The linear relationship between retention volume (if elution temperature is programmed) or the logarithm of retention volume (if elution is isothermal) and the carbon number of a member of an homologous series is the basis for the method described here. The situation is illustrated in Figure 9-1. The plots of adjusted retention volume vs. carbon number for several homologous series of functional group compounds are shown. Two compounds having the same retention time are represented by the dashed line. They could not be identified from retention data alone. If the functionality is known, interpolation on the corresponding retention volume vs. carbon number plot then permits identification of the compound as a three-carbon member of the normal thiol series, i.e., propanethiol, or the six-carbon member of the normal alkane series, i.e., normal hexane.

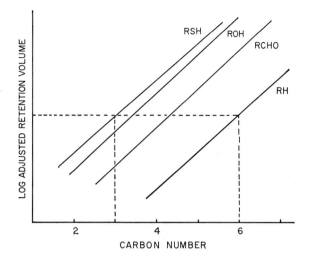

Fig. 9-1. Graph showing the adjusted retention volume vs. carbon number relationship for a set of homologous series of compounds of various functional group types. R = normal alkyl group.

B. The Use of Retention Volume Plots in Combination with Classification Reactions

An example of the analysis of a separated mixture of compounds having different functionalities is shown in Figures 9-2 and 9-3. The chromatogram is shown in Figure 9-2. Each of the peaks, as it is eluted from the column, is passed through a set of classification reagents and its functionality is determined by the resulting change in one of the reagents. The individual compounds are then identified by comparing the adjusted retention volume of the eluted peak with the corresponding adjusted retention volumes of the individual members of the appropriate homologous series.

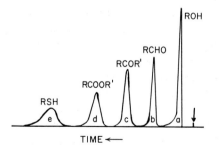

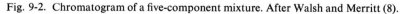

Fig. 9-2. Chromatogram of a five-component mixture. After Walsh and Merritt (8).

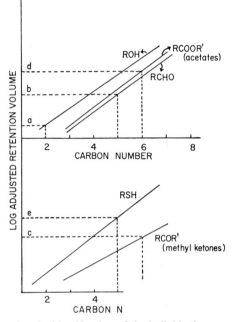

Fig. 9-3. Graph showing the identification of the individual compounds corresponding to peaks shown in Figure 9-2. After Walsh and Merritt (8).

This may be accomplished most conveniently by a graphical procedure. As shown in Figure 9-3, a plot of the logarithm of the adjusted retention volume vs. carbon number is prepared for each homologous series which has been shown to be present by means of the functional group tests. Interpolation on this graph then permits identification of the various components as individual compounds. The alcohol is accordingly identified as ethanol, the aldehyde as pentanal, the ketone as 2-hexanone, the ester as butyl acetate, and the mercaptan as pentanethiol.

Frequently, in the analysis of unknown mixtures, neither the types of components nor the total number of components is known, and the chromatogram of such a mixture may show peaks which result from two or more components having the same retention time. With the aid of functional group classification reagents, this situation can be resolved quite readily. This is illustrated by means of the chromatogram shown in Figure 9-4. The chromatogram appears to be that of a mixture having five components. Functional group tests show, however, that the first peak, *a*, is due to an alcohol and a ketone; the second, *b*, a ketone and an ester; the third, *c*, a mercaptan; the fourth, *d*, an aldehyde, an alcohol, and an

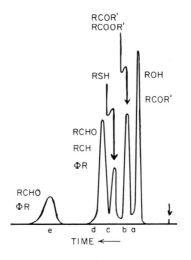

Fig. 9-4. Chromatogram of a partly resolved ten-component mixture. After Walsh and Merritt (8).

aromatic hydrocarbon; and the last, *e*, an aldehyde and an aromatic hydrocarbon. Ten components, therefore, are present rather than the five indicated by the chromatogram. The individual compounds may be subsequently identified from the appropriate log adjusted retention volume vs. carbon number plots as shown in Figure. 9-5.

Three special situations which require further clarification arise in the use of plots of log adjusted retention volume vs. carbon number. The first case arises when the same reagent is used to detect more than one type of functional class. For example, the LeRosen test may be used to detect both aromatic and aliphatic unsaturation. Little difficulty arises, however, in identifying the individual components from retention volume once the compound has been classified as either aromatic or aliphatic unsaturated. This is illustrated in Figure 9-6 where the log adjusted retention volume vs. carbon number plots for both alkylbenzenes and olefins are shown. The log adjusted retention volume for a given compound will intersect the plot corresponding to an integral carbon number only for the appropriate compound. In this example, a compound having the retention volume shown must be toluene and could not be octene.

A second situation arises from the fact that the classification tests usually fail to distinguish among isomers within the functional group class. Fortunately, however, log adjusted retention volume vs. carbon number plots are valid for any given series of isomers within a functional

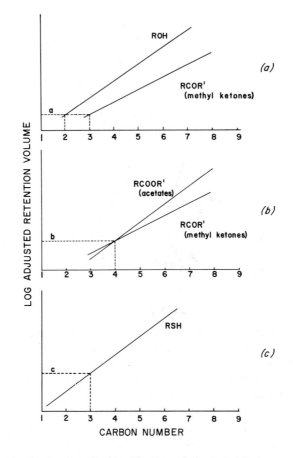

Fig. 9-5. Graph showing the identification of the individual compounds corresponding to the peaks in Figure 9-4. After Walsh and Merritt (8). (a) Ethanol + 2-propanone; (b) ethyl acetate + 2-butanone; (c) propanethiol.

class. Thus, one can obtain suitable plots for groupings such as normal alcohols, iso-alcohols, or secondary alcohols; methyl ketones or ethyl ketones; primary or secondary amines, etc. The application of this principle is illustrated in Figure 9-7. The log adjusted retention volume vs. carbon number plot is shown for both the normal alcohols and the isomeric alcohols. The graph indicates that the alcohol having the retention volume shown would have to be isobutyl alcohol, and could not be normal butyl alcohol or normal propyl alcohol.

Finally, when two classification reagents respond for a given peak, the question arises whether the peak is due to two components or a bifunc-

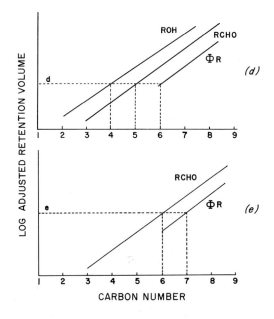

Fig. 9-5 *Continued.* (*d*) butanol + pentanal + benzene; (*e*) hexanal + toluene.

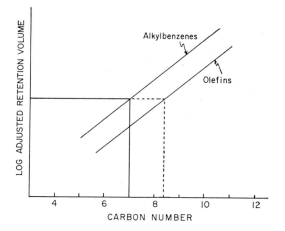

Fig. 9-6. Graph illustrating the identification of a peak showing a positive LeRosen reaction as toluene. After Walsh and Merritt (8).

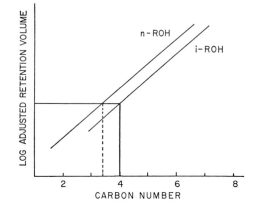

Fig. 9-7. Graph illustrating the identification of a peak showing a positive nitrochromic acid reaction as 2-methylpropanol. After Walsh and Merritt (8).

tional compound. For example, if positive results are obtained from the reaction of an eluent peak with both Schiff's and the LeRosen reagents, the possibility of whether the peak is due to an unsaturated aldehyde or a combination of an olefin and a saturated aldehyde must be considered. Figure 9-8 shows, however, from the plots for the corresponding saturated aldehydes, unsaturated aldehydes, and olefins, that the compound having the retention volume shown would have to be pentenal, the unsaturated five-carbon aldehyde, and could not be a mixture of an olefin and a

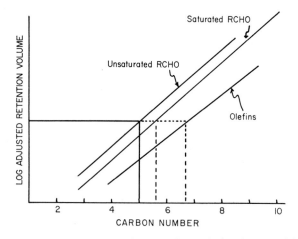

Fig. 9-8. Graph illustrating the identification of a peak showing a positive LeRosen and Schiff's reactions as pentenal. After Walsh and Merritt (8).

saturated aldehyde. Of course, many problems may occur in longer chained branched systems.

III. Experimental

A. Apparatus and Procedure

The use of classification reagents with conventional packed column GC apparatus with a thermal conductivity detector may be easily accomplished in the manner described by Walsh and Merritt (8). The effluent from the gas chromatograph is passed through a stream-splitting device, as shown in Figure 9-9, which may be attached to the exit tube from the thermal conductivity cell. The splitter is constructed of a piece of 0.25-in. o.d. stainless steel tubing, with a rubber serum cap with five hypodermic needles of the same gauge inserted through the cap and into the open end of the stainless tubing. The chromatographic effluent is divided thereby into five equal streams, each of which is allowed to bubble through a vial containing an appropriate classification reagent.

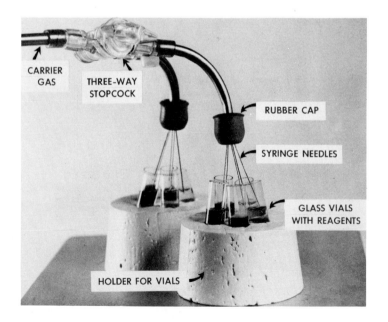

Fig. 9-9. Stream-splitting device for functional group testing of eluted peaks. After Walsh and Merritt (8).

334   CHARLES MERRITT, JR.

The number of individual effluent streams in the splitter and the number of reagents employed, will depend, of course, on the number of functional group compounds expected to be found in the mixture. A five-way split of the chromatographic effluent, however, has been found most convenient. When a larger number of reagents is used, an even and continuous flow in all the vials may not be obtained and handling may become rather cumbersome. It is generally found more convenient, if more reagents are required, to run a second sample with another set of reagents. Another factor which affects the choice of the number of functional group tests to be performed at one time is the sensitivity of the reagent. It is undesirable to decrease excessively the amount of component bubbling through the reagent by using more than four or five splitters.

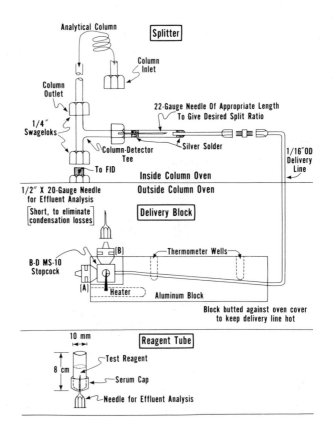

Fig. 9-10. Schematic of a stream splitter and testing apparatus for flame ionization detectors. After Hoffmann and Evans (9).

As each chromatographic peak passes, it is split and allowed to enter the set of the various reagents. To make a rapid change in the stream splitter and reagent vials, when successive peaks are closely adjacent, the exit tube is provided with a three-way stopcock so that as soon as one peak has passed, the stopcock may be turned and the next peak passes through a second splitter. While the second peak is passing, a new set of reagent vials may be positioned to receive the succeeding peak.

When a flame ionization detector is used with the gas chromatograph the column effluent is combusted in the detector. If classification tests are to be performed, a stream splitter is required upstream from the detector. Such a device has been described by Hoffmann and Evans (9) and a schematic diagram of it is reproduced in Figure 9-10.

Another ingenious device for conducting spot tests on column eluates is the automatic reagent-impregnated thin-layer strip tester described by Casu and Cavallotti (10). A drawing of the assembly is shown in Figure 9-11. Although this device affords nearly automatic and synchronized testing of the eluting components, it does not allow for simultaneous testing of the effluent with several reagents. The chromatogram must be repeated employing a strip impregnated with a different reagent for each

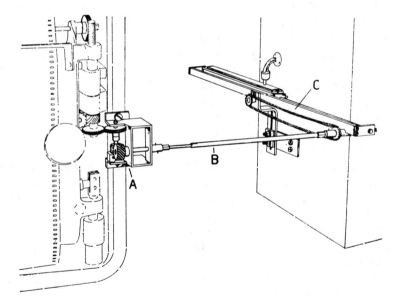

Fig. 9-11. Schematic of a synchronous automatic testing device. After Casu and Cavallotti (10). *A*, gear system attached to recorder chart drive; *B*, transmission shaft; *C*, slide support for thin-layer strip.

pass. Casu and Cavallotti have used two strips cut in half longitudinally, but to employ a greater number of strips would be cumbersome without modifying the delivery procedure for the eluate.

B. Reagents

1. Reagents Used in Group Classification

Several reagents which have been employed to test for the functional group classes are summarized in Table 9-1. The sensitivities of the various reagents have all been determined experimentally by Walsh and Merritt (8). These values are those reported in their original paper. Subsequent reports (10,11) have indicated that the first estimates were quite conservative, and sensitivities of at least one order of magnitude better may be expected.

Each of the classification reagents listed in Table 9-1 was selected if 0.5 mg of a pure component would give a positive test. However, this does not represent the limit of sensitivity, since in many instances the reagent will detect smaller amounts. The approximate minimum amount that could be detected by a given reagent was determined by injecting a known amount of a typical component onto the column, and after splitting, observing the response of the reagent. The values for minimum detectable amounts given in the table correspond to the estimated amount in the effluent stream passing through the reagent and represent approximately one-fifth of the amount of sample injected on the column. Since the smallest sample which could be conveniently measured was 0.1 μl, the maximum sensitivity of any of the reagents was taken as 20 μg. Some of the reagents are actually even more sensitive, as indicated by a strong response of the reagent at the 20-μg level.

The sensitivity tests were made with appropriately chosen compounds containing two or three carbon atoms. The sensitivity of the reagents decreases somewhat with increasing molecular weight of a component. All the reagents were checked for response over a fairly wide range of carbon number within each homologous series, and the response was generally found to be adequate. The compounds tested are also given in Table 9-1.

The usual cautions must be observed, of course, in the general application and interpretation of the color tests because of the variations in sensitivity, unusual responses of the reagents, etc. In the selection and application of the various reagents, the worker should consult freely with standard references on qualitative functional group reagents (12,13).

TABLE 9-1

Functional Group Classification Tests

Compound type	Reagent	Type of positive test	Minimum Detectable Amount μg	Compounds tested
Alcohols	$K_2Cr_2O_7$–HNO_3	Blue color	20	C_1–C_8
	Ceric nitrate	Amber color	100	C_1–C_8
Aldehydes	2,4-DNP	Yellow ppt.	20	C_1–C_6
	Schiff's	Pink color	50	C_1–C_6
Ketones	2,4-DNP	Yellow ppt.	20	C_3–C_8 (methyl ketones)
Esters	Ferric hydroxamate	Red color	40	C_1–C_5 acetates
Mercaptans	Sodium nitroprusside	Red color	50	C_1–C_9
	Isatin	Green color	100	C_1–C_9
	Pb(OAc)$_2$	Yellow ppt.	100	C_1–C_9
Sulfides	Sodium nitroprusside	Red color	50	C_2–C_{12}
Disulfides	Sodium nitroprusside	Red color	50	C_2–C_6
	Isatin	Green color	100	C_2–C_6
Amines	Hinsberg	Orange color	100	C_1–C_4
	Sodium nitroprusside	Red color, 1°; Blue color, 2°	50	C_1–C_4 diethyl and diamyl
Nitriles	Ferric hydroxamate–propylene glycol	Red color	40	C_2–C_5
Aromatics	HCHO–H_2SO_4	Red-wine color	20	ϕH–ϕC_4
Aliphatic unsaturation	HCHO–H_2SO_4	Red-wine color	40	C_2–C_8
Alkyl halide	Alc. AgNO$_3$	White ppt.	20	C_1–C_5

337

The reagents used will depend on the types of compounds expected and on the desired level of response of the reagent. For present purposes, nine functional groups are considered: alcohols, aldehydes, ketones, esters, unsaturated aliphatic and aromatic hydrocarbons, amines, alkyl halides, sulfur compounds, and nitriles.

The potassium dichromate–nitric acid (nitrochromic acid) test is found to be most suitable for alcohols. This reagent has proved to be more sensitive than the more conventional ceric nitrate test. However, the nitrochromic acid reagent fails to respond to tertiary alcohols, for which ceric nitrate reagent may be used.

In some cases, the response of two or more reagents must be considered in combination. For example, since both aldehydes and ketones give a positive reaction with 2,4-dinitrophenylhydrazine, Schiff's reagent was used to distinguish aldehydes from ketones. Mercaptans, sulfides, and disulfides all give a positive test with sodium nitroprusside. This reagent, therefore, may be used as a general test for the presence of these sulfur compounds. If one wishes to distinguish among mercaptans, sulfides, and disulfides, lead acetate may be used to detect mercaptans, and the isatin test may be used to distinguish between sulfides and disulfides.

The ferric hydroxamate test is used for the classification of both esters and nitriles, but nitriles may be distinguished from esters, since the development of the color for nitriles must take place in propylene glycol solution. In either case, the color reaction is not given immediately, as the effluent stream passes through the reagent, but must be subsequently developed by the addition of ferric chloride.

Similarly, the formaldehyde–sulfuric acid reagent is used to detect both aromatic compounds and aliphatic unsaturation. However, these types may be differentiated on the basis of retention time, as described above, once the presence of either type is established.

Primary and secondary amines can be detected by either the Hinsberg reaction or by the Rimini and Simon test with sodium nitroprusside. The sodium nitroprusside test for amines employs different reaction conditions than the sodium nitroprusside test for sulfur compounds. It is this difference in reaction conditions which permits the identification of an amine or a sulfur compound with the same reagent. The sodium nitroprusside reagent is preferable to benzenesulfonyl chloride used in the Hinsberg reaction, since the color reaction immediately distinguishes between a primary and a secondary amine. On the other hand, tertiary amines may be distinguished from primary and secondary amines with benzenesulfonyl chloride if modified reaction conditions (12) are employed.

2. Preparation of Reagents

The reagents used in the test vials may be prepared as follows.

Alcohols. *Nitrochromic acid* (12): 10 drops (ca. 0.5 ml) of $7.5N$ HNO_3 plus 1 drop of 1% $K_2Cr_2O_7$. Turns from bright yellow to blue-gray. Good for primary and secondary alcohols.

Ceric nitrate (21,13): 5 drops of reagent plus 5 drops of H_2O. Yellow to amber. Good for all aliphatic alcohols.

Aldehydes. *Dinitrophenylhydrazine* (12,13): 10 drops. Yellow or orange precipitate.

Schiff's reagent (12,13) (must be freshly prepared): 10 drops. Colorless to pink or purple.

Alkyl Sulfides. *Isatin* (12): 10 drops of 1% isatin in concentrated H_2SO_4 gives green color.

Sodium nitroprusside: 10 drops of 95% ethyl alcohol plus 2 drops of a 5% KCN–1% $NaOH$ solution. Two to three minutes after sample passes, add 5 drops of 1% sodium nitroprusside solution. Red color results.

Alkyl Halides. *Alcoholic $AgNO_3$* (13): 10 drops of 2% alcoholic $AgNO_3$. White precipitate.

Mercurous nitrate: 10 drops of a $7.5N$ HNO_3–5% mercurous nitrate solution. Iodides, yellow to orange precipitate; chlorides, white precipitate; bromides, white or gray precipitate.

Alkyl Nitriles. *Ferric hydroxamate* (12): 10 drops of $1N$ NH_2OH. HCl in propylene glycol plus 2 drops of $1N$ KOH in propylene glycol. After sample passes, heat to boiling and cool. Solution is clear and colorless. Add 1 to 2 drops of 10% $FeCl_3$. Red-wine color is positive test.

Amines. *Benzenesulfonyl chloride* (Hinsberg test) (12,13): 5 drops of pyridine, 1 drop of 5% $NaOH$. After sample passes, add 1 to 2 drops of benzenesulfonyl chloride. Colorless to yellow for primary or secondary aliphatic amines. Tertiary amines give a rose to deep purple color.

Sodium nitroprusside (Rimini and Simon test for primary and secondary amines) (12): 10 drops of H_2O plus 2 drops of acetone plus 1 drop of 1% sodium nitroprusside. Primary amine gives red color. Add 1 to 2 drops of acetaldehyde. Secondary amine gives blue color. Omitting acetone permits test for secondary amine directly.

Aromatic Nucleus and Aliphatic Unsaturation. *Formaldehyde–sulfuric acid* (Le Rosen test) (12): 10 drops of concentrated H_2SO_4 plus 1 drop of 37% HCHO gives wine color.

Esters. *Ferric hydroxamate* (12): 10 drops of $1N$ NH_2OH. HCl in methanol plus 3 to 4 drops of $2N$ alcoholic KOH or until solution turns blue. After passing sample through solution add 5 to 6 drops of $2N$ HCl

until solution is clear and colorless. Add 1 to 2 drops of 10% $FeCl_3$. Colorless to red.

Ketones. *Dinitrophenylhydrazine* (12,13): 10 drops. Yellow or orange precipitate.

Mercaptans. *Alcoholic silver nitrate* (13): 10 drops of 2% alcoholic $AgNO_3$ gives white precipitate. (H_2S gives black precipitate.)

Lead acetate (12): 10 drops of saturated alcoholic PbOAc gives yellow precipitate. (H_2S gives black precipitate.)

Isatin: Same as test for isatin under Alkyl Sulfides above.

Sodium nitroprusside: Same as test with sodium nitroprusside under Alkyl Sulfides above.

References

1. R. L. Pecsok, Ed. *Principles and Practice of Gas Chromatography*, Wiley, New York, 1959, pp. 137–138.
2. L. Dubois and J. L. Monkman, in *Gas Chromatograpy*, H. J. Noebels, R. F. Wall, and N. Brenner, Eds., Academic Press, New York, 1961, pp. 237–246.
3. J. A. Attaway, R. W. Wolford, and G. J. Edwards, *J. Agr. Food Chem.*, **10**, 102 (1962).
4. P. P. Cherevka and T. Malyutina, *Zh., Kim. Prom.*, **40**, 582 (1964).
5. C. E. Döring and H. G. Hauthal, *J. Prakt. Chem.*, **22**, 59 (1960).
6. Y. R. Naves and A. V. Frampoloff, *Bull. Soc. Chim. France*, **1960** (1), 37.
7. R. Rowan, *Anal. Chem.*, **33**, 658 (1961).
8. J. T. Walsh and C. Merritt, Jr., *Anal. Chem.*, **32**, 1378 (1960).
9. R. L. Hoffmann and C. D. Evans, *J. Chromatog.*, **26**, 491 (1967).
10. B. Casu and L. Cavallotti, *Anal. Chem.*, **34**, 1514 (1962).
11. J. L. Monkman, private communication.
12. N. D. Cheronis and J. B. Entrikin, *Semimicro Qualitative Organic Analysis*, 2nd ed., Interscience, New York, 1957.
13. R. L. Shriner, R. C. Fuson, and D. Y. Curtin., *The Systematic Identification of Organic Compounds*, 4th ed., Wiley, New York, 1956.

CHAPTER 10

Special Identification Detectors

Charles T. Malone, *Corporate Research Department,*
The Coca-Cola Company, Atlanta, Georgia, and
William H. McFadden, *International Flavors & Fragrances (U.S.),*
Union Beach, New Jersey

I. Introduction

If a genie promised a chemist any detector he could describe, the chemist would probably ask for a system that would give him functionality, molecular weight, structural features, conformation, specificity, etc., and of course, he would also ask for ultrahigh sensitivity and an unlimited linear dynamic range. Surprisingly, most of the features that might be requested are available! But, unfortunately, not in any one magic genie box.

Modern separating techniques and modern instrumentation have had a fabulous impact on qualitative organic analysis. Results obtainable today with mass, NMR, and infrared spectrometers, and so forth, especially coupled with gas chromatography, probably surpass the wildest dreams of the organic chemist two decades ago. These wonders have been so well described in the previous chapters that it would seem impossible for the chemist to have an unknown sample that could not be easily determined. Lamentably, this is not yet so.

The limitations encountered in the more widely used instrumental systems are generally due to some combination of (a) lack of sufficient sensitivity (at least not comparable to the flame ionization detector), (b) lack of specificity, or (c) lack of understanding (no prior investigations of that chemical system). Unfortunately, today, the old tricks or the simple tricks are often overlooked. A well-equipped high-resolution mass spectrometry laboratory is very impressive, particularly since it will often cost as much as $200,000. Yet, this laboratory would be hard pressed to determine the existence of sulfur or phosphorus in a chromatographic peak separated in, say, nanogram quantities. Such limitations must be forgiven, considering the tremendous value of such a laboratory in other projects. But it must also be remembered that several relatively simple analytical systems do exist that can solve such problems.

It is to these often unpraised techniques that this chapter is devoted. Two common characteristics will be noted. First, they generally possess the advantage of being economical (by today's standards). Second, they usually possess the mixed blessing of being highly specific. Such a feature is necessary in some applications but, of course, also limits the general usefulness.

All the systems discussed in this chapter fall within the definition of ancillary techniques (see Chapter 1). Either they are based on some chemical reactions and thus may be considered as belonging, in a broader sense, to the field of reaction gas chromatography; or they utilize for detection purposes analytical methods which can be, and are, also used independently of the GC system.

II. Detectors Specific for Elements

One of the important applications of specific detectors is in the determination of elements. Frequently, a peak may be encountered in such small amounts as to preclude infrared or NMR measurements and the expensive high-resolution mass spectrometers are not always available. When the existence of heteroatoms is probable, one should resort to a specific elemental detector.

In general, elemental detectors are used for detection of halides, phosphorus, sulfur, and nitrogen, although some can cover a much wider range of elements when desired.

A. Microcoulometric Titrating System

The microcoulometric titrating system is a specific detection device for qualitative and quantitative analyses of nitrogen-, sulfur-, halogen-, and phosphorus-containing compounds (1,2). Although these elements can be determined by a variety of other detectors, the coulometric system has been found to give reliable results combined with ease of operation. It is rugged, and the sensitivity is not seriously affected by improper operation.

Suitable commercial equipment is available in the price range of $5500–6000. The system consists of the following parts: (A) pyrolysis furnace, (B) microcoulometer, and (C) interchangeable titration cells (see Figs. 10-1 and 10-2).

As individual components are eluted from the column, they are mixed with reactant gases (oxygen or hydrogen) and pass into the pyrolysis chamber. Here they are oxidized or reduced, and eluted into the titration cell. The inorganic ions formed from the sample proportionately change the concentration of titrant in the electrolyte. Ion imbalance is sensed by the sensor/reference pair of electrodes and this imbalance is used as the signal input to the microcoulometric amplifier. The amplifier supplies a balancing voltage to the generator electrodes which regenerate titrant to

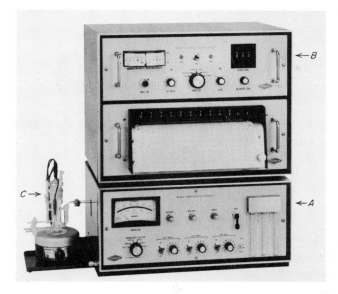

Fig. 10-1. Microcoulometric detection system. (Dohrmann Instrument Company) (2). (A) Pyrolysis furnace. (B) Microcoulometer. (C) Interchangeable titration cells.

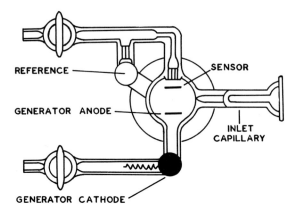

Fig. 10-2. Microcoulometric titration cell (3).

bring the electrolyte back to its original concentration. The area beneath the peak represents the coulombs of electricity required to carry out the titration.

Rapid microcombustion or microreduction is performed in the pyrolysis furnace between 600 and 1100°C. If oxygen is used as the combustion gas (the pyrolysis temperature is generally 800°C), the following reactions take place:

$$RCl + O_2 \xrightarrow{800°C} CO_2 + H_2O + HCl$$
$$R_2S + O_2 \longrightarrow CO_2 + H_2O + SO_2$$
$$RNH_2 + O_2 \longrightarrow \text{No inorganic electrolyte}$$
$$R_3PO_4 + O_2 \longrightarrow P_2O_5, \text{involatile}$$

If hydrogen is used (pyrolysis temperatures to 950°C), the following reactions take place:

$$RCl + H_2 \xrightarrow{950°C} RH + HCl$$
$$R_2S + H_2 \longrightarrow 2HR + H_2S$$
$$RNH_2 + H_2 \longrightarrow RH + NH_3$$
$$R_3PO_4 + 7H_2 \longrightarrow 3RH + 4H_2O + PH_3$$

Most of the titration cells employ a four-electrode system (see Fig. 10-2). The reference sensor pair detects changes in ion concentration due to acid–base reaction, and the generator anode–cathode pair re-establishes ion concentration by generating sufficient ions to replace those lost by the reaction. At present, there are three types of titration cells in common use (Table 10-1) (3–5).

TABLE 10-1

Cell Specifications—Dohrmann Instrument Company

	T300-S (Silver cell) (3)	T300-P (Platinum cell) (4)	T400-H (Hydrogen cell) (5)
Electrolyte	70% Acetic acid in water	0.04% Acetic acid; 0.05% potassium iodide	0.04% Sodium sulfate
Sensor electrode	Silver	Platinum	Platinum black
Reference electrode	Silver–silver acetate; saturated	Platinum triiodide; saturated	Lead–lead sulfate; saturated
Generator anode	Silver	Platinum	Platinum
Generator cathode	Platinum	Platinum	Platinum

The T300-S silver cell is available for the determination of halide ions (except fluoride) or materials which react to form a stable complex or precipitate with Ag^+ in acetic acid solution. The cell reaction in the reference/sensor pair is:

$$Ag^+ + X^- \longrightarrow AgX \downarrow$$

and the generator electrode reaction is:

$$Ag^\circ \longrightarrow Ag^+ + e^-$$

Another titration cell, the T-300-P platinum cell is used to titrate SO_2, or any other material which will react with iodine (triiodide) in solution. The cell reaction for SO_2 is:

$$I_3^- + SO_2 + H_2O \longrightarrow SO_3 + 3I^- + 2H^+$$

The I_3^- depletion is detected by the reference/sensor electrode and replaced electrically at the generator electrode by a current from the microcoulometer. The reaction at the generator anode is:

$$3I^- \longrightarrow I_3^- + 2e^-$$

The third type of cell presently available is used to titrate nitrogen (as ammonia) as well as other basic or acid constituents which will react to change the pH of a slightly acidic Na_2SO_4 solution. Typical cell reaction, using ammonia as an example is:

$$NH_3 + H_2O \longrightarrow NH_4OH$$
$$NH_4OH + H^+ \longrightarrow NH_4^+ + H_2O$$

The hydrogen ion removal in the electrolyte is detected by the reference electrode and replaced by the generator electrodes, where the following reactions take place:

$$\tfrac{1}{2}H_2 \longrightarrow H^+ + e^-$$
$$H_2O + e^- \longrightarrow OH^- + \tfrac{1}{2}H_2$$

Phosphorus compounds present a special problem in that they are oxidized to phosphorus pentoxide, which is so involatile it is not carried from the oxidation chamber to the titration cell. Burchfield has recommended that hydrogen be used as a reducing agent (6), and organic compounds containing phosphorus are thus reduced to phosphine. Phosphine can then be detected using the silver titration cell where the precipitate formed is reported to be Ag_2PH. Some problems can be encountered in quantitative determinations of phosphine, and it is necessary to reduce the flow to increase trapping efficiency. Burchfield et al. (7) also reported that the addition of 0.2% Triton X-35 (Rohm & Haas Co., Philadelphia, Pa.) to the electrolyte reduced gas–liquid interfacial tension. These modifications resulted in quantitative recovery of phosphine.

The sensitivity of the microcoulometric systems is reported to be 0.1 ppm in a 10-mg sample or 1 ng for halogen-, sulfur-, and nitrogen-containing compounds.

An interesting example of the use of the microcoulometric detector has been recently reported by Johnson and Burchfield (8). The separation of metabolites derived from individual phenothiazine tranquilizers has been difficult. Some workers report that these drugs are metabolized in different ways according to various psychopharmacological responses exhibited by mental patients. A chromatogram shown in Figure 10-3 is the result obtained from a dichloromethane extract of urine of a patient on this drug. The chromatogram shows a complex mixture of metabolites from chlorpromazine [2-chloro-10-(3-dimethylaminopropyl)-phenothiazine], but the specificity of the detector readily permits their identification even in the presence of many artifact materials.

B. Flame Photometric Detector

Another device for the selective determination of specific elements is the flame photometric detector. As the name implies, this system is essentially a flame photometer that has been modified for use in a flow system such as a gas chromatograph. Quite obviously, a detector based on these principles can have essentially the same specificity as a flame spectrophotometer, and the system has been used for studies of phosphorus, sulfur, and chlorine compounds as well as for detection of metals in coordination complexes.

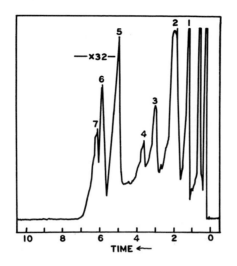

Fig. 10-3. Microcoulogram of chlorpromazine metabolites (8). Peaks: *1*, chlor-promazine-*N*-oxide. *2*, chlorpromazine. *3*, demethylchlorpromazine. *4*, 2-chloro-phenothiazine. *5*, chlorpromazine-*S*-oxide. *6*, demethylchlorpromazine-*S*-oxide. *7*, didemethylchlorpromazine-*S*-oxide.

The flame photometric detector system can be designed so that a rela-tively narrow wavelength band is selected, thus affording a high degree of specificity (9). Such applications are generally free of interference, but of course, each element must be detected with a separate gas chromatographic run. Winefordner and Glenn (10) summarize the detection limits of various compounds using different flame photometer systems and these are presented in Table 10-2. Use of these data is illustrated in Figure 10-4, which shows three chromatograms of a mixture of chromium, iron, and rhenium chelates (11). By selection of the appropriate wavelength, each chelate is successively identified with minimal interference from the others.

For many elements the sensitivity of the flame photometer is comparable to the flame ionization detector. For example, phosphorus compounds are reported to have a sensitivity in the range of 10^{-12} g/sec with linear response over four orders of magnitude. On the other hand, sulfur compounds have a sensitivity limit of only 10^{-10} g/sec (12). Another advantage is that the sensitivity is not seriously affected by changes in carrier gas flow rates, and up to 160 ml/min can be used without loss of response.

In one experimental setup, a Beckman DU spectrophotometer flame attachment was joined directly to the capillary of the burner assembly

TABLE 10-2

Detection Limits of Flame Emission Detection (10)

Sample	Band or line, Å	Limit of detection with DU,[a] g	Limit of detection with Spectronic 20,[b] g	Limit of detection nonselective, g
$TiCl_4$	5449[c]	4×10^{-11}	—	—
$AsCl_4$	AsO,5000	2×10^{-9}	—	—
$ZrCl_4$	5640[c]	2×10^{-8}	—	—
$Rh(hfa)_3$[d]	Rh,3692,3701	1×10^{-11}	7×10^{-10}	3×10^{-10}
$Cr(hfa)_3$[d]	Cr,4254,4275,4290	1×10^{-10}	6×10^{-10}	2×10^{-10}
$Al(tfa)_3$[e]	AlO,4866	—	7×10^{-9}	2×10^{-9}

[a] Beckman DU flame photometer.
[b] Bausch & Lomb Spectronic 20.
[c] Bands unassigned.
[d] 1,1,1,5,5,5-Hexafluoro-2,4-pentanedione.
[e] 1,1,1-Trifluoro-2,4-pentanedione.

of a Beckman Model 4030 gas chromatograph (13). Alternatively, a flame photometric detector can be attached directly to the burner jet of a flame ionization gas chromatograph (14). Nitrogen is generally used as a carrier gas and oxygen is mixed in at the column exit. Hydrogen enters the burner directly and the mixture of the three gases is burned in a hollow tip which shields the flame from view of the photomultiplier tube (Fig. 10-5). When components containing a suitable heteroatom are eluted, optical emission occurs above the shielded flame. The light transmitted to the photocell gives a qualitative and quantitative measure of the atom for which the system is tuned.

C. Thermionic Detector

The thermionic detector is a sensitive, highly selective GC detector used primarily for the analysis of organophosphorus compounds. The sodium thermionic detector is a modified flame ionization detector in which the electrode is coated with a sodium salt (sodium hydroxide, sodium sulfate, etc.) (15). In some modifications, other alkali metals such as potassium, cesium, and rubidium are used (16,17). The addition of the alkali metal to

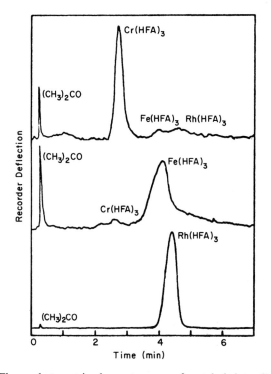

Fig. 10-4. Flame photometric chromatograms of metal chelates (11). Cr, 3.9 μg; Fe, 39.9 μg; Rh, 12.8 μg. Slit, 0.3 mm. Top: λ = 425.4 mμ. Middle: λ = 372.0 mμ. Bottom: λ = 369.2 mμ.

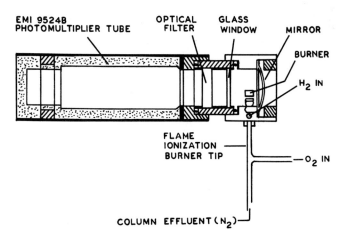

Fig. 10-5. Flame photometric detector (14).

the flame electrode greatly increases the ionization efficiency of the system for organophosphates. The sensitivity to halides is also increased, but to a lesser extent.

A typical flame ionization detector modified for parallel use as a thermionic detector is shown in Figure 10-6. The column effluent enters through the standard hydrogen flame jet and then diffuses into the second flame chamber in which the wire has been coated with alkali metal. This tandem operation permits the response ratio to be obtained for ease of peak identification. Voltage is applied across the polarized electrodes and produces a small standing current through the cell. When an organic compound is burned in the flame, some of the molecules are ionized, which thus increases the cell current. According to Johnson (18), efficiency in the flame ionization detector is only about 10 ppm. Addition of alkali metal to the electrode does not alter this efficiency for most organic compounds but does increase it for the organophosphates and organohalides.

The choice of one alkali metal versus another is generally a matter of trading off between such parameters as sensitivity, selectivity, stability, and ease of fabrication. For example, according to Hoffman (19), when sodium sulfate is coated on the electrode, phosphates respond around 600 times greater and halides 20 times greater than when a regular flame ionization detector is used. However, purified potassium chloride gives an additional threefold increase in sensitivity for phosphates while the response for halides is less. If an investigation required determination of

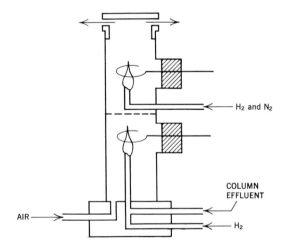

Fig. 10-6. Thermionic detector (15).

both phosphorus and halide compounds, the best choice would be to use the sodium salt and compare the results with those obtained from an unmodified flame detector. On the other hand, if phosphates were the sole interest, one might prefer the higher sensitivity obtained by using KCl. One disadvantage of the KCl coating is that the salt volatilizes rather quickly from the tip and causes a gradual decrease in sensitivity.

The Varian Aerograph phosphorus detector (20) is a flame ionization detector modified according to the method of Giuffrida (17). The standard flame ionization detector is converted to a phosphorus detector by inserting a cesium bromide tip over the flame tip. The conversion tip consists of cesium bromide and a filler pressed under high pressure to form a rugged ceramic material. The flow rates of air and hydrogen must be controlled closely for optimum performance. A typical phosphorus detector should have an air flow of 170 ± 0.1 ml/min and a hydrogen flow of 16 ± 0.01 ml/min. Under these conditions, the sensitivity of the detector permits minimum detectability of phosphorus in common pesticides of about 10^{-13} g/sec. The range of linearity of this detector is about 100 above the minimum detectable quantity.

The selectivity of the sodium thermionic detector is illustrated in the chromatogram of Figure 10-7. Lindane, parathion, and methyl stearate, representing halogenated, phosphorus, and standard organic compounds, were chromatographed on both the flame ionization detector, A, and the sodium thermionic detector, B. In A, full-scale deflection corresponds to 1×10^{-9} A, whereas in B, full-scale deflection is 3×10^{-7} A. The results indicate a 300-fold increase in the parathion signal.

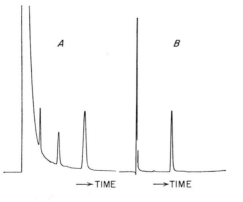

Fig. 10-7. Analysis of 1 μg lindane, 1 μg parathion, and 2 μg methyl stearate (15). (A) By flame ionization detector, 10^{-9} A full scale. (B) By sodium thermionic detector, 3×10^{-7} A full scale.

D. Microwave Emission Detector

The selectivity obtainable by spectral emission (as described for flame photometric detector) is also utilized in the microwave emission detector (21–23). In this system, the effluent from the GC column passes through a quartz capillary tube positioned in the cavity of a microwave generator (Fig. 10-8). A discharge, which must be initiated by a spark coil, is maintained by the microwave generator as a sustained plasma. Light from this discharge is examined by a suitable scanning monochrometer and the phototube output of the selected wavelength is recorded as a chromatographic signal.

Both helium and argon are suitable carrier gases. Helium has a simpler background emission spectrum, but the plasma discharge can only be sustained under low pressures. Hence, argon is usually preferred.

This system cannot be obtained commercially as a complete package at the present time, and the components (a 2450 Hz power oscillator, scan-

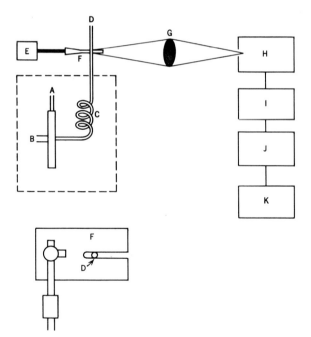

Fig. 10-8. Schematic of microwave emission detector (21). *A*, Carrier gas inlet; *B*, injection port; *C*, column; *D*, quartz discharge tube; *E*, microwave generator; *F*, microwave cavity; *G*, quartz lens; *H*, monochrometer; *I*, multiplier phototube; *J*. amplifier; *K*, recorder.

ning monochrometer, phototube, amplifier, recorder) cost approximately $8000. This is sufficiently expensive to discourage casual use, and some of the other specific detectors, though less definitive, are often preferred.

The plasma discharge is sufficiently energetic to cause virtually complete fragmentation of effluent compounds, and in addition, some diatomic

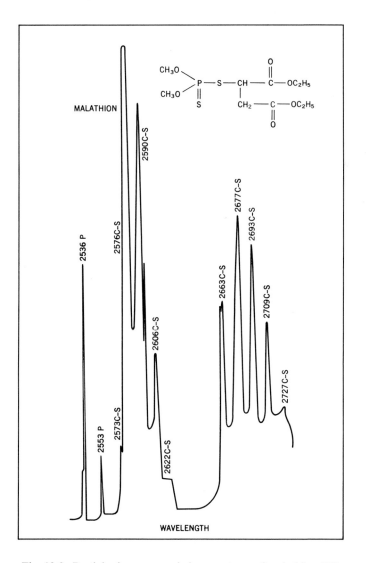

Fig. 10-9. Partial microwave emission spectrum of malathion (21).

TABLE 10-3

Sensitivity Data for the Microwave Emission Detector (21)

Element	Compound	Wavelength, Å	Assignment	Detection limit, g/sec
Carbon	—	3883	CN	2×10^{-16}
		5165	C_2	3×10^{-14}
Fluorine	C_6F_6	5166	?	3×10^{-12}
Chlorine	$CHCl_3$	2788	CCl	8×10^{-10}
Bromine	$CHBr_3$	2985	?	2×10^{-7}
Iodine	CH_3I^3	2062	I	7×10^{-14}
Phosphorus	$(C_2H_5O)_3PO$	2555	P	1×10^{-11}
Sulfur	CS_2	2575	CS	1×10^{-9}

recombinations are observed. Thus, the emitted spectrum will contain lines due to any and all of the atoms present, and one observes spectral lines due to the species C_2, CN, CH, CS, P, PO, I, CCl, SiF_4, etc., depending on what atoms are present in the sample or background. This is partially illustrated in Figure 10-9, which shows a small portion of the microwave emission spectrum of malathion. (Test spectra of this type are obtained by scanning the plasma spectrum during passage of carrier gas which contains a constant concentration of the test compound.)

With a suitable monochromer, selection of almost any atomic or diatomic species that might be encountered is feasible. As an example, if a compound is suspected of containing phosphorus and/or sulfur, the monochromer can be set for the P peak at 2536 Å in the first chromatographic run. If the phosphorus test is positive, then the second run would be set to select the PS peak at 4957 Å. If the phosphorus test is negative, then the CS peak at 2576 Å would be selected. In this way, the presence or absence of the specific atomic species can be established.

The sensitivity of this detector depends, of course, on the band selected for the run. A summary of a few sensitivities is given in Table 10-3.

III. Detectors for Specific Organic Functions

A. Electron Capture Detector

The electron capture detector is a detector system that is highly sensitive to organic compounds which possess a high electron affinity (24). In general, the system may be used for alkyl halides, conjugated carbonyls, nitriles, nitrates, organic metal compounds, and many sulfur-containing compounds, particularly disulfides and trisulfides. This detector has been used for the analysis of pesticide residues in food products and water pollutants, and it is also frequently applied in the drug and medical field because of its high selectivity and sensitivity to certain elements in the presence of other compounds. It is sometimes used in flavor and aroma chemistry on peaks separated in subnanogram quantities. In such cases, the ratio of the hydrogen flame signal to the electron capture signal provides information on the type of functionality.

Many commercial GC instruments presently available are modular in construction, and electron capture attachments (cells and electrometers) can be purchased for approximately $1300 (more if the ^{63}Ni source is required). Two typical cells are shown in Figures 10-10 and 10-11. The cell shown in Figure 10-10 is the Aerograph concentric tube detector (25) which operates using a constant potential. Figure 10-11 is a high-temperature detector of coaxial design (26) which uses a ^{63}Ni radiation source in the pulsed mode.

The electron capture detector works on the principle that when carrier gas flows through the detector, a radioactive source ionizes the carrier gas molecules, thus generating electrons and positive ions. A voltage is applied across two electrodes which produces a standing electron current of approximately 10^{-8} A. This voltage is of such a value (nominally, in the range of 50 V), and/or is pulsed at such a duration (about 1 μsec duration, 15 μsec repetition rate) that collection of all the electrons at the anode is

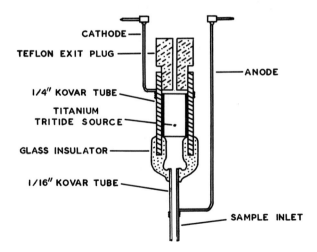

Fig. 10-10. Concentric tube electron capture detector (25).

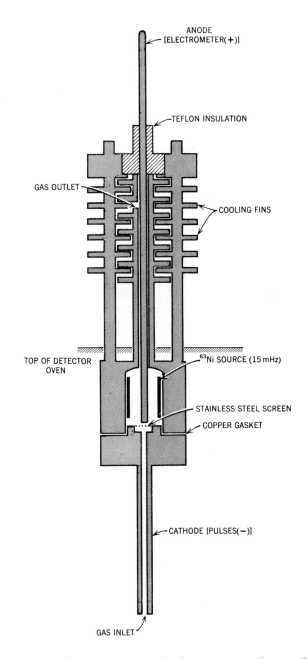

Fig. 10-11. High temperature coaxial electron capture detector (26).

barely attained. When a sample of high electron affinity enters the chamber, negative ions are formed by electron capture by the dissociative process (*a*), or the nondissociative process (*b*):

(*a*) $AB + e^- \longrightarrow A + B^-$
(*b*) $AB + e^- \longrightarrow AB^-$

In either event, the greatly reduced mobility of the negative ion, as compared with the electron mobility, results in a significant decrease in standing current. This fractional decrease in standing current is sensed by the electrometer and recorded as the electron capture signal.

The linearity of a device based on such a principle is necessarily poor. Typically, the response of the electron capture detector may be considered as linear for about one or two orders of magnitude and over a greater range may be considered to approximate an exponential function. However, the high selectivity and the very high sensitivity render it extremely valuable for detection of electron-capturing materials, especially when they cannot be properly separated from other compounds.

The ionization source, and hence source of electrons, is usually tritium or ^{63}Ni. Both provide soft β-particles of a suitable range. In general, a higher sensitivity can be attained by the tritium source, but if temperatures in the range of 210–350°C are required, the ^{63}Ni source must be used. The carrier gas can be hydrogen or nitrogen, but most commonly, argon mixed with 5% methane is employed. The methane prevents formation of metastable argon which would then convert the system to an argon ionization detector (27).

The simultaneous use of both the flame ionization and electron capture detectors provides information on the type of compounds present in the mixture. Figure 10-12 (28) shows a chromatogram of a head space analysis on macerated garlic cloves using the combination electron capture and flame ionization detector. In Table 10-4 the response ratio (response electron capture/response hydrogen flame) is given for each peak which, along with known retention values, permits identification of compounds.

Many fused ring polycyclic aromatic compounds do not give suitable response to a hydrogen flame detector. An additional application of the electron capture device for identification of such compounds is shown in Figure 10-13 (29). As is noted, benz[*b*]-fluoranthene and benz[*a*]pyrene gave essentially no response in the flame detector, but were easily observed by the electron capture device.

An interesting example on the use of the electron capture detector occurred during 1967 (30). In Mexico and Colombia, hundreds of people were taken ill and many died from eating contaminated food. Within two

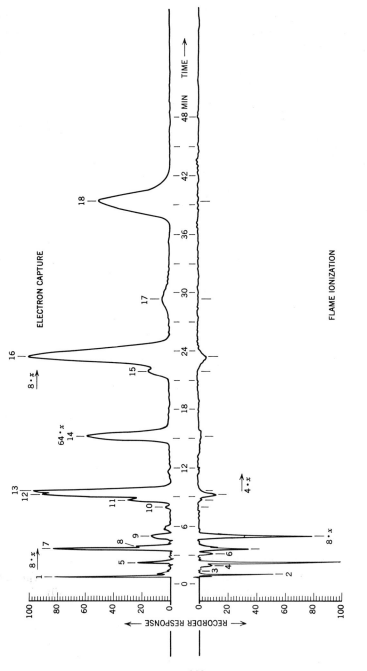

Fig. 10-12. Chromatogram of head space analysis of garlic cloves (28).

358

TABLE 10-4

Response Ratios and Compound Identification for
Chromatograms of Figure 10-12

Peak No.	Response ratio	Compound
1	—	Oxygen from air
2	0.4	Methyl mercaptan
3	0.01	Dimethyl sulfide
4	—	Unknown
5	0.05	Methyl allyl sulfide (tentative)
6	—	Unknown
7	2.4	Dimethyl disulfide
8	1.7	Unknown
9	0.09	Diallyl sulfide
10	3.0	Unknown
11	2.5	Unknown
12	10.0	Methyl allyl disulfide
13	50.0	Unknown
14	300.0	Dimethyl trisulfide
15	—	Unknown
16	20.0	Diallyl disulfide
17	—	Methyl *n*-propyl trisulfide
18	200.0	Methyl allyl trisulfide

days after the outbreak, food samples were extracted and sent to the California Department of Agriculture for analysis. Two hours after the samples were received, the pesticide was identified as parathion. Doctors were able immediately to stop further deaths by applying special drugs. The simplicity of the analyses and interpretation is illustrated in Figure 10-14, which shows that the chromatograms obtained are free from interfering peaks. Combined with a reasonable knowledge of possible contaminants, the analytical data provided unequivocal identification on subnanogram quantities of the unknown.

B. The Acid–Base Titration Cell

The first detection system employed in gas chromatography was the automatic titrator used by James and Martin in their investigation of volatile acids and bases (31). Figure 10-15 illustrates the principle of this detector. The effluent from the gas chromatographic column is eluted to the titration cell which contains appropriate pH color indicator. Phenol

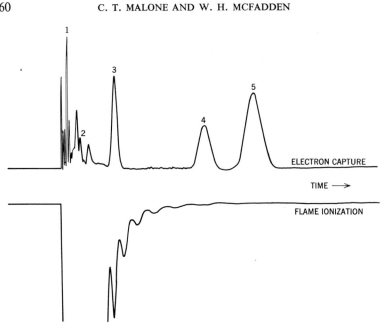

Fig. 10-13. Application of electron capture detector for identification of polycyclic aromatic compounds (29). Peaks: *1*, anthracene; *2*, pyrene; *3*, 1,2-benzanthra-
cene; *4*, benz[*b*]fluoranthene; *5*, benz[*a*]pyrene. Peaks *3*, *4*, and *5* each correspond to
1 ng substance.

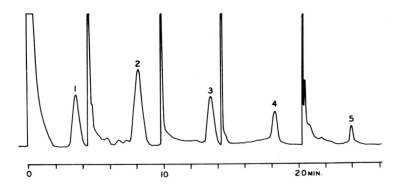

Fig. 10-14. Chromatograms obtained on extracts from contaminated food (30).
1, 2 ng Parathion standard; *2*, 0.24 mg stomach content, 14 ppm; *3*, 0.12 mg stomach
content, 16 ppm; *4*, 0.04 mg stomach content, 31 ppm; *5*, extract of 0.1 mg bread,
8 ppm.

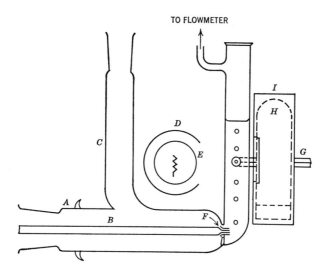

Fig. 10-15. Automatic acid–base titration detector (31). *A*, Vapor jacket; *B*, chromatographic tube; *C*, air-condenser socket; *D*, lamp house; *E*, lamp; *F*, rubber gasket on capillary; *G*, capillary lead from buret; *H*, photoelectric cell; *I*, photoelectric cell housing with adjustable window.

red is used for fatty acids and methyl red for volatile bases. When the vapors enter the cell, the pH is changed, and the system is automatically titrated by means of a photocell relay. Alkali or acid is added from the buret until the original pH is restored as indicated by the photocell signal.

The original work of James and Martin used an integral detection system rather than the differential system which is most commonly used today. Figure 10-16 shows the original chromatograms obtained on the separation and analysis of fatty acids, and Figure 10-17 shows the first chromatogram recorded on aliphatic amines. The integral curves represent the number of equivalents of titrant that have been added to the cell during the chromatographic run.

Several authors have modified the original apparatus described by James and Martin (32,33). Langan (34) found that a glass electrode pH-stat assembly was a suitable recording detector for acids. Acids are neutralized as they emerge from the column and the volume of alkali consumed is recorded by the pH-stat assembly. The limit of readability on the titration curve is ± 0.01 μmole. Since this generally corresponds to about 1 μg, it indicates a considerable limitation for problems dealing with small samples. Nevertheless, the system is still very valuable because of its complete selectivity for acid–base systems.

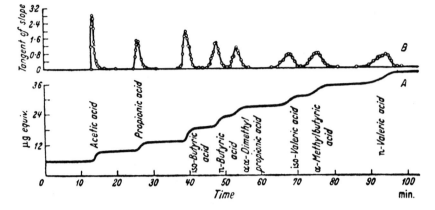

Fig. 10-16. Original gas chromatogram of fatty acids obtained by James and Martin (31).

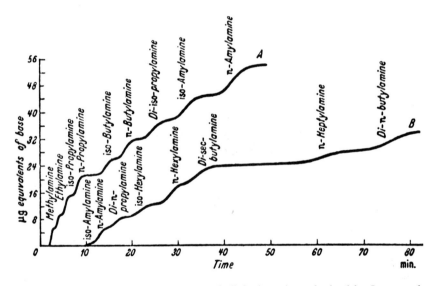

Fig. 10-17. Original gas chromatogram of aliphatic amines obtained by James and Martin (31).

C. The Spectrofluorimetric Detector

One type of functionality that can frequently be determined by a specific detector is aromaticity. Almost all compounds classified as aromatic are fluorescent or possess ultraviolet absorption, and both these properties have been used for specific detection with modest success.

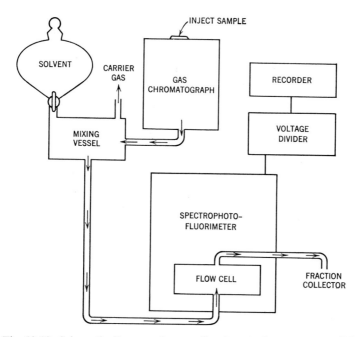

Fig. 10-18. Schematic diagram of spectrofluorimetric detector system (35).

Recently, a modified mixing cell and flow system has been developed that provides easy manipulation of eluted samples (35) when used with a spectrofluorimetric detector.

A schematic diagram of the spectrofluorimetric system is given in Figure 10-18. The mixing vessel (Fig. 10-19) provides a convenient method of dissolving an eluted vapor into a solvent (such as ethanol) for transport to the detector. Carrier gas exits through the open tube in the cell and except for samples of high volatility or low solubility, reasonably efficient trapping into solution is possible. The exit port of the GC column is kept at 200°C, which effectively prevents holdup of sample. Transfer of heat to the ethanol solution at this point was not sufficient to be troublesome.

For spectrofluorimetric detection, it is necessary to preset the instrument for both the excitation wavelength and the emission wavelength. This presents an annoying obstacle even when considerable is known about the sample; and it is inevitable that in a single run, the most efficient settings for one material will not be the most efficient settings for another. Nevertheless, Bowman and Beroza report (35) minimal detectable quantities down to 0.2 ng and have demonstrated rapid insecticide analysis of

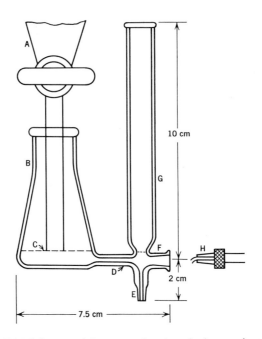

Fig. 10-19. Mixing vessel for spectrofluorimetric detector system (35).

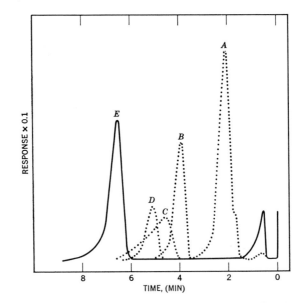

Fig. 10-20. Chromatographic peaks obtained by spectrofluorimetry (35). *A*, 0.2 μg fluorene; *B*, 0.05 μg *p*-terphenyl; *C*, 0.5 μg anthracene; *D*, 0.5 μg chrysene; *E*, 0.1 μg benzo[*a*]pyrene.

364

0.5 ppm of carbaryl (1-naphthol-*N*-methyl carbamate) and 1-naphthol in milk. A few typical chromatographic peaks are shown in Figure 10-20.

Adsorption of radiation in the ultraviolet and far ultraviolet spectral regions has also been used to detect aromaticity in GC flow systems (36,37). Adsorption has one advantage over fluorescence in that only the observed wavelength must be preselected. However, in general, sensitivities for UV adsorption are lower and minimal detection limits are in the microgram range. For the far-UV detector, minimal quantities down to 10^{-8} g have been reported for naphthalene adsorbing at 2108 Å. This is considerably more sensitive than a typical thermal conductivity detector as seen in the chromatographic peaks of Figure 10-21, which shows the thermal conductivity and far-UV response for equivalent amounts of naphthalene (38). However, the necessity of carefully selecting the most favorable wavelength is evident from the UV spectrum of naphthalene

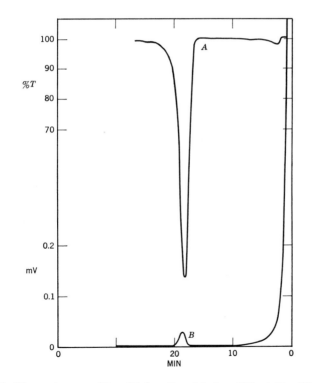

Fig. 10-21. Chromatogram of 2×10^{-5} g of naphthalene (38). *A*, Far-UV detector; *B*, thermal conductivity detector.

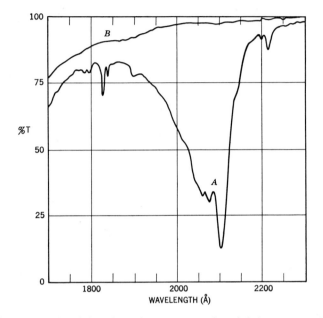

Fig. 10-22. Ultraviolet adsorption spectrum of naphthalene vapors (36).

(Fig. 10-22). If the signals from a thermal conductivity detector and far-UV detector were being simultaneously used in a ratio mode to qualitatively (and quantitatively) determine naphthalene, any small shift of the wavelength selection would make the results invalid.

Techniques reported for UV or far-UV absorption in gas chromatography have used a heated gas cell (35). In some applications, the solution system described by Morton and Beroza would be more convenient.

IV. The Biological Detector

A. The Human Sensor

One of the most interesting and specific properties that has been employed as a GC detector is the sense of smell. It is well known that animals —particularly lower forms—possess extremely high sensitivity for detection of certain chemicals or pheromones. In many instances, the sensitivity of olfactory detection greatly exceeds the most sensitive instrumental system, and many materials can be detected in amounts of 10^{-15} g or less.

When this particular sensory excitation is the parameter of interest, then it alone can be used as the detector. All instrumental methods are supplemental. The purpose of this section is to correlate the requirements for coordination of instrumental (GC) and sensory techniques and to discuss some of the current applications of this method of GC detection.

Living cells are composed of membranes which separate biological solutions that differ in chemical and ionic concentrations, and this unequal distribution of ions plays a major role in propagation of nerve impulse. The concept of smell can be schematically described (39):

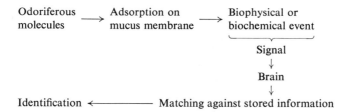

Odorous chemicals which enter the nose absorb in the mucus and a biophysicochemical event produces a signal. The signal is matched in the brain against known information. Since this final step is usually the recall from a human experience (the smell of steak frying, the aroma of fresh oranges, etc.), this stage cannot be duplicated by instrumental processes.

In its simplest application, the nose is used to sniff effluent from the chromatographic column (Fig. 10-23) without any modification of the

Fig. 10-23. Simplified nasal detector technique.

GC apparatus (40,41). In other cases, a hood or cage is constructed to isolate the human sensor (42). In this way, reasonable appraisals can be made on the olfactory properties of separated GC components.

Figure 10-24 shows the results of an odor evaluation of the effluent from a concentrated flavor sample (40). Two very interesting features are noted. First, it is not possible to ascertain from the nasal appraisal what material was being analyzed. (In this case, commercial security forbade revealing the substance.) Second, the actual determinations cover a considerable spectrum of aromas that frequently repeat, presumably because of the occurrence of homologous materials.

A similar evaluation was made on the chromatographic peaks from a Delicious apple essence (43). In this case, the results shown in Figure 10-25 directed the research to the peaks of importance. A subsequent mass spectral study gave chemical identification and the materials have been used with remarkable success to enhance the flavor of apple juice.

In the case above, the active materials were determined to be hexanal, 2-hexanal, and ethyl 2-methyl butyrate. This points to a possible pitfall of this methodology. When the synergistic effects of two or more components are required to give the desired natural aroma, sniffing the effluent from a column of high separating efficiency may not reveal any aroma which resembles the starting material. The investigator must then revert to a column of *poorer* separation in the hope of bringing the necessary combination together again.

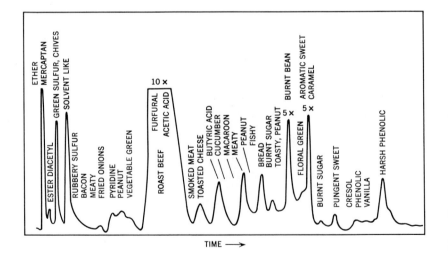

Fig. 10-24. Odor evaluation of a chromatogram from a flavor concentrate (40).

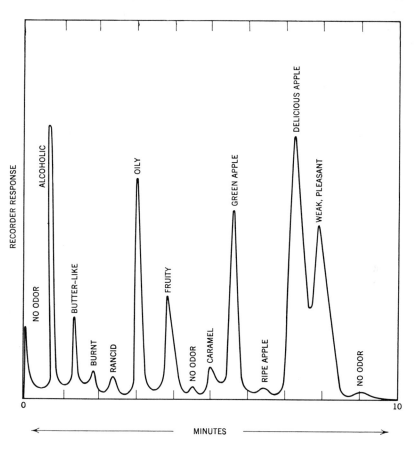

Fig. 10-25. Odor evaluation of a chromatogram from a Delicious apple extract (43).

Of course, a very important corollary of the above principle is to be careful that chemical identification of two or more components in the odorous peak does not cause one to overlook a more important odor component. In preliminary studies on the apple essence discussed above, the ethyl 2-methyl butyrate was overlooked. The threshold value for this compound in water is 0.1 ppb, which is about 100 times less than the other two components. At the early stage of the work, identification of the aldehydes seemed sufficient.

Sometimes odor analysis is performed by untrained laboratory personnel; sometimes a selected trained panel is used. In one rather esoteric refinement of the human sensor, trained, highly skilled perfumers have been used (42). With very modest training, a perfumer can use his excellent memory of some 1000 compounds to identify chemically the GC effluents, without recourse to ancillary instrumental aids. The results of such an investigation are shown in Figure 10-26. The caption gives the perfumer's identification and when this did not correspond to the known identity, the correct information is given in parentheses.

Use of a perfumer in this manner has some curious advantages, not the least of which is to expose the perfumer to highly purified chemicals. As a consequence, in some instances the perfumer has indicated extreme interest in certain chemicals thus encountered. Unfortunately, the use of a full-time perfumer is possibly the most expensive detector ever proposed for GC and in a 10-year period, the cost would rival the cost of a well-equipped high-resolution mass spectrometry laboratory.

B. The Insect

Although the human model of the nasal detector is desired in certain applications, it is well known that lower animals possess a very much higher level of nasal sensory perception. Unfortunately, it is not generally possible for the animal to communicate back to the GC operator and until this channel of communication can be developed, use of animals as GC detectors must be limited.

One important and very specific exception exists to this rule, namely, when the chemical of interest excites a particular reaction in that animal. This type of response has been rather widely used in entomological studies, particularly those studies that seek a sex attractant or colonization pheromone (44–46).

Results obtained in research on the sex pheromone of the cotton leafworm illustrate the techniques and advantages of this type of biological detector (45). The abdominal segments from female moths were extracted in methylene chloride. Male cotton leafworm moths were placed in a small cage at the exit of a chromatograph and the extract was injected. The elution of an active peak was characterized by "the lifting of antennae, vibrating of the wings, flying in a hovering position, extending of the claspers, and attempting to copulate with one another." Similar results have been observed in studies of other insects (44,46).

One important aspect of this type of detection is that the researcher is able to direct his attention to the GC peak of interest and can ignore the many inactive compounds that are generally coextracted. Furthermore, the quantities of material obtainable are always small and, if they are not detected by the standard gas chromatographic detector, biological activity observed at a particular elution time will alert the investigator to the fact

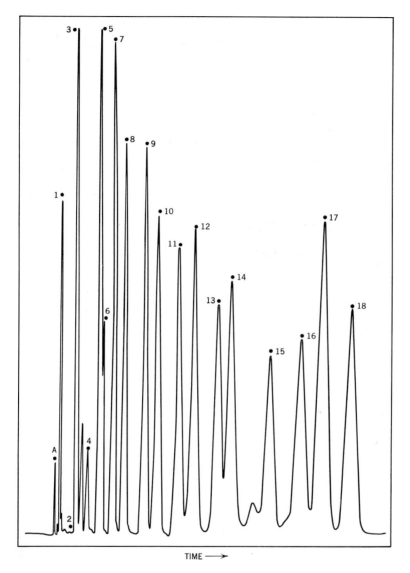

Fig. 10-26. Perfumer identification of GC effluents (42). *A*, air; *1*, dimethyl sulfide; *2*, dipentene (unknown); *3*, α-pinene; *4*, myrcene (unknown); *5*, limonene; *6*, *p*-cymene (unknown); *7*, C_9-aldehyde (C_8 aldehyde); *8*, methyl heptenone; *9*, dihydromyrcenol (tetrahydrolinalool); *10*, dihydromyrcenyl acetate (dihydromyrcenol); *11*, citronellal; *12*, linalool; *13*, borneol (camphor); *14*, isopulegol; *15*, methyl heptinecarbonate; *16*, α-terpineol; *17*, recognize but cannot identify (C_{12} aldehyde); *18*, carvone.

that a very potent pheromone is being eluted. Another somewhat surprising bonus has been that some extracts do not show activity as a mixture (presumably the active ingredient is masked or neutralized), but do have effluent peaks that excite the expected sexual behavior (44).

References

1. D. M. Coulson, L. A. Cavanagh, E. deVries, and B. Walther, *J. Agr. Food Chem.*, **8**, 399 (1960).
2. Dohrmann Instrument Co., Mountain View, California, Technical Bulletin Number 508.
3. Dohrmann Instrument Co., Mountain View, California, Technical Bulletin Number 520.
4. Dohrmann Instrument Co., Mountain View, California, Technical Bulletin Number 521.
5. Dohrmann Instrument Co., Mountain View, California, Technical Bulletin Number 522.
6. H. P. Burchfield, J. W. Rhoades, and R. J. Wheeler, in *Lectures in Gas Chromatography 1964*, L. R. Mattick and H. A. Szymanski, Eds., Plenum Press, New York, 1965, p. 59.
7. Ref. 6, p. 66.
8. D. E. Johnson and H. P. Burchfield, in *Lectures in Gas Chromatography 1964*, L. R. Mattick and H. A. Szymanski, Eds., Plenum Press, New York, 1965, p. 109.
9. R. W. Moshier and R. E. Sievers, *Gas Chromatography of Metal Chelates*, Pergamon Press, New York, 1965, p. 69.
10. J. D. Winefordner and T. H. Glenn, in *Advances in Gas Chromatography*, Vol. 5, J. C. Giddings and R. A. Keller, Eds., Dekker, New York, 1968, p. 263.
11. R. S. Juvet and R. P. Durbin, *J. Gas Chromatog.*, **1** (12), 14 (1963).
12. J. C. Winefordner and T. H. Glenn, in *Advances in Gas Chromatography*, Vol. 5, J. C. Giddings and R. A. Keller, Eds., Dekker, New York, 1968, p. 268.
13. R. S. Juvet and R. P. Durbin, *J. Gas Chromatog.*, **1** (12), 14 (1963).
14. S. S. Brody and J. E. Chaney, *J. Gas Chromatog.*, **4**, 42 (1966).
15. L. Giuffrida, *J. Assoc. Offic. Anal. Chemists*, **47**, 293 (1964).
16. L. Giuffrida and N. Ives, *J. Assoc. Offic. Anal. Chemists*, **47**, 1112 (1964).
17. L. Giuffrida and J. Bostwick, *J. Assoc. Offic. Anal. Chemists*, **49**, 8 (1966).
18. R. E. Johnson, in *Lectures in Gas Chromatography 1966*, L. R. Mattick and H. A. Szymanski, Eds., Plenum Press, New York, 1967, p. 1.
19. R. V. Hoffman, in *Lectures in Gas Chromatography 1966*, L. R. Mattick and H. A. Szymanski, Eds., Plenum Press, New York, 1967, p. 137.
20. Varian Aerograph, Walnut Creek, California, Accessory Catalog (1968).
21. A. J. McCormack, S. S. C. Tong, and W. D. Cooke, *Anal. Chem.*, **37**, 1470 (1965).
22. C. A. Bache and D. J. Lisk, in *Lectures in Gas Chromatography 1966*, L. R. Mattick and H. A. Szymanski, Eds., Plenum Press, New York, 1967, p. 17.
23. J. D. Winefordner and T. H. Glenn, in *Advances in Gas Chromatography*, Vol. 5, J. C. Giddings and R. A. Keller, Eds., Dekker, New York, 1968, p. 264.
24. J. E. Lovelock and S. R. Lipsky, *J. Am. Chem. Soc.*, **82**, 431 (1960).

25. E. J. Bonelli and K. P. Dimick, in *Lectures in Gas Chromatography 1964*, L. R. Mattick and H. A. Szymanski, Eds., Plenum Press, New York, 1965, p. 15.
26. P. C. Simmonds, D. C. Fenimore, B. C. Pettitt, J. E. Lovelock, and A. Zlatkis, *Anal. Chem.*, **39**, 1429 (1967).
27. Varian Aerograph Research Notes, Walnut Creek, California, Winter 1965.
28. D. E. Oaks, H. Hartmann, and K. P. Dimick, *Anal. Chem.*, **36**, 1563 (1964).
29. Aerograph Research Notes "Previews and Reviews," Walnut Creek, California, April, 1965.
30. E. J. Bonelli, Varian Aerograph Research Notes, Walnut Creek, California, February, 1968.
31. A. T. James and A. J. P. Martin, *Biochem. J.*, **50**, 679 (1952).
32. Sisters V. Heines, R. Jukasy, and M. A. O'Leary, and Mr. G. Schramm, *Trans. Kentucky Acad. Sci.*, **18**, 1 (1957).
33. K. Aunstrup and R. Djurtoft, *Brygmesteren*, **6**, 137 (1960).
34. G. W. Langan and R. B. Jackson, *J. Chromatog.*, **17**, 238 (1965).
35. M. Bowman and M. Boroza, *Anal. Chem.*, **40**, 535 (1968).
36. W. Kaye, *Anal. Chem.*, **34**, 287 (1962).
37. J. Merritt, F. Comendant, S. Abrams, and V. Smith, *Anal. Chem.*, **35**, 1461 (1963).
38. J. D. McCallum, in *Lectures in Gas Chromatography 1962*, H. A. Szymanski, Ed., Plenum Press, New York, 1963, p. 213.
39. H. L. Rosano, in *Flavor Chemistry*, I. Hornstein, Ed., American Chemical Society Publications, Washington, D. C., 1966, pp. 29–52.
40. W. S. Ryder, in *Flavor Chemistry*, I. Hornstein, Ed., American Chemical Society Publications, Washington, D. C., 1966, pp. 70–93.
41. D. G. Guadogni, S. Okano, R. G. Buttery, and H. K. Burr, *Food Technol.*, **20**, 166 (1966).
42. G. H. Fuller, R. Steltenkamp, and G. A. Tisserand, *Ann. N. Y. Acad. Sci.*, **116**, 711 (1964).
43. R. A. Flath, D. R. Black, D. G. Guadogni, W. H. McFadden, and T. H. Schultz, *J. Agr. Food Chem.*, **15**, 29 (1967).
44. R. S. Berger, J. M. McGough, and D. F. Martin, *J. Econ. Entomol.*, **58**, 1023 (1965).
45. R. S. Berger, *J. Econ. Entomol.*, **61**, 326 (1968).
46. G. B. Pitman, J. A. A. Renwick, and J. P. Vite, *Contribs. Boyce Thompson Inst.*, **23**, 243 (1966).

Author Index

Numbers in parentheses are reference numbers and indicate that the author's work is referred to although his name is not mentioned in the text. Numbers in italics show the pages on which the complete references are listed.

Subject Index

Acetaldehyde, nuclear magnetic resonance, 274-276

Acetone, infrared spectrum, 259
photolysis, 17-18

Acetylene subtractive processes, 130

Acid-base titration detector, 359-362

Acid derivatives, carbon-skeleton chromatography, 99-100

Acids, carbon-skeleton chromatography, 98, 99-100
identification, 134
precolumn reactions, 98, 99-100, 128, 130, 131-132, 136
subtractive processes, 128, 130, 131-132
see also Fatty Acids

Active sites of catalysts, 50

Adsorption-diffusion separator, 174-178

Adsorption equilibria of catalysts, 49-50

Aerosol propellents, infrared spectrum, 249

Alcohols, carbon-skeleton chromatography, 98, 107, 108
catalytic reactions, 38, 47, 48
dehydration, 47, 48
fragmentation reactions, 137
functional group tests, 337, 338, 339
hydrogenation, 114
ozonolysis, 119
precolumn reactions, 128, 129, 130-131, 134, 136
subtractive processes, 128, 129, 130-131, 134

Aldehydes, carbon-skeleton chromatography, 98
fragmentation reactions, 137
functional group tests, 337, 338, 339
hydrogenation, 114
identification, 134

precolumn reactions, 129, 130, 132-133, 134, 135
subtractive processes, 129, 130, 132-133

Alkaloids, mass spectrometry, 151-154, 212-216
pyrolysis gas chromatography, 56

Alkimino group, precolumn reactions, 135

Alkoxy group, precolumn reactions, 135

Alkylbenzenes, carbon-skeleton chromatography, 101
dealkylation, 36

Alkylbenzene sulfonates, pyrolysis gas chromatography, 63

Alkyl halides, *see* Halides

Allyl compounds, subtractive processes, 128

Aluminum, detection, 348

Amides, carbon-skeleton chromatography, 98, 99
hydrogenation, 114
precolumn reactions, 98, 99, 114, 136

Amines, acid-base titration, 362
carbon-skeleton chromatography, 96, 97, 98
functional group tests, 337, 338, 339
hydrogenation, 114
tertiary, precolumn reactions, 136

Amino acids, precolumn reactions, 136

Amino group, primary, precolumn reactions, 135

Ancillary techniques, classification, 2
connecting the systems, 10-11
definition, 2
gas flow rates, 11
sample size, 11-12
utilization, 7-12

Anthracene, spectrofluorimetry, 364
subtractive processes, 128, 130

387